Renewable

Energy

Resources

*John W. Twidell and
Anthony D. Weir*

London   New York
E. & F. N. Spon

First published in 1986 by
E. & F. N. Spon Ltd.
11 New Fetter Lane, London EC4P 4EE
Published in the USA by
E. & F. N. Spon
29 West 35th Street, New York NY 10001

Printed in Great Britain at
The University Press, Cambridge

ISBN 0 419 12000 9 (hardback)
         0 419 12010 6 (paperback)

British Library Cataloguing in Publication Data

Twidell, John
    Renewable energy resources.
    1. Power resources   2. Renewable energy
    sources
    I. Title II. Weir, Anthony D.
    333.79   TJ163.2

    ISBN 0-419-12000-9
    ISBN 0-419-12010-6 Pbk

Library of Congress Cataloging in Publication Data

Twidell, John.
    Renewable energy resources.

    Includes bibliographies and index.
    1. Renewable energy sources. I. Weir, Anthony D.
II. Title.
TJ808.T95   1985   333.79'2   85-10861
ISBN 0-419-12000-9
ISBN 0-419-12010-6 (pbk.)

Contents

Preface

Our aim

We have written this book to cover a subject of increasing technical and economic importance worldwide. It is primarily intended to support courses for undergraduates in physical science and engineering, beyond first year level. However since many practicing scientists and engineers will not have had a general training in renewable energy, the book is also intended for wider use beyond colleges and universities. Each chapter begins with fundamental theory from a physical science perspective, then considers applied examples and developments, and finally concludes with a set of problems and solutions. The whole book is structured to share common material and to relate aspects together. After each chapter, reading material is reviewed for further study.

Therefore the book is intended both for basic study and for application. Throughout the book and in the appendices, we include essential and useful reference material.

The subject

Renewable energy supplies are of steadily increasing importance in all countries. Most governments have substantial plans directed towards commercial development, and World agencies, such as the United Nations, have large programs to encourage the technology. In this book we stress the scientific understanding and analysis of renewable energy, since we believe these are distinctive and require specialist attention. The subject is not easy, mainly because of the spread of disciplines involved, and it certainly cannot be fully covered in just one book. This book is intended to bridge the gap between descriptive reviews and specialized engineering treatises on particular aspects. It centres on demonstrating how fundamental physical processes govern renewable energy resources and their application. Although the applications are being updated continually, the fundamental principles remain the same and we are confident that it will continue to provide a useful platform for those advancing the subject in the future. We have been encouraged in this approach by the rapidly increasing commercial importance of renewable energy technologies.

Readership

We expect our readers to have a basic understanding of science, especially of physical science, and of mathematics including calculus. It is not necessary to read or refer to chapters consecutively, as each aspect of the subject is treated, in the main, as independent of the other aspects. However some common elements, especially heat transfer, will have to be studied seriously if the reader is to progress to any depth of understanding. The disciplines behind a proper understanding and application of renewable energy include both biology and engineering. We are aware that our readers with a physical science background will usually be unfamiliar with life science and agricultural science, but we stress the importance of these subjects with obvious application for biofuels and for developments akin to photosynthesis.

Ourselves

We would like our readers to *enjoy* the subject of renewable energy, as we do, and to be stimulated to *apply* the energy sources for the benefit of their societies. Our own interest and commitment has evolved from work in both hemispheres and in a range of countries. Common elements have been teaching and application in the South Pacific and Scotland. We do not see the world as divided sharply between developed industrialized countries and developing countries of the Third World. Rural and remote communites of all countries share common opportunities, including opportunities associated with renewable energy. Developments from these circumstances will be of increasing industrial im-- portance. This is meaningful to us personally, since we wish our own energies to be directed for a just and sustainable society, increasingly free of poverty and the threat of cataclysmic war. We sincerely believe the development and application of renewable energy technology will favor these aspirations, and we therefore entirely endorse the sentiments of those who have promulgated a smaller scale of resource use, such as the late E. F. Schumacher. Our readers may not share these views, and this fortunately does not affect the content of the book. One thing they will have to share, however, is contact with the outdoors. Renewable energy is drawn from the environment, and practitioners must put on their rubber boots or their sun hat and move from the closed environment of buildings to the outside. This is no great hardship however, as anyone who knows the beauties of the South Pacific and the Scottish islands will affirm.

Suggestions for using the book in teaching

How a book is used in teaching depends mainly on how much time is devoted to its subject. For example, at the University of the South Pacific, we have taught the one-semester course on'Energy Resources and Distribution' to senior undergraduates in Physics. About half this course was on energy use in society, fossil fuels and their limitations, and other similar matters adequately and accessibly covered in many existing books. The remaining lecture hours were devoted to the analysis of those renewable energy supplies that seemed most applicable in that part of the world. A similar course has been taught at the University of Strathclyde. We found no existing books covering this range of topics at a suitable level for students in science or engineering. We have also taught other lecture and laboratory courses, and have found many of the subjects in renewable energy can be incorporated with great benefit into conventional teaching.

This book deliberately contains more material than could be covered in one specialist course. This enables the instructor and reader to concentrate on those particular energy supplies of benefit in their situation. To assist in this selection, most chapters start with a preliminary estimate of each resource, and its geographical variation.

The chapters are broadly grouped into similar areas. Chapter 1 introduces renewable energy supplies in general, and in particular the characteristics that distinguish their application from that for fossil or nuclear fuels. Chapter 2 (Fluid Mechanics) and Chapter 3 (Heat Transfer) are background material for later chapters. They contain nothing that a senior student in Mechanical Engineering will not already know, but are included for reference because much of this classical material has disappeared from Physics courses. Chapters 4–7 deal with various aspects of direct solar energy. Readers interested in this area are advised to start with the early sections of Chapter 5 (Solar Water Heating) or Chapter 7 (Photovoltaics), and review Chapters 3 and 4 as required. Chapters 8 (Hydro), 9 (Wind), 12 (Waves), 13 (Tides) present applications of fluid mechanics. Again the reader is advised to start with an applications chapter, and review the elements from Chapter 2 as required. Chapters 10, 11 deal with biomass as an energy source – respectively how the energy is stored and how it can be used. Chapters 14 (OTEC) and 15 (Geothermal) treat sources that are, like those in Chapters 12 and 13, important only in fairly limited geographical areas. Chapter 16, like Chapter 1, treats matters of importance to all renewable energy sources, namely the storage and distribution of energy and the integration of energy sources into energy systems. Appendices A (units), B (data) and C (heat transfer formulas) are referred to either implicitly or explicitly throughout the book. Suggestions for further reading and problems (mostly numerical in nature) are included with most chapters. Answers are provided.

Acknowledgments

As authors we bear responsibility for all interpretations, opinions and errors in this work. However, many have helped us, and we express our gratitude to them. Successive drafts of all, or portions of, this book have been used with undergraduate physics classes mainly at the University of the South Pacific and the University of Strathclyde, and also at a number of Universities where we have corresponded or visited. Our first debt is to these classes, whose reactions have helped shape the choice and presentation of material. We have also benefited from detailed comment from our close colleagues (Ken Taylor, Mahendra Kumar, Bob Lloyd, Surendra Prasad, Clifford Yee, Bill Grainger, Fiona Riddoch and Charles Giles) and from staff at our own and other institutions. At the University of New South Wales, these include Charles Sapsford, Graham Bowden, Hugh Outhred and Peter Barker; at the University of Queensland, Neville Jones; at the Ministry of Energy, Fiji, Peter Johnston, Jerry Ricolson; at the University of Khartoum, Farouk Habbani, Tony Egram; at the University of the South Pacific, Philip Whitney, Dick Solly; at the University of Strathclyde, Malcolm Slesser, Chris Lewis. Many others have helped us, including John Huthnance, Peter Giddens, Norman Bellamy, Norman Lipman, Jim Halliday, David Barbour, Hilary Wyper, Herick Othieno, Jerry Bass and Adam Pinney.

We also acknowledge with thanks the financial and administrative help given to us in the preparation of this book by: the University of the South Pacific (and especially successive Heads of the School of Natural Resources), the University of Strathclyde (especially Professor Edward Eisner, who has encouraged our interest and collaboration), the Nuffield Foundation (who financed the trials of several chapters as 'teaching units'), the University of Queensland, the United Nations Development Advisory Team for the Pacific, the Fiji Ministry of Energy, and the South Pacific Bureau for Economic Co-operation.

We have also to thank many support staff who have worked on this book (especially Sobha Narayan, Jean Lindores, Ann Clark, Elsie MacVarish, Muni Raj Deo, and Joan Finch).

And last, but not least, we have to thank the publishers, and our families, for

their patience with what must have seemed to them – as it did sometimes to us –
to be a never-ending project.

JOHN TWIDELL MA DPhil
Energy Studies Unit,
University of Strathclyde,
Glasgow,
Scotland.

TONY WEIR BSc PhD
South Pacific Bureau for
Economic Co-operation,
Suva,
Fiji.

List of symbols

Symbol	Main use	Other use or comment
Capitals		
A	area (m^2)	acceptor; ideality factor
A_R	Richardson's constant ($A\,m^{-3}\,K^{-2}$)	
AM	air mass ratio	
C	thermal capacitance (JK^{-1})	electrical capacitance (F); constant
C_D	drag coefficient	
C_F	thrust, force, coefficient	
C_L	lift coefficient	
C_P	power coefficient	
C_r	concentration ratio	
C_Γ	torque coefficient	
D	distance (m)	pipe or blade diameter (m); donor
E	energy (J)	
E_F	Fermi level	
E_g	band gap (eV)	
E_k	kinetic energy (J)	
E_p	potential energy (J)	
EMF	electromotive force	
F	force (N)	Faraday constant
F'_{i-j}		radiation exchange factor i to j
G	solar irradiance ($W\,m^{-2}$)	gravitational constant ($Nm^2\,kg^{-2}$); temperature gradient ($K\,m^{-1}$); Gibbs Energy
$G_{b,d,h}$	irradiance; beam, diffuse, on horizontal	
H	enthalpy (J), heat of combustion	pressure height (head) of fluids (m); wave crest/trough height (m); insolation ($J\,m^{-2}\,day^{-1}$)
I	electric current (A)	moment of inertia ($kg\,m^2$); integral
J	current density ($m^{-2}\,s^{-1}$ or $A\,m^{-2}$)	recombination current (A)
K	extinction coefficient (m^{-1})	clearness index (K_T); constant
L	distance, length (m)	diffusion length (m)
M	mass (kg)	molecular weight
M	momentum ($kg\,m\,s^{-1}$)	
N	concentration (m^{-3})	hours of daylight
N_o	Avogadro number	
P	power (W)	
P'	power per unit length ($W\,m^{-1}$)	
PS	photosystem	

Symbol	Main use	Other use or comment
Q	volume flow rate ($m^3\,s^{-1}$)	
R	thermal resistance (KW^{-1})	radius (m); electrical resistance (Ω); reduction level; gas constant; tidal range (m)
R_n	thermal resistance (co*n*duction)	
R_r	thermal resistance (*r*adiation)	
R_v	thermal resistance (con*v*ection)	
R_m	thermal resistance (*m*ass transfer)	
RFD	radiant flux density ($W\,m^{-2}$) (see ϕ)	
S	surface area (m^2)	entropy
S_v	surface recombination velocity (ms^{-1})	
STP	standard temperature & pressure	
T	absolute temperature (K)	period (s)
U	potential energy (J)	
V	volume (m^3)	electrical potential (V)
W	width (m)	mass flow rate per unit area ($kg\,s^{-1}\,m^{-2}$); energy density
X	characteristic dimension (m)	concentration ratio

Script capitals	(Non dimensional)	
$\mathscr{A}$	Rayleigh number	
$\mathscr{G}$	Grashof number	
$\mathscr{N}$	Nusselt number	
$\mathscr{P}$	Prandtl number, ν/κ	
$\mathscr{R}$	Reynolds number	
$\mathscr{S}$	Shape number of turbine	

Lower case		
a	amplitude (m)	wind interference factor, area (m^2)
b	width, of channel, (m)	wind profile exponent
c	specific heat capacity ($J\,kg^{-1}\,K^{-1}$)	velocity of electromagnetic radiation in vacuum ($m\,s^{-1}$); phase velocity of wave ($m\,s^{-1}$); chord length (m); Weibull speed factor ($m\,s^{-1}$)
d	distance (m)	zero plane displacement (wind) (m), depth
e	electron charge (C)	base natural logarithm (2.718)
f	frequency of cycles ($Hz = s^{-1}$)	pipe friction coefficient; fraction
g	acceleration of gravity ($m\,s^{-2}$)	
h	heat transfer coefficient ($Wm^{-2}\,K^{-1}$)	vertical displacement (m); Planck constant (6.63×10^{-34} Js); hole
$\hbar$	$h/(2\pi) =$ Planck constant$/(2\pi)$	
i	$\sqrt{-1}$	integer
j		integer
k	thermal conductivity ($Wm^{-1}\,K^{-1}$)	wave vector, $2\pi/\lambda$, (m^{-1});
k		Boltzmann constant $1.38 \times 10^{-23}\,JK^{-1}$
l	distance (m)	
m	mass (kg)	air mass ratio (see AM)
n	number	number of wind turbine blades; hours of bright sunshine; concentration (m^{-3}); number of nozzles
$\hat{n}$	unit vector normal to plane	
p	pressure ($Nm^{-2} = Pa$)	concentration (m^{-3}); (porosity p')

Symbol	Main use	Other use or comment
q	power per unit area (Wm^{-2})	
r	thermal resistivity of unit area ($m^2\,K\,W^{-1}$) ($r = h^{-1} = RA$)	radius (m); distance (m)
s	angle of slope (of a collector) (deg)	
t	time (s)	thickness (m)
u	velocity along stream ($m\,s^{-1}$)	group velocity ($m\,s^{-1}$)
v	velocity (not along stream) ($m\,s^{-1}$)	
w	distance (m)	moisture content, w dry basis, w' wet basis.
x	coordinate (along stream) (m)	
y	coordinate (across stream) (m)	
z	coordinate (upwards) (m)	z (downwards) depth (m)

Greek symbol

Symbol	Main use	Other use or comment
Γ gamma	torque (Nm)	gamma function
Δ delta	increment of ... (other symbol)	
Λ lambda	latent heat ($J\,kg^{-1}$)	
Σ sigma	summation sign	
Φ phi	radiant flux W	probability function
Φ_u	probability distribution of wind speed (($m\,s^{-1})^{-1}$)	
Ω omega	solid angle (deg)	phonon frequency (s^{-1}) angular velocity of blade ($rad\,s^{-1}$)
α alpha	absorptance	angle of attack, (deg)
α_λ	monochromatic absorptance	
β beta	angle (deg)	expansion coefficient (K^{-1})
γ gamma	angle (deg)	blade setting angle of wind turbine (deg)
δ delta	boundary layer thickness (m)	angle of declination (deg)
ϵ epsilon	emittance	wave 'spectral' width; permittivity; dielectric constant
ϵ_λ	monochromatic emittance	
ζ zeta	angle (deg)	
η eta	efficiency	
θ theta	temperature difference (°C)	angle of incidence (deg)
κ kappa	thermal diffusivity ($m^2\,s^{-1}$)	angle (deg)
λ lambda	wavelength (m)	tip speed ratio of wind turbine
μ mu	dynamic viscosity ($N\,m^{-2}\,s$)	
ν nu	kinematic viscosity ($m^2\,s^{-1}$) (NB $\nu = \mu/\rho$)	
π pi	3.1416	
ξ xi	electric potential (V)	roughness height (m); angle (deg)
ρ rho	density ($kg\,m^{-3}$)	reflectance; electrical resistivity ($\Omega\,m$)
ρ_λ	monochromatic reflectance	
σ sigma	Stefan constant ($= 5.67 \times 10^{-8}\,Wm^{-2}K^{-4}$)	
τ tau	transmittance	relaxation time (s); duration (s); sheer stress ($N\,m^{-2}$)
τ_λ	monochromatic transmittance	
ϕ phi	radiant flux density RFD ($W\,m^{-2}$)	wind/blade relative velocity angle (deg) potential difference (V), latitude (deg)
ϕ_λ	spectral distribution of RFD ($W\,m^{-3}$)	
χ chi	absolute humidity ($kg\,m^{-3}$)	
ψ psi	angle (deg)	longitude (deg)
ω omega	frequency ($= 2\pi f$) ($radian\,s^{-1}$)	solid angle (steradian); hour angle (deg)

Symbol	Main use	Other use or comment
Subscript		
B	black body	band
D	drag	dark
E	earth	
F	force	
G	generator	
L	lift	load, loss
M	Moon	
P	power	
R	rated	
S	Sun	
T	tangential	turbine
a	ambient	aperture, available head, aquifer
abs	absorbed	
b	beam	blade, bottom, base, biogas
c	collector	cell (photoelectric), cold
ci	cut in	
co	cut out	
cov	cover	
d	diffuse	dopant, digester
e	electrical	equilibrium, energy
f	fluid	forced, friction, flow
g	glass	generation current, band gap
h	horizontal	hot
i	integer	intrinsic
in	incident	
int	internal	
j	integer	
m	*m*ass transfer	mean (average), methane
max	maximum	
n	co*n*duction	negative charge carriers (electrons)
net	heat flow across surface	
o	(see numeral zero)	free space, dry
oc	open circuit	
p	plate	peak, positive charge carriers (holes)
r	*r*adiation	relative, recombination current, room, resonant, rock
rad	radiated	
refl	reflected	
rms	root mean square	
oc	open circuit	
s	surface	significant, saturated, sun
sc	short circuit	
t	tip	total
th	thermal	
trans	transmitted	
u	useful	
v	con*v*ection	vapor
w	wind	water
z	zenith	
λ	monochromatic e.g. α_λ, spectral distribution w.r.t. wavelength, ϕ_λ.	
0	distant approach	ambient, extra terrestrial, dry matter, saturated, ground level
1	entry to device	first

Symbol	Main use	Other use or comment
2	exit from device	second
3	output	third
Superscript		
m or max	maximum	
*	measured perpendicular to direction of propagation, e.g. G_b^*	sidereal day
· (dot)	rate of . . . , e.g. $\dot{m}\ (kg\ s^{-1})$	
Other Symbols		
bold face	vector	
$\hat{x}$	unit vector	Same symbol in ordinary type indicates the magnitude of the vector) e.g. $|\mathbf{u}| = u$
=	mathematical equality	
≈	approximate equality (within a few %)	
~	equality in 'order of magnitude' (with a factor of 2 to 10)	
≡	mathematical identity (or definition), equivalent	

1 *Principles of renewable energy*

1.1 Introduction

The aim of this text is to analyze the full range of renewable energy supplies needed for modern economies. Subjects will include power from wind, water, biomass, sunshine and other sources. Although the scale of application ranges from hundreds to many millions of watts, three questions have to be asked when considering any practical application of renewable energy:

(1) How much energy is available in the environment?
(2) For what purposes can the harnessed energy be used?
(3) What is the price of the energy as compared with other supplies – is it 'economic'?

The first two technical questions will be tackled chapter by chapter through this text. The last question understandably dominates in the mind of the energy consumer and always becomes the final question for practical installations. It is of the greatest importance to realize that harnessing renewable energy is only likely to be 'economic' if two conditions have been met:

(1) *The distinctive scientific principles* of renewable energy have been understood and applied (we outline these in Section 1.3).
(2) Each stage of the energy harnessing process is made *efficient* in terms of both minimizing losses and maximizing economic and social benefits.

When these conditions have been met, it is possible to calculate the costs and benefits of a particular scheme and compare these with alternatives for an economic assessment.

Failure to understand the distinctive scientific principles for harnessing renewable energy will almost certainly lead to poor engineering and uneconomic operation. Frequently there will be a marked contrast between the methods developed for renewable supplies and those used for the nonrenewable fossil fuel and nuclear supplies.

The *need* for harnessing renewable energy supplies is apparent as fossil fuels (especially oil) become increasingly expensive, as world population increases, and as each individual presses for a higher standard of living in terms of material goods, especially in rural and developing regions.

Consider the following simple model describing the need for commercial and noncommercial energy resources:

$$R = EN \qquad (1.1)$$

Here R is the total yearly energy requirement for a population of N people. E is the per caput energy resource use averaged over one year, related closely to provision of food and manufactured goods. Standard of living relates in a complex and an ill-defined way to E. Thus per caput gross national product S (a crude measure of standard of living) may be related to E by:

$$S = fE \qquad (1.2)$$

Here f is a complex and nonlinear coefficient that is itself a function of many factors. It may be considered an efficiency for transforming energy into wealth and should be as large as possible. Obviously unnecessary waste of energy leads to a lower value of f than would otherwise be possible. Substituting for E in (1.1),

$$R = SN/f \qquad (1.3)$$

The world population is now over 4200 million and increasing at approximately 2 to 3% per year so as to double every 20 to 30 years. Tragically high infant mortality and low life expectancy tend to hide the intrinsic pressures of population growth in many countries. In terms of total energy use, E on a world level is low at 0.8 kW, but national values range widely, from the USA at 10 kW and Europe at 4 kW to Central Africa at 0.1 kW. Economic growth implies exponential increase of S, say at 2 to 5% per year. Thus in (1.3), at constant efficiency f, the growth of world energy supply should be 4 to 8% per year. Without new and alternative supplies such growth cannot be maintained.

Yet at the same time as more energy is required, fossil fuels are being depleted. Whatever the status of nuclear energy, all national energy plans include two vital factors for improving or maintaining social benefit from energy:

(1) The harnessing of renewable supplies.
(2) The increase in efficiency of energy use.

Fortunately renewable energy sources are available in many remote and less privileged areas, as well as developed countries.

Consider the following elementary calculation for supplying 2 kW of power per person, which should be satisfactory for a developed lifestyle with due regard for energy conservation and efficiency. Each square meter of the earth's habitable surface is crossed by or accessible to an average energy flux of about 500 W. This includes solar, wind or other renewable energy forms in an overall estimate. If this flux is harnessed at just 4% efficiency, 2 kW of power can be drawn from an area of 10 m × 10 m, assuming suitable methods. Suburban areas of residential

1.2.3 *Environmental energy*

The currents of energy passing continuously as renewable energy through the earth are shown in Fig. 1.2. For instance, total solar flux incident at sea level is 1.2×10^{17} W. Thus the solar flux per person on earth (with more than 4×10^{9} people) is nearly 30 MW – the power of ten very large diesel electric generators. The maximum solar flux density perpendicular to the solar beam is about $1 \, \text{kW m}^{-2}$

However, the global data of Fig. 1.2 are of little value for practical engineering applications, since particular sites can have remarkably different environments

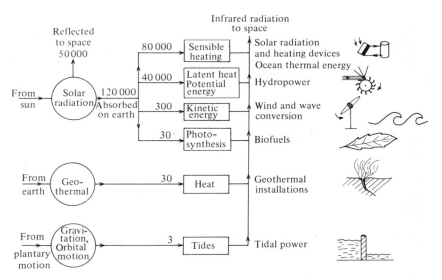

Fig. 1.2 Natural energy currents on earth, showing renewable energy systems. Note the great range of energy flux $(1:10^{5})$ and the dominance of solar radiation and heat. Units terawatts (10^{12}W)

and possibilities for harnessing renewable energy. Obviously flat regions, such as Denmark, have little opportunity for hydro-power but may have wind power. Yet neighboring regions, for example Norway, may have vast hydro potential. Tropical rain forests may have biomass energy sources, but deserts at the same latitude have none (moreover, forests must not be destroyed for energy so making more deserts). Thus practical renewable energy systems have to be matched to particular local environmental energy flows occurring in a particular region.

1.2.4 *Primary supply to end-use*

All energy systems can be visualized as a series of pipes or circuits through which the energy currents are channeled and transformed to become useful in domestic, industrial and agricultural circumstances. Fig. 1.3(a) is a spaghetti diagram of

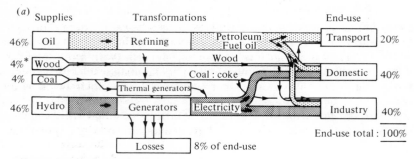

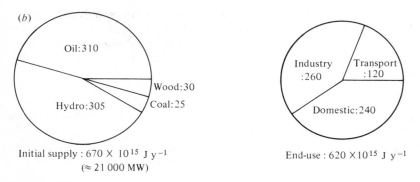

Fig. 1.3 Energy flow diagrams for Norway, 1980: (a) spaghetti (b) pie. Data and presentation simplified from official statistics. Population 4.0 million. Note importance of hydro-power. In contrast to most industrialized countries, there is little generation of electricity and hence little loss

energy supply. Sections across such a diagram can be drawn as pie diagrams for primary delivered and end-use supply (Fig. 1.3(b)).

1.2.5 *Energy planning*

(1) *Complete energy systems* must be analyzed, and supply must not be separated from end-use. Unfortunately precise *needs* for energy are too frequently forgotten, and supplies are not well matched to end-use. Energy losses and uneconomic operation therefore frequently result. For instance if a dominant domestic energy requirement is heat for warmth and hot water, it is probably not sensible to generate grid quality electricity from a fuel, waste the majority of the energy as thermal emission from the boiler and turbine, distribute the electricity and then dissipate this electricity as heat. Direct heat production with local distribution could be more energy and financially efficient. A variation of this principle is combined heat and electricity production.

(2) *System efficiency* calculations can be most revealing and can pinpoint

unnecessary losses. Here we define 'efficiency' as the ratio of the useful energy output from a process to the total energy input to that process. Consider mains electric lighting from thermally generated electricity. Successive energy efficiencies might be: electricity generation 30%, distribution 90%, and incandescent lighting (energy in visible radiation) 5%. The total efficiency is 1.4%. Contrast this with cogeneration of useful heat and electricity (efficiency ~80%), distribution (90%) and lighting in modern low consumption lamps (~20%). The total efficiency is now 14% – a tenfold improvement. The total cost of the more efficient system can be lowered, despite high initial unit capital costs, because less capital equipment is needed, fuel supplies are reduced, and equipment (especially lamps) will last longer.

(3) *Energy management* is always important to improve overall efficiency and reduce economic losses. No energy supply is free, and renewable supplies are usually more expensive in practice than might be assumed. Thus there is no excuse for wasting energy of any form unnecessarily.

1.3 Scientific principles of renewable energy

The definitions of renewable and finite energy supplies (Section 1.2.1) indicate the fundamental differences between the two forms of supply. As a consequence the efficient use of renewable energy can only follow the correct application of certain principles.

1.3.1 *Energy currents*

It is essential that a sufficient renewable current is *already present* in the local environment. It is not good practice to try to create this energy current especially for a particular system. In a public meeting in Suva, the capital of Fiji, renewable energy was once ridiculed by calculating the number of pigs required to produce dung for sufficient methane generation to power the whole city. It is obvious, however, that biogas (methane) production should only be contemplated as a *byproduct* of an animal industry already established, and not vice versa. The practical implication of this principle is that the local environment has to be monitored and analyzed over a long period to establish precisely what energy flows are present. In Fig. 1.1 the direct current ABC must be assessed before the diverted flow through DEF is established.

1.3.2 *Dynamic characteristics*

End-use requirements for energy vary with time. For example electricity demand on a power network often peaks in the morning and evening, and reaches a minimum through the night.

If power is provided from a finite source, such as oil, the input can be adjusted in response to demand. Unused energy is not wasted, but remains with the source fuel. However, with renewable energy systems, not only does end-use vary uncontrollably with time but so too does the natural supply in the environment. Thus a renewable energy device must be matched dynamically at both D and E of Fig. 1.1; the characteristics will probably be quite different at both interfaces. Examples of these dynamic effects will appear in most of the following chapters.

The major periodic variations of renewable sources are listed in Table 1.2, but precise dynamic behavior may well be greatly affected by irregularities. Systems range from the highly unpredictable (e.g. wind power) to the highly predictable (e.g. tidal power). Sunshine may be highly predictable in some regions (e.g. Khartoum) but elusive in others (e.g. Glasgow).

1.3.3 *Quality of supply*

The quality of an energy supply or store is often discussed, but usually remains undefined. We define *quality* as the proportion of an energy source that can be converted to mechanical work. Thus power from an electric motor has high quality ($\gtrsim 95\%$) because the shaft power can be efficiently harnessed as mechanical work, say to lift a weight. The quality of thermal power from burning fuel in a conventional power station is moderately low, because only about 30% of the calorific value of the fuel can be made to appear as mechanical work.

Renewable energy supply systems divide into three broad divisions by quality:

(1) *Mechanical supplies*, such as hydro, wind, wave and tidal power. In general the quality of the supply is high and mechanical work is usually extracted for electricity generation at quite high efficiency. The proportion of power in the environment extracted by the devices is determined by the mechanics of the process, as explained in later chapters. The proportions are, commonly, wind 30%, hydro 60%, wave 75% and tidal 75%.

(2) *Heat supplies*, such as biomass combustion and solar collectors. The maximum proportion of heat energy extractable as mechanical work is given by the second law of thermodynamics, which assumes reversible, infinitely long transformations. In practice maximum mechanical power produced in a dynamic process is about half that predicted by the second law. For thermal boiler heat engines, maximum quality is about 35%.

(3) *Photon processes*, such as photosynthesis (Chapter 10) and photovoltaic conversion (Chapter 7). For example, solar photons of a single frequency may be transformed into mechanical work with high efficiency using a matched solar cell. In practice the broad band of frequencies in the solar spectrum makes matching difficult and photon conversion efficiencies of 15% are considered good.

1.3.4 *Dispersed versus centralized energy*

A pronounced difference between renewable and finite energy supplies is the energy flux density at the initial transformation. Renewable energy commonly arrives at about $1 \, \mathrm{kW \, m^{-2}}$ (e.g. solar beam irradiance, energy in the wind at $10 \, \mathrm{m \, s^{-1}}$), whereas finite centralized sources have energy flux densities that are orders of magnitude greater. For instance, boiler tubes in gas furnaces easily transfer $100 \, \mathrm{kW \, m^{-2}}$, and in a nuclear reactor the first wall heat exchanger must transmit several $\mathrm{MW \, m^{-2}}$. At end-use after distribution, however, supplies from

Table 1.2 Intensity and frequency properties of renewable sources

System	Major periods	Major variables	Power relationship	Comment	Text reference
Direct sunshine	24 h, 1 y	Solar beam irradiance G_b^* (W m^{-2}) Angle of beam from vertical θ_z	$P \propto G_b^* \cos \theta_z$, max. 1 kW m^{-2}	Daytime only!	(4.2)
Diffuse sunshine	24 h, 1 y	Cloud cover	$P << G$; $P \lesssim 300$ W m^{-2}	Significant energy, however Very many variations	(4.3)
Biofuels	1 y	Soil condition, solar irradiation, water, species type, wastes	Stored energy 10 MJ kg^{-1}	Link with agriculture and forestry	Table 11.1
Wind	1 y	Wind speed u_0 Height above ground z Height of meteorological mast h	$P \propto u_0^3$ $u_z/u_h = (z/h)^b$	Highly fluctuating $b \sim 0.15$	(9.2) (9.54)
Wave	1 y	Significant wave height H_s Wave period T	$P \propto H_s^2 T$	High power density ~ 50 kW m^{-1}	(12.46)
Hydro	1 y	Reservoir height H Water flow volume rate Q	$P \propto HQ$	Established resource	(8.1)
Tidal	12 h 25 min	Tidal range R; contained area A; length of estuary L; depth of estuary h	$P \propto R^2 A$	Enhanced tidal range if $L/\sqrt{h}$ a critical value (36400 m$^{0.5}$)	(13.35) (13.28)
Ocean thermal energy conversion	Constant	Temperature difference between sea surface and deep water, ΔT	$P \propto (\Delta T)^2$	Some tropical locations Large energy fluxes harnessed at low efficiency	(14.4)

finite sources must be greatly reduced in flux density. Thus apart from major exceptions such as metal refining, end-use loads for both renewable and finite supplies are similar. In summary *finite energy is most easily harnessed centrally and is expensive to distribute. Renewable energy is most easily harnessed in dispersed locations and is expensive to concentrate.*

A practical consequence of renewable energy application is development and increased cash flow in the rural economy. Thus the use of renewable energy favors rural development and not urbanization.

1.3.5 *Complex systems*

Renewable energy supplies are intimately linked to the natural environment, which is not the preserve of any one academic discipline such as physics or electrical engineering. Frequently it is necessary to cross disciplinary boundaries from as far apart as, say, plant physiology to electronic control engineering.

An outstanding example is the energy planning of integrated farming as in the Philippine Islands (Section 11.8.1). Animal and plant wastes may be used to generate methane, liquid and solid fuels, and the whole system integrated with fertilizer production and nutrient cycling for optimum agricultural yields.

1.3.6 *Situation dependence*

No one renewable energy system is universally applicable, since the ability of the local environment to supply the energy and the suitability of society to accept the energy vary greatly. It is as necessary to 'prospect' the environment for renewable energy as it is to prospect geological formations for oil. It is also necessary to conduct energy surveys of the domestic, agricultural and industrial needs of the local community. Particular end-use needs and local renewable energy supplies can then be matched, subject to economic constraints. In this respect renewable energy is similar to agriculture. Particular environments and soils are suitable for some crops and not others, and the market pull for selling the produce will depend on particular needs.

The main consequence of the situation dependence of renewable energy is the impossibilty of making simplistic international or national energy plans. Solar energy systems in southern Italy should be quite different from those in Belgium and indeed in northern Italy. Corn alcohol fuels might be suitable for farmers in Missouri but not in New England. A suitable scale for renewable energy planning might be 250 km, but certainly not 2500 km. Unfortunately large urban and industrialized societies are not well suited for such flexibility and variation.

1.4 Technical implications

1.4.1 *Prospecting the environment*

Usually monitoring is needed for several years at the site in question. Ongoing analysis must insure that useful data are being recorded, particularly with respect to dynamic characteristics of the energy systems planned. Meteorological data are always important, but unfortunately the sites of official stations are often different from the energy generating sites, and the methods of recording and analysis are not ideal for energy prospecting. However an important use of the long term data from official monitoring stations is as a base for comparison with local site variations. Thus wind velocity may be monitored for several months at a prospective generating site and compared with data from the nearest official base station. Extrapolation from many years of base station data may then be possible.

Data unrelated to normal meteorological measurements may be difficult to obtain. In particular flows of biomass and waste materials will often not have been previously assessed, and will not have been considered for energy generation. In general prospecting for supplies of renewable energy requires specialized methods and equipment that demand significant resources of finance and manpower. Fortunately the links with meteorology, agriculture and marine science give rise to much basic information.

1.4.2 *End-use requirements and efficiency*

Energy generation should always follow quantitative and comprehensive assessment of energy end-use requirements. Since no energy supply is cheap or occurs without some form of environmental disruption, it is also important to use the energy efficiently with good methods of energy conservation. With electrical systems the end-use requirement is called the *load*, and the size and dynamic characteristics of the load will greatly affect the type of generating supply. Money spent on energy conservation and improvements in end-use efficiency usually gives better benefit than money spent on increased generation and supply capacity.

The largest energy requirements are usually for heat and transport. Both uses are associated with energy storage capacity in thermal mass, batteries or fuel tanks, and the inclusion of these uses in energy systems can greatly improve overall efficiency.

1.4.3 *Matching supply and demand*

After quantification and analysis of the separate dynamic characteristics of end-use demands and environmental supply options, the total demand and supply have to be brought together. This may be explained as follows:

(1) The maximum amount of environmental energy must be utilized within the capability of the renewable energy devices and systems. In Fig. 1.4(a), the resistance to energy flow at D, E and F should be low. The main benefit of this is to reduce the size and amount of generating equipment.
(2) Negative feedback control from demand to supply is *not* beneficial since the result is to waste or spill harnessable energy (Fig. 1.4(b)). Such control should only be used at times of emergency or when all conceivable end-uses have been satisfied. Note that the disadvantage of feedback control is a consequence of *renewable* energy being flow or current sources that can never be stopped. With *finite* energy sources, feedback control to the energy source is beneficial since less fuel is used.

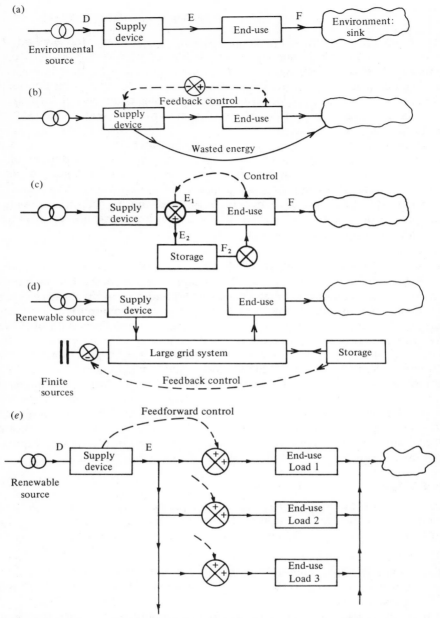

Fig. 1.4 Matching renewable energy supply to end-use. (a) Maximum energy flow: for minimum size of device or system requires low resistance to flow at D, E and F. (b) Negative feedback control: wastes energy. (c) Energy storage: allows the dynamic characteristics of end-use to be decoupled from the supply characteristics. (d) Decoupling with a large grid system. (e) Feedforward load management control of the supply; arguably the most efficient way to use renewable energy. Total load at E may be matched to the available supply at D at all times and so control the supply device

(3) The natural periods and dynamic properties of end-use will never be exactly the same as those of the supply, as discussed in Section 1.3.2. The only way to match supply and demand with different dynamic characteristics, and yet not to waste harnessable energy, is to incorporate storage (Fig. 1.4(c)). Satisfactory energy storage is expensive (see Chapter 16), especially if not incorporated at the earliest stages of planning.

(4) The difficulties of matching renewable energy supplies to end-use are so great that one common approach is to completely decouple supply from local demand (Fig. 1.4(d)). Here the renewable supply feeds a large and usually complex energy grid system having input from finite sources which benefit from feedback control. Such systems imply large scale operation and usually electricity grids. As in (3) the addition of substantial energy storage in the system, say pumped hydro or thermal capacity for heating, can improve efficiency and allow the proportion of renewable supply to increase.

(5) The most efficient way to use renewable energy is shown in Fig. 1.4(e). Here a range of end-uses is available and can be switched or adjusted so that the total load equals the supply at any one time. Some of the end-use blocks could themselves be adjustable (e.g. variable voltage water heating, pumped water storage). Such systems require *feedforward control* (see Section 1.4.4).

1.4.4 *Control options*

Good matching of renewable energy supply to end-use demand is accomplished by control of machines, devices and systems. The previous discussion shows that there are three possible categories of control: (1) spill the excess energy, (2) incorporate storage and (3) operate load management control. These categories may be applied in different ways, separately or together, to all renewable energy systems, and will be illustrated here with a few examples (see Fig. 1.5).

(1) *Spill excess energy* Since renewable energy derives from energy flow sources, energy not used is energy wasted. Nevertheless spilling excess energy provides easy control and may be the cheapest option. Examples occur with run-of-the-river hydroelectric systems (Fig. 1.5(a)), shades and blinds with passive solar heating of buildings, and wind turbines with adjustable blade pitch.

(2) *Incorporate storage* Storage before transformation allows a maximum amount of energy to be trapped from the environment and eventually harnessed or used. Control methods are then similar to conventional methods with finite sources with the store equivalent to fuel. The main disadvantages of such systems are the high relative costs of storage, and the difficulty of reducing conventional control methods to small scale and remote operation. In the example of Fig. 1.5(b), hydro storage is usually only contemplated for generation above $\sim 10\,MW$. The mechanical flow control devices become unwieldy and expensive at a microhydro scale of $\sim 10\,kW$. A disadvantage of hydro storage may be the environmental damage caused by reservoirs.

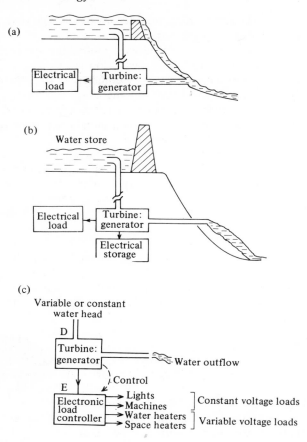

Fig. 1.5 Examples of control. (a) Control by spilling excess energy: constant pressure maintained for the turbine. (b) Control incorporating storage in hydroelectric catchment dam. (c) Control by load variation: feedforward control. Load controller automatically shunts power between end-uses, maintaining constant total generator load at E. Turbine also has constant load and hence constant frequency: only rudimentary mechanical control of turbine necessary

Storage after energy transformation, e.g. battery charging or hydrogen production, is also possible and may become increasingly important especially in small systems. Thermal storage is already common.

(3) *Load control* Parallel arrangements of end-uses may be switched and controlled so as to present optimum total load to the supply. An example of a microhydro load controller for household power is shown in Fig. 1.5(c) (see also Section 8.6). The principle may be applied on a small or a large scale, but is perhaps most advantageous when many varied end-uses are

available locally. There are considerable advantages if load control is applied to renewable energy systems:

(a) No environmental energy need be wasted if parallel outputs are opened and closed to take whatever input energy flow is available.

(b) Priorities and requirements for different types of end-use can be incorporated in many varied control modes (e.g. low priority uses can receive energy at low cost, provided that they can be switched off by feedforward control; electrical resistive heaters may receive variable voltage and hence variable power).

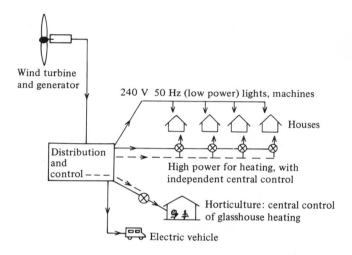

Fig. 1.6 Wind energy conversion system for Fair Isle, Scotland. Electrical loads are switched by small changes in the supply frequency, so presenting a matched load to the generator over a wide range of wind speeds

(c) End-uses having storage capability (e.g. water heating and building space conditioning) can be switched to give the benefits of storage in the system at no extra cost.

(d) Electronic and microprocessor based control may be used with benefits of low cost, reliability, and extremely fast and accurate operation.

Feedforward load control may be particularly advantageous for autonomous wind energy systems (see Chapter 9, especially Section 9.10). Wind fluctuates greatly in speed and the wind turbine should change rotational frequency to maintain optimum output. Rapid accurate control is necessary without adding greatly to the cost or mechanical complexity, and so electronically based feedforward control into several parallel electrical loads is most useful. An example is shown in Fig. 1.6.

1.5 Social implications

The Industrial Revolution in Europe and North America, and industrial development in all countries, have profoundly affected social structures and patterns of living. The influence of changing and new energy sources, especially coal and oil, has been the driving function for much of this change. Thus there is a historic relationship between coal mining and the development of industrialized countries which will continue for several hundred years. In the non-industrialized countries there has been an equally close relationship between development and the international availability of relatively cheap oil. These oil supplies became available worldwide in the 1950s at the same time as many countries obtained independence from colonialism. This development in many respects has coincided with the availability of cheap oil. Thus in all countries the use of fossil fuels has led to profound changes in lifestyle.

Likewise changes in social patterns will occur as renewable energy systems become widespread. The influence of modern science and technology will insure that there are considerable improvements to older methods, and subsequently standards of living can be expected to rise, especially in the rural sector. It is impossible to predict the long term effect of such changes in energy supply, but the sustainable nature of renewable energy should produce greater stability than has been the case with fossil fuels, especially oil. In particular the great diversity of renewable energy supplies will insure a similar diversity in local economic and social characteristics.

1.5.1 *Dispersed living*

In Sections 1.1 and 1.3.4 the dispersed and low energy flux density of renewable sources was discussed. Renewable energy arrives dispersed in the environment and is difficult and expensive to concentrate. By contrast finite energy sources are energy stores that are easily concentrated at source and expensive to disperse. Thus electrical distribution grids from fossil fuel and nuclear sources have tended to radiate from central, intensive distribution points, typically with $\sim 1000\,MW$ capacity. Industry has developed on these grids, with heavy industry closest to the points of intensive supply. Domestic populations have grown in response to the employment opportunities of industry and commerce. Similar effects have occurred with the relationships between coal mining and steel production, oil refining and chemical engineering and the availability of gas supplies with urban complexes.

This physical review of the effect of the primary flux density of energy sources suggests that widespread application of renewable energy will favor dispersed rather than concentrated communities. In Section 1.1 an approximate estimate of 500 people km^{-2} was made of maximum population density for communities relying on renewable sources. This is considerably higher than for rural communities (≤ 100 people km^{-2}) and corresponds with the population densities of the main administration and commercial towns of rural regions. Thus the gradual acceptance of significant supplies of renewable energy could allow relief from the concentrated metropolises of excessive urbanization, yet would not require unacceptably low population densities. Sensible and efficient use of remaining fossil fuel sources should allow ample time for such changes.

1.5.2 *Pollution*

Renewable energy is always extracted from a flow of energy already occurring in the environment (Fig. 1.1). The energy is then returned to the environment, so no thermal pollution can occur on anything but a small scale. Likewise material and chemical aspects of pollution in air, water and refuse tend to be minimal. These effects favor the use of renewable energy for power generation against the finite sources of fossil and nuclear fuels. An exception is air pollution from incomplete combustion of wood (see Chapter 11). Environmental pollution can of course occur if the devices are manufactured using finite supplies, and from materials processing.

A main environmental objection to renewable energy developments is likely to be their visual impact. Every development must penetrate the natural environment in order to extract energy, and may be apparent in otherwise open scenery. The most serious objections often relate to the ecological impact of large scale or concentrated renewable energy developments. Large hydro-power installations occur where water is naturally concentrated by mountain or hill formations, usually in regions of outstanding beauty. Large catchment dams cause most definite and possibly unacceptable interference. Soil erosion and loss of soil quality can all too easily result from ignorant agricultural practice. Such abuses must be prevented, and are most likely to arise from large scale developments having the aim of producing concentrated or intense energy supplies.

1.5.3 *Long term influences*

The physical factors of energy sources constrain technology and society to move in certain directions. Renewable sources lead to more involvement with the natural environment, whereas fossil fuel and nuclear sources produce methods and scales of operation that isolate technology from the environment. A society deciding to depend significantly on renewable energy supplies for all forms of end-use must allocate considerable resources towards these technologies. Educational, planning, financial and industrial policies must be adjusted for the new strategy. The overall effect will be to make the society more knowledgeable and conscious of its environment. Since much energy is used within commercial and domestic buildings, a move towards environmental and solar sources will lead to a significant change in architectural styles. For instance, in cold and temperate regions it becomes sensible to place a glass covered enclosure (a conservatory) in the sun-facing sides of houses for passive solar heating. On a widespread scale such a development would have marked effects in living patterns and styles of buildings.

Another significant effect might be on education and research. In physics, for instance, the classical studies of heat transfer, mechanics, optics and fluid dynamics would have to be given fresh emphasis. Methods to monitor and analyze the properties of the natural environment would require advances in measurement and instrumentation. The electronic and quantum effects of photovoltaic and photosynthetic processes would be essential parts of a syllabus and are scientific challenges of great depth. Such influences on school and college physics education would relate to distinctive practical applications within the experience of the pupils and reinforce practical work.

Similar influences would also occur in chemistry and biology, and in the applied sciences. Hopefully one common feature of those changes at all levels would be to dissociate science, and particularly physics, from close involvement with military weapons. Many physicists have experienced serious unease at the historical and present links of nuclear physics with nuclear weapons. It is our hope that renewable energy development will also have a strong influence, but one that will lead to an open exchange of information for the benefit of all mankind.

Bibliography

Refer to the bibliographies at the end of each chapter for particular subjects.

Comprehensive study of renewable energy
Sørensen, B. (1979) *Renewable Energy*, Academic Press, London.
    Outstandingly the best general text at postgraduate level, considering energy from the environment to final use. Useful for final year undergraduates.
Advances in Energy, series of review publications by Academic Press, London.
    This series is setting excellent standards at postgraduate and professional level.
McMullan, J. T., Morgan, R. and Murray, R. B. (1977) *Energy Resources and Supply*, John Wiley and Sons, London.
    Considers all energy supplies, not just renewables. Clearly presented, useful text for undergraduates and general energy study.

Official publications
United Nations Agencies produce a wide range of essential publications regarding energy. These are especially important for data. For instance we recommend:
United Nations *World Energy Supplies*, UN document no. ST/ESA/STAT/SER.J/19, annual. Gives statistics of energy consumption around the world, classified by source, country, continent etc., but counts only 'commercial energy' (i.e. excludes firewood etc.).

Government publications are always important. For instance:
UK Department of Energy series of Energy Papers. Such publications are usually clearly written and include economic factors at the time of writing. Basic principles are covered, but usually without the details required for serious study.

National studies are made by both government agencies and specialist institutes, for instance:
Ford Foundation (1974) *A Time to Choose: America's Energy Future*, Ballinger.
    Summary of an extensive project, which spawned some 20 specialized reports (e.g. on agriculture, industry, nuclear safety). Makes the important distinction between the 'near term' (up to, say, AD 2000) and 'long term' beyond: energy choices now have little immediate effect, but may determine the long term future of society.

General energy studies and the 'energy debate'

Cook, E. (1976) *Man, Energy, Society*, Freeman, Oxford.
  Good historical and social perspective.
Darmstadter, J., Dunkerley, J. and Alterman, J. (1977) *How Industrial Societies Use Energy: A Comparative Analysis*, Resources for the Future/Johns Hopkins University Press, Baltimore.
Foley, G. (1976) *The Energy Question*, Penguin, London.
Lovins, A. B. (1977) *Soft Energy Paths: Towards a Durable Peace*, Penguin, London.
  Explores the wider social and political issues of energy supply, especially those associated with renewable and nuclear supplies.
O'Toole, J. (1976) *Energy and Social Change*, MIT Press, Boston.
  Like Lovins (1977), this lays out the choice for the USA between the scenarios of 'historical growth', 'technical fix' and 'zero energy growth'.
Ruedisili, L. C. and Firebaugh, M. W. (eds) (1978) *Perspectives on Energy*, 2nd edn, Oxford University Press.
  Well-chosen collection of reprints, often of contrasting views. Includes a good review of fusion by Rose and Feirtas.
Smil, V. and Knowland, W. E. (eds) (1980) *Energy in the Developing World: The Real Energy Crisis*, Oxford University Press.

Journals and Magazines

We urge readers to scan the serious scientific and engineering journals, e.g. *Nature, Annual Review of Energy*, and magazines, e.g. *Electrical Review, Modern Power Systems*. These publications regularly cover renewable energy projects among the general articles. There are also specialist journals, e.g. *Solar Energy, Wind Engineering*.

There are many publications, journals and newsletters by specialist groups, e.g. *Soft Energy Paths* by Friends of the Earth. These do not attempt to compete with professional journals, but are excellent for general awareness and news of developments.

Do-it-yourself publications

There are many publications for the general public and for enthusiasts. Do not despise these, but take care if the tasks are made to look easy. Many of these publications give stimulating ideas and are attractive to read, e.g. Merrill, R. and Gage, T. (eds) *Energy Primer*, Dell, New York (several editions).

2 *Essentials of fluid mechanics*

2.1 Introduction

The transfer of energy from a moving fluid is the basis of hydro, wind, wave and some solar power systems. To understand these energy sources we must start with the basic laws of mechanics: the conservation of mass, energy and momentum. We shall consider mainly incompressible flow. The term *fluid* includes both liquids and gases. The feature that distinguishes a fluid from a solid is the inability of a fluid, while remaining in a state of equilibrium, to resist shearing forces. The distinction between a liquid and a gas is that a gas is much more compressible. A certain mass of a particular liquid will have a definite volume which will vary only slightly with temperature and pressure. In a gas the volume varies with temperature and pressure, approximately in accordance with the perfect gas law ($pV = n\mathrm{R}T$). Fortunately, for gas flowing at speeds $<100\,\mathrm{m\,s}^{-1}$ and not subject to large changes in pressure or temperature, the changes in density are negligible. This is the case in most applications in this book.

Many important fluid flows are also *steady*, in the sense that the flow *pattern* does not vary with time. In this case it is useful to picture a set of lines drawn so as to coincide with the velocity vectors at each point. These are known as *streamlines*.

A further important distinction is between laminar and turbulent flow. This is discussed in more detail in Section 2.5, but as an example watch the smoke rising from a smoldering taper in still air. Near the taper the smoke rises in an orderly stream. The paths followed by neighboring particles of smoke remain more or less parallel. Such an orderly, smooth flow is called *laminar*. Further from the taper the flow becomes more chaotic, with individual smoke particles intermingling in three dimensions. This flow is called *turbulent*. Although it is very difficult to calculate the details of a turbulent flow, it is often sufficient to treat it as a steady mean flow subject to friction caused by the velocity fluctuations.

2.2 Conservation of energy: Bernoulli's equation

Consider the most important case of steady, incompressible flow. As discussed in Section 2.1, the flow stays within well-defined (though imaginary) stream-tubes, i.e. tubes bounded by streamlines.

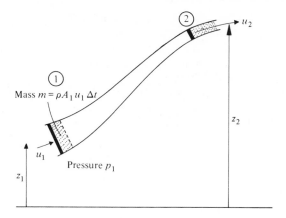

Fig. 2.1 Illustrating conservation of energy: a streamtube rises from height z_1 to z_2

Fig. 2.1 shows a streamtube which rises from a height of z_1 to a height of z_2. The tube is narrow enough that z may be supposed constant over each section of the tube. We consider a control volume bounded by the streamlines and two perpendicular slices across the streamtube at 1 and 2.

A mass $m = \rho A_1 u_1 \Delta t$ enters the control volume at 1, and an equal mass $m = \rho A_2 u_2 \Delta t$ leaves at 2. Then the energy balance on the fluid within the control volume is

potential energy lost + work done by pressure forces
  = gain in kinetic energy + heat losses due to friction

and may be written as

$$mg(z_1 - z_2) + [(p_1 A_1)(u_1 \Delta t) - (p_2 A_2)(u_2 \Delta t)] = \tfrac{1}{2}m(u_2^2 - u_1^2) + E_f \qquad (2.1)$$

where the pressure force $p_1 A_1$ acts through a distance $u_1 \Delta t$, and similarly for $p_2 A_2$, and E_f is the heat energy generated by friction.

We neglect fluid friction for the moment, but we will examine some of its effects in Section 2.5. In this ideal, frictionless case (2.1) reduces to

$$(p_1/\rho) + gz_1 + \tfrac{1}{2}u_1^2 = (p_2/\rho) + gz_2 + \tfrac{1}{2}u_2^2 \qquad (2.2)$$

or, equivalently,

$$\frac{p}{\rho g} + z + \frac{u^2}{2g} = \text{constant along a streamline} \qquad (2.3)$$

Either of these forms of the energy equation is called *Bernoulli's equation*.

The sum of the terms on the left hand side of (2.3) is usually called the total *head* of fluid. It represents the total energy of a unit mass of fluid, and may vary from point to point in a flow. In particular, the constant in (2.3) may vary from streamline to streamline. Head has the dimensions of length, and is visualized as the height of a column of the fluid.

The main limitation of (2.2) and (2.3) is that they apply only to an ideal fluid, with zero viscosity, zero compressibility and zero thermal conductivity. The

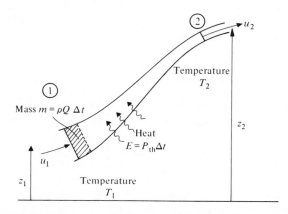

Fig. 2.2 As for Fig. 2.1, but with heat sources present. Thermal power P_{th} is added to the flow

energy equation can however be modified to overcome each of these limitations. The form for compressible fluids is given in any of the references: it reduces to (2.3) for the low speeds we are concerned with here. The assumption that the fluid does not conduct heat implies that no heat from external sources reaches the streamtube under consideration. Equation (2.3) will therefore be valid if there are no external heat sources to be considered, as in most wind or hydro-power systems.

In solar heating systems and heat exchangers power P_{th} is added to the fluid as it flows past a heat source (Fig. 2.2). Heat $E = P_{th}\Delta t$ is added to the energy inputs on the left hand side of (2.1). The mass m coming into the control volume at temperature T_1 has heat content mcT_1 (where c is the specific heat capacity of the fluid), and that going out has heat content mcT_2. Thus we have to add to the right hand side of (2.1) the net heat carried out of the control volume in time Δt, namely $mc(T_2 - T_1)$. This gives an equation corresponding to (2.2), namely

$$(p_1/\rho) + gz_1 + \tfrac{1}{2}u_1^2 + cT_1 + (P_{th}/\rho Q) = (p_2/\rho) + gz_2 + \tfrac{1}{2}u_2^2 + cT_2 \tag{2.4}$$

where the volume flow rate

$$Q = Au \tag{2.5}$$

In practice, heating systems are designed so that the thermal contributions dominate the energy balance, so that (2.4) reduces to

$$P_{th} = \rho c Q (T_2 - T_1) \qquad (2.6)$$

2.3 Conservation of momentum

In standard textbooks Newton's second law of motion is generalized from particles to fluids: 'At any instant in steady flow the resultant force acting on the moving fluid within a fixed volume of space equals the net rate of outflow of momentum from the closed surface bounding that volume'. This is known as the *momentum theorem*.

As an example, consider fluid passing across a turbine in a pipe. In Fig. 2.3, fluid flowing at speed u_1 into the left of the control surface carries momentum $\rho u_1 \hat{x}$ per unit volume, where $\hat{x}$ is the unit vector in the direction of flow. In time Δt, the

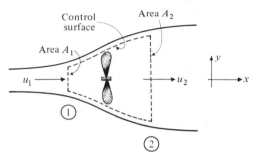

Fig. 2.3 A turbine in a pipe. The dotted line shows the control surface over which the momentum theorem is applied

volume entering the surface is $A_1 u_1 \Delta t$. Therefore the *rate* at which the momentum is entering the control surface is

$$(A_1 u_1 \Delta t)(\rho u_1 \hat{x})/\Delta t = \rho A_1 u_1^2 \hat{x}$$

Similarly the rate at which momentum is leaving the control volume is $\rho A_2 u_2^2 \hat{x}$. The net rate of outflow of momentum is

$$\mathbf{F}_1 = \rho (A_2 u_2^2 - A_1 u_1^2)\hat{x} = (\dot{m} u_2 - \dot{m} u_1)\hat{x} \qquad (2.7)$$

where $\dot{m} = \rho A_1 u_1 = \rho A_2 u_2$ is the mass flow. The momentum theorem tells us that $\mathbf{F}_1$ is the force on the fluid and so, by Newton's third law, $-\mathbf{F}_1$ is the force exerted on the turbine and pipe by the fluid. Normally $u_2 < u_1$, so that $\mathbf{F}_1$ points in the negative x direction and $-\mathbf{F}_1$ (i.e. the force on the turbine) is in the direction of the flow, as expected.

This illustrates two points to note in applying the momentum theorem: (1) Momentum is a *vector*, (2) Expressions for flow of momentum (e.g. $\rho A_1 u_1^2 \hat{x}$) typically involve *products* of velocity.

2.4 Viscosity

Suppose we have two parallel plates with fluid filling the gap between them, and the top plate moving at a velocity u_1 relative to the bottom one, as shown in Fig. 2.4. We choose axes as shown, with x in the direction of motion, and y across the gap between the plates. It is found experimentally that *fluid does not slip at a solid surface*. Thus the fluid immediately adjacent to each plate moves with the same velocity as the plate.

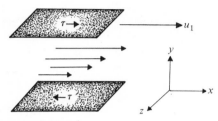

Fig. 2.4 Flow between two parallel plates

Now the molecules of the fluid are in random (thermal) motion, and the effect of this is to carry fluid with high x velocity (acquired from the top plate) downwards, and fluid with low x velocity (acquired from the bottom plate) upwards. This diffusion of momentum limits the velocity gradient that the fluid can sustain, i.e. it gives rise to an internal friction opposing the horizontal slip in the flow.

It is found that the shear stress (i.e. the force per unit area, in the direction indicated in Fig. 2.4) is

$$\tau = \mu(\partial u/\partial y) \tag{2.8}$$

where μ is a property of the fluid called its *dynamic viscosity* (unit $N\,s\,m^{-2}$). This viscosity is independent of τ and $\partial u/\partial y$; it depends only on the composition and temperature of the fluid.

A closely related property of the fluid is the *kinematic viscosity*

$$\nu = \mu/\rho \tag{2.9}$$

In incompressible fluids the flow pattern often depends more directly on ν than on μ. By combining (2.8) and (2.9) we find that the units of ν are

$$\frac{(kg\,m\,s^{-2})m^{-2}}{kg\,m^{-3}}\,\frac{m}{m\,s^{-1}} = m^2 s^{-1}$$

Thus ν has the character of a diffusivity. That is to say, the time taken for a change in momentum to diffuse a distance x is x^2/ν (cf. thermal diffusivity κ defined in Section 3.3). Typical values of ν are given in Appendix B.

2.5 Turbulence

Turbulent flow motions arise because, in general, rapid fluid motion is unstable. Suppose fluid is flowing through a pipe in the orderly manner illustrated by the pathlines in Fig. 2.5(a) and something disturbs the motion (e.g. a knock on the pipe). Fluid particles will alter course and, if they are moving rapidly enough, fluid friction will not be able to restore them to their original paths. Moreover the disturbed particles disturb other particles of fluid from their original path, and soon the entire flow is in the semi-chaotic state called *turbulence*, illustrated in Fig. 2.5(b).

Thus it is the ratio of fluid momentum ('inertia forces') to viscous friction which determines whether the flow is smooth (i.e. laminar) or turbulent. This ratio is usually measured by the *Reynolds number*.

$$\mathcal{R} = uX/\nu \tag{2.10}$$

in which u is the mean velocity of the flow, X is a characteristic length of the system (in this case the diameter of the pipe), and ν is the kinematic viscosity of the fluid. It is found experimentally that flow in a pipe is turbulent if $\mathcal{R} \gtrsim 2300$.

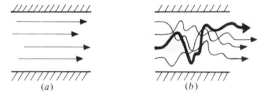

Fig. 2.5 Pathlines of flow in a pipe (a) laminar (b) turbulent

In turbulent flow large, more or less random fluctuations in velocity are imposed on the mean flow. These velocity fluctuations are three dimensional and not just variations in the flow speed. Thus a blob of fluid in the pipe, while moving along the pipe, also moves rapidly inwards and outwards across the pipe, as illustrated by Fig. 2.5(b). Since fluid does not slip at the pipe surface (Section 2.4), the mean speed near the surface is low and the mean speed near the center of the pipe is correspondingly high. Therefore the effect of the sideways motions of the fluid is to carry fluid of high velocity outwards, and fluid of low velocity inwards. This transfer of momentum by blobs of fluid is much more effective than the corresponding transfer by molecular motions described in Section 2.4, because a blob of fluid may move all the way across the pipe in a single jump, whereas the mean free path of a molecule is of the order of nanometers.

This transfer of momentum from the fluid to the walls constitutes a sizeable friction force opposing the motion of the fluid.

If the walls of the pipe are hotter than the incoming fluid, then these rapid inward and outward motions transfer heat rapidly to the bulk of the fluid. A blob of cold fluid can jump from the center of the pipe, pick up heat by conduction from the hot wall, and then carry it much more rapidly back into the center of the pipe than could molecular conduction. This mechanism of heat transfer is discussed in more detail in Section 3.4.

2.6 Friction in pipe flow

Suppose we have a pipe of length L and uniform diameter D. What pressure Δp has to be maintained between the ends to overcome fluid friction in the pipe, so that the fluid will flow through the pipe at a mean speed u?

Effectively, Δp is the work done against fluid friction in moving a unit mass of fluid through the length L. Since the flow is statistically uniform along the pipe, each meter of pipe will contribute the same friction. Therefore Δp increases with L. Since much of the resistance to the flow originates from the no-slip condition at the walls (Section 2.4), moving the walls closer to the bulk of the fluid will increase the friction. Therefore Δp increases as D decreases. Equation (2.8) implies that fluid friction increases with flow speed, so that Δp increases with u. Bernoulli's equation (2.3) shows that the quantity $\frac{1}{2}\rho u^2$ has the same dimensions as Δp. All these facts can be expressed in the single equation

$$\Delta p = 2f(L/D)(\rho u^2) \tag{2.11}$$

in which f is a dimensionless pipe *friction coefficient* that changes with experimental conditions.

Since f is dimensionless it can depend only on the pattern of flow, i.e. the shape of the streamlines. This is because the factor $\frac{1}{2}\rho u^2$ in (2.11) represents a natural unit of pressure drop in the pipe. The factor f tells what *proportion* of the kinetic energy ($\frac{1}{2}\rho u^2$) entering unit area of the pipe has to be applied as external work (Δp) to overcome frictional forces. This will depend on the time that a typical fluid particle spends in contact with the pipe wall, expressed as a *proportion* of the time the particle takes to move a unit length along the pipe. This proportion is much larger for the turbulent paths (Fig. 2.5(b)) than for the laminar paths (Fig. 2.5(a)).

Fluid flow depends mainly on the dimensionless Reynolds number (2.10). A plot of f against $\mathcal{R}$ should give a single curve applying to pipes of any length and diameter, carrying any fluid at any speed. There is no particular reason why this curve should be a straight line, or even continuous. Indeed we might expect a discontinuity at $\mathcal{R} \approx 2000$, where the flow pattern changes from laminar to turbulent.

It is found experimentally that friction data for pipes follow a single curve only if the pipes are *smooth* inside. This curve, shown in Fig. 2.6, does indeed have a discontinuity at $\mathcal{R} \approx 2000$.

If we consider real pipes with rough walls it is reasonable to suppose that the pattern of the flow will depend on the ratio of the height ξ of the surface bumps to the diameter of the pipe D. Plotting the experimental data on this basis, we obtain a series of curves in Fig. 2.6 with one curve for each roughness ratio ξ/D.

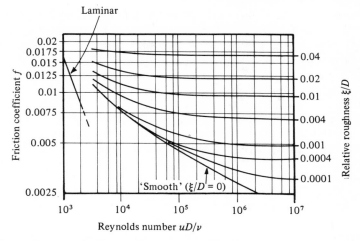

Fig. 2.6 Friction coefficient f for pipe flow (see (2.11))

Provided the appropriate value of ξ is used, these curves give a reasonable estimate of pipe friction. Typical values of ξ are given in Table 2.1, but it should be realized that the roughness of a pipe tends to increase with age.

Table 2.1 Approximate pipe roughness ξ

Material	ξ/mm
PVC	0 ('smooth')
Asbestos cement	0.012
New steel	0.1
'Smooth' concrete	0.4

Example 2.1
What is the pressure head required to force $0.1\,\mathrm{m^3\,s^{-1}}$ of water through a concrete pipe of length $200\,\mathrm{m}$ and diameter $0.3\,\mathrm{m}$, at $20°\mathrm{C}$?

Solution
Mean speed

$$u = Q/A = \frac{0.1\,\mathrm{m^3\,s^{-1}}}{\pi(0.15\,\mathrm{m})^2} = 1.4\,\mathrm{m\,s^{-1}}$$

From (2.10), the Reynolds number

$$\mathscr{R} = \frac{uD}{\nu} = \frac{(1.4\,\mathrm{m\,s^{-1}})(0.3\,\mathrm{m})}{1.0 \times 10^{-6}\,\mathrm{m^2\,s^{-1}}} = 0.4 \times 10^6 \geq 2000$$

Therefore flow is turbulent.

For concrete (from Table 2.1), $\xi = 0.4\,\text{mm}$. Thus the ratio

$$\xi/D = \frac{0.4\,\text{mm}}{300\,\text{mm}} = 0.0013$$

For this $\mathcal{R}$ and ξ/D, Fig. 2.6 gives

$$f = 0.0050$$

Expressing (2.11) in terms of the head loss due to friction,

$$H_f = \Delta p/\rho g = 2fLu^2/Dg \tag{2.12}$$

Hence

$$H_f = \frac{(2)(5.0 \times 10^{-3})(200\,\text{m})(1.4\,\text{m}\,\text{s}^{-1})^2}{(0.3\,\text{m})(9.8\,\text{m}\,\text{s}^{-2})}$$

$$= 1.3\,\text{m}$$

Note that the units used in such a calculation must always be consistent (e.g. SI).

Fig. 2.6 shows only one curve for $\mathcal{R} < 2000$, indicating that the friction coefficient f is independent of pipe roughness ξ in this range. This is because the flow is laminar, and the bumps hardly disturb the smoothness of the flow. In this laminar case the pressure drop Δp can be explicitly calculated from (2.8) for viscous shear stress, as indicated in Problem 2.4. The corresponding expression for the friction coefficient is

$$f = 16v/(uD) \quad \text{(laminar)} \tag{2.13}$$

Problems

2.1 Fig. 2.7(a) shows an ideal *Venturi meter* for measuring flow in a pipe.

(a) Use the equations expressing conservation of mass and conservation of energy to show that the volume of fluid flowing past cross-section 1 of the pipe per unit time is

$$Q = u_1 A_1 = A_1 \{(A_1/A_2)^2 - 1\}^{-1/2} \left\{ 2g \left[\frac{p_1 - p_2}{\rho g} + (z_1 - z_2) \right] \right\}^{1/2}$$

(b) What is the volume per second flowing past cross-section 2 in Fig. 2.7(a)?

(c) The pressures p_1 and p_2 are measured by the heights of columns of fluid rising up the side tappings, as shown in Fig. 2.7(a). Indicate on the diagram a distance corresponding to the factor in square brackets in the expression in Problem 2.1(a).

(d) Fig. 2.7(b) shows a similar system for measuring flow in a pipe of nominally uniform cross-section. A plate with a sharp edged hole in it is inserted as shown, and the pressure is measured on either side. Will the flow rate calculated from the equation of Problem 2.1(a) be correct, too high or too low? Why?
Hint: consider the full energy equation (2.1).

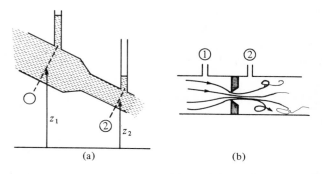

(a)   (b)

Fig. 2.7 For Problem 2.1. Venturi meters for measuring flow in a pipe. (a) Smooth contraction in diameter along the pipe. (b) Sharp edged orifice

2.2 A two-dimensional nozzle discharges water as a plane horizontal jet of thickness t and width b. The jet then strikes a large inclined flat surface as shown in Fig. 2.8.

(a) Neglecting the effects of gravity and viscosity, apply Bernoulli's equation to show that $u_1 = u_2 = u_j$.
Hint: the pressure change across a thin layer is negligible.

(b) What is the component parallel to the plane of the force acting on the fluid? By considering the change in momentum of the fluid in a suitable control volume, and applying conservation of matter, show that the flow rates up and down are, respectively,

$$Q_1 = \tfrac{1}{2}(1 + \cos\alpha)Q_j$$
$$Q_2 = \tfrac{1}{2}(1 - \cos\alpha)Q_j$$

(c) Find also an expression for the force on the plane, and evaluate it for the case $b = 10\,\text{cm}$, $t = 1\,\text{cm}$, $Q_j = 10\,\text{ls}^{-1}$, $\alpha = 60°$.

2.3 Consider a steady *laminar flow between two fixed parallel plates* at $y = 0$ and $y = D$ (cf. Fig. 2.4). The fluid is pushed to the right (x increasing) by a constant pressure gradient $\partial p/\partial x$ (<0).

(a) What are the forces acting on an element of fluid of length Δx and width Δz, lying between y and $y + \Delta y$?
     Show that the balance of forces on this fluid element is given by

$$\frac{\partial}{\partial y}\left(\mu\,\frac{\partial u}{\partial y}\right) = \frac{\partial p}{\partial x}$$

(b) By integrating the last expression, show that the velocity at a distance y above the lower plate is

$$u(y) = \frac{y}{2\mu}\left(\frac{\partial p}{\partial x}\right)(D - y)$$

(c) The plates have width $W \gg D$, so that edge effects are negligible. Show that the volumetric flow rate between the plates is

$$Q = \frac{1}{12}W D^3\left|\frac{\partial p}{\partial x}\right|$$

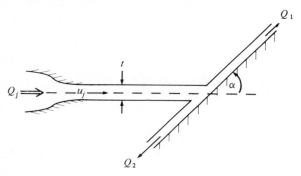

Fig. 2.8 For Problem 2.2. A plane jet strikes an inclined wall

2.4 *Laminar flow in a pipe*

A constant pressure gradient $(\partial p/\partial x)$ between the ends pushes fluid straight down a circular pipe of diameter $D = 2R$. Let x be the distance along the pipe and r the distance from the axis.

(a) As in Problem 2.3, show that the balance of forces on an annulus of length Δx lying between r and $r + \Delta r$ leads to the equation

$$\frac{\partial}{\partial r}\left(r\mu\,\frac{\partial u}{\partial r}\right) = -r\,\frac{\partial p}{\partial x}$$

Hint: the shear force at r is $(2\pi r\,\Delta x)\tau(r)$, and r varies across the pipe.

(b) By integration, show that the velocity at a distance r from the axis is

$$u(r) = \frac{1}{4\mu}\left(\frac{\partial p}{\partial x}\right)(R^2 - r^2)$$

(c) Hence show that the volume of fluid flowing out of the pipe per unit time is

$$Q = \frac{\pi a^4}{8\mu}\left(\frac{\partial p}{\partial x}\right)$$

(d) The mean velocity in the pipe is $\bar{u} = Q/A$. By putting the results so far into (2.11) show that $\bar{u}f = 16\nu/D$, thus verifying (2.13).

2.5 Having been informed of the results of Example 2.1, the accountant of the engineering firm concerned suggests that it would be cheaper to use narrow PVC pipe to carry the flow.

(a) Disillusion the accountant by calculating the theoretical friction head incurred in passing $0.1\,\mathrm{m^3 s^{-1}}$ of water through 200 m of PVC pipe of diameter 5 cm, and show that it exceeds the available head.
(b) If gravity is the only force available to move the water, calculate an upper limit to the flow which could in fact be pushed through this pipe. *Hint:* take $H_f = L$ (vertical pipe), estimate f, and calculate the corresponding u.

2.6 A steel pipe of diameter D and length L is to carry a flow Q. Assuming that the pipe friction coefficient f varies only slowly with Reynolds number, show that the head loss due to friction is proportional to D^{-5} (for fixed L and Q).

Solutions

2.1 (a) Mass: $u_2 = u_1 A_1/A_2$. Bernoulli equation (2.2) as written. Simple algebra.
(b) Q
(c) Geometric difference in levels.
(d) Calculated flow too high because some kinetic energy at 1 is degraded to turbulence and heat at 2.

2.2 (a) All p equal in (2.4) and $g = 0$.
(b) Zero, because without viscosity fluid cannot apply shear force. Mass reduces to $A_j = A_1 + A_2$. Momentum (*vector*):

$$\rho A_j u_j^2 \cos\alpha = \rho A_1 u_1^2 - \rho A_2 u_2^2$$

Simple algebra.
(c) $F = \rho Q_j u_j \sin\alpha = 86\,\mathrm{N}$, pushing plane down and to the right.

2.3 (a) Difference in pressure forces at x, $x + \Delta x$ balances difference between forward (for $y < \frac{1}{2}D$) shear force on top and backward shear force on bottom. Use (2.8).
(b) $u = 0$ at $y = 0$, $y = D$. Simple algebra.
(c) $Q = \int u(W\,dy)$

2.4 (a) $[(\partial p/\partial x)\,\Delta x][\pi(r+\Delta r)^2 - \pi r^2] = \tau(r)\,2\pi r\,\Delta x - \tau(r+\Delta r)\,2\pi(r+\Delta r)\,\Delta x$
    (b) Symmetry: $\partial u/\partial r = 0$ at $r = 0$, since $u = 0$ at $r = R$.
    (c) $Q = \int u\,2\pi r\,dr$
    (d) Simple algebra.

2.5 (a) Following Example 2.1 gives $H_f = 12\,\text{km} \gg L$.
    (b) $f = 0.006$ (more or less independent of $\mathcal{R}$ in this range) gives $u = 4.4\,\text{m s}^{-1}$, $Q = 8\,\text{ls}^{-1}$.

2.6 Note $u = Q/(\tfrac{1}{4}\pi D^2)$.

Bibliography

The following selection from the many books on fluid mechanics may prove useful. There are many other good books besides those listed. For work on turbomachinery books written for engineers are usually more useful than those written for mathematicians, who often ignore friction and forces.

Batchelor, G. K. (1967) *An Introduction to Fluid Dynamics*, Cambridge University Press.
    A most precise statement of the foundations (see especially Chapter 3), with many examples. Repays careful reading, but perhaps unsuitable for beginners.
Francis, J. R. (1974) *A Textbook of Fluid Mechanics*, 4th edn, Edward Arnold, London.
    Clear writing makes easy reading for beginners. More engineering detail than Kay and Nedderman (1974).
Hughes, W. F. and Brighton, J. A. (1967) *Theory and Problems of Fluid Dynamics*, Schaum's Outline Series.
    Multitude of worked examples, but beware of units.
Kay, J. M. and Nedderman, R. M. (1974) *Fluid Mechanics and Heat Transfer*, 3rd edn, Cambridge University Press.
    A concise and wide ranging introduction.
Webber, N. (1971) *Fluid Mechanics for Civil Engineers*, Chapman and Hall, London.
    Delightfully simple but useful introduction for students with little knowledge of physics or engineering.

3 Heat transfer

3.1 Introduction

With direct solar, geothermal and biomass sources, most energy transfer is by heat rather than by mechanical or electrical processes. Heat transfer is a well-established subject, and we shall not go into the detail of specialized engineering texts because such detail is rarely needed in moderate or small scale renewable energy applications. For instance temperature differences are often smaller, geometric configurations less complicated and (most importantly) energy fluxes much lower.

This book uses a unified approach to heat transfer processes, so that several interrelated processes can be analyzed as one 'heat circuit'. For instance the solar heater of Fig. 3.1 receives heat by solar radiation at about 1kW m^{-2} maximum intensity, producing surfaces about 50°C higher than the environment. Heat is lost from these surfaces by long wavelength radiation, by conduction and by convection. The useful heat is removed by mass transport. Our recommended method of analysis is to set up a heat transfer circuit of the interconnected processes (e.g. Fig. 3.2(c)) and calculate each transfer process to

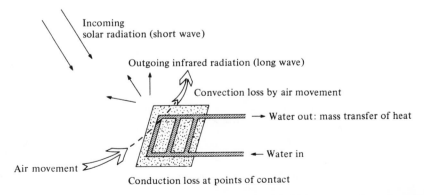

Fig. 3.1 Solar heater (absorbing plate fitted with water pipes): demonstrating types of heat transfer

an accuracy of about 50%. At this stage insignificant processes can be neglected, and then important transfers can be analyzed to greater accuracy. Even so, it is unlikely that final accuracies will be better than $\pm 10\%$ of actual performance.

This chapter is supportive to subsequent chapters on individual renewable energy systems. The main formulas needed for practical calculations are summarized in Appendix C.

3.2 Heat circuit analysis and terminology

We introduce our method of heat transfer and circuit analysis with a simple example (which will in reality be more complicated). At night a large tank of hot water stands in a cool enclosed room with an even colder environment outside, so the direction of net heat transfer is down a temperature gradient from the (hot) tank to the (cold) outside environment (Fig. 3.2(a)). For simplicity, the floor and ceiling are so well insulated that heat leaks only through the walls. Heat is transferred from the tank by radiation and convection to the room walls, by conduction through the walls, and then by radiation and convection to the environment (Fig. 3.2(b)). This complex transfer of parallel and series connections is described in the heat circuit of Fig. 3.2(c).

Each process can be described by an equation of the form

$$P_{ij} = (T_i - T_j)/R_{ij} \tag{3.1}$$

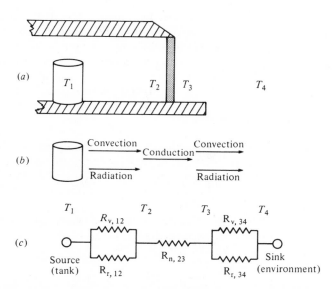

Fig. 3.2 Example of a heat circuit. (a) Physical situation. A hot tank is in a cool room with cold air outside. The roof and the floor are well insulated. T_1, T_2, T_3 and T_4 are the temperatures of the tank surface, the wall surfaces and the outside environment respectively. (b) Energy flow mechanisms. (c) Analog circuit.

where the power P_{ij} is the *heat flow* between surfaces at temperatures T_i (hotter) and T_j (colder), and R_{ij} is called the *thermal resistance*. (See Appendix A for units.)

In general R_{ij} is *not* constant with respect to the temperatures, and may well change rapidly and nonlinearly with change in surface temperature and temperature difference. However for our purposes the heat flows P_{ij} will depend most strongly on the term $(T_i - T_j)$, and the variation of R_{ij} with temperature is always weaker. Thus the concept of thermal resistance is useful. If the direction of heat flow is obvious, then (3.1) may be written

$$P_{ij} = \Delta T / R_{ij} \qquad (3.2)$$

R is called a 'resistance' by analogy with Ohm's law in electricity.

The thermal resistance method allows each step of a complex of heat transfers to be added together as a set of series and parallel connections. In our example of Fig. 3.2

$$P_{14} = (T_1 - T_4)/R_{14} \qquad (3.3)$$

where

$$R_{14} = R_{12} + R_{23} + R_{34}$$

and

$$\frac{1}{R_{12}} = \frac{1}{R_{12}\,(\text{convection})} + \frac{1}{R_{12}\,(\text{radiation})}$$

$$R_{23} = R_{23}\,(\text{conduction})$$

$$\frac{1}{R_{34}} = \frac{1}{R_{34}\,(\text{convection})} + \frac{1}{R_{34}\,(\text{radiation})}$$

Even if we use only approximate values of the temperatures, we shall find that the individual resistances can be calculated to obtain the overall resistance R_{14}. The heat loss from the tank P_{14} is now calculated in terms of T_1 and T_4 only. This simplification, together with the diagrammatic quantification of the heat flows, makes thermal resistance a powerful concept.

For general considerations it is often useful to consider heat flow q per unit area of surface. Then across a surface of area A,

$$q = \Delta T / r \qquad (3.4)$$

$$P = qA = \Delta T / (r/A) \qquad (3.5)$$

so

$$R = r/A \ (\text{unit of } R \text{ is } K\,W^{-1})$$

and

$$r = RA \text{ (unit of } r \text{ is m}^2 \text{ K W}^{-1})$$ (3.6)

Here we call r the *thermal resistivity* of unit area. (r is *not* the resistance *per* unit area, since we multiply R by the area to calculate r, as with electrical resistivity.)

A common expression for the heat flow per unit area is

$$q = h \Delta T$$ (3.7)

where h is the *heat transfer coefficient* $(\text{W m}^{-2} \text{K}^{-1})$.

Note that

$$h = 1/r$$ (3.8)

In this text we use the following subscripts on R, r and h to distinguish the various heat transfer mechanisms: R_n for conduction, R_v for convection, R_r for radiation and R_m for mass transfer.

3.3 Conduction

Thermal conduction is the transfer of heat by the vibrations of atoms, molecules and electrons without bulk movement. It is the only mechanism of heat transfer in opaque solids, but transparent media also pass heat energy by radiation. Conduction occurs in liquids and gases, but is usually dominated by convection, in which heat is carried by the fluid circulating or moving in bulk.

Consider the heat flow P by conduction through a slab of material, area A, thickness Δx, surface temperature difference ΔT:

$$P = -kA \Delta T / \Delta x$$ (3.9)

k is the *thermal conductivity* $(\text{unit W m}^{-1} \text{K}^{-1})$, and the negative sign indicates that heat flows in the direction of decreasing temperature. By comparison of (3.9) and (3.2), the thermal resistance of conduction is

$$R_n = \frac{\Delta x}{kA}$$ (3.10)

and the corresponding thermal resistivity of unit area is

$$r_n = R_n A = \Delta x / k$$ (3.11)

The thermal conductivity of a dry solid is constant over a wide range of temperature, and so the thermal resistance R_n of dry opaque solids may be considered constant. This is in marked contrast with liquids, gases and vapors, whose thermal resistance varies distinctly with temperature owing to convection. Values of thermal conductivity are tabulated in Appendix B.

Note that the thermal conductance (often called the *U* value of a material) is given by

$$U = 1/r \tag{3.11'}$$

Example 3.1
Some values of conductive thermal resistance

(1) $1\,m^2$ of window glass:

$$R_n = \frac{5\,mm}{(1\,W\,m^{-1}K^{-1})(1\,m^2)} = 0.005\,K\,W^{-1}$$

(2) $5\,m^2$ of the same glass:

$$R_n = 0.001\,K\,W^{-1}$$

(3) $1\,m^2$ of continuous brick wall 220 mm thick:

$$R_n = \frac{220\,mm}{(1\,W\,m^{-1}K^{-1})(1\,m^2)} = 0.022\,K\,W^{-1}$$

and $U = 45\,W\,m^{-2}K^{-1}$
(4) $1\,m^2$ of loosely packed glass fibers 80 mm deep as used for ceiling insulation:

$$R_n = \frac{80\,mm}{(0.04\,W\,m^{-1}K^{-1})(1\,m^2)} = 2\,K\,W^{-1}$$

and $U = 5\,W\,m^{-2}K^{-1}$.

Note the following:

(1) The conductive resistance of a glass sheet is much less than the overall resistance of a window made from that glass sheet, because of convection in the air near the window (see Problem 3.7).
(2) Since the thermal conductivity of metals is high ($k \sim 100\,W\,m^{-1}K^{-1}$), the conductive thermal resistance of metal components in series with other components is often negligible.
(3) Loosely packed glass fibers have a much higher thermal resistance than pure glass sheet, because the packed fibers incorporate small pockets of still air. *Still air* is one of the best insulators available ($k \approx 0.03\,W\,m^{-1}K^{-1}$), and all natural and commercial insulating materials rely on it. The thermal resistance of such materials drops drastically if the material becomes wet or if the air pockets are too big (in which case the air carries heat by convection).
(4) There are two mechanisms that lower the apparent conductive thermal resistance of wet or damp materials:

(a) Liquid water provides a thermal short circuit within the structure of the material. This is particularly so for biological material, e.g. wood.
(b) Evaporation may occur so that vapor diffuses within the structure to regions of condensation (see Section 3.7).

Another property closely related to the conductivity is the *thermal diffusivity* κ, which indicates how quickly changes in temperature diffuse through a material:

$$\kappa = k/\rho c \tag{3.12}$$

where ρ is the density and c is the specific heat capacity at constant pressure. κ has the unit of $m^2 s^{-1}$ as with kinematic viscosity ν (see (2.9)). The temperature will change quickly only if heat can move easily through the material (large k in the numerator of (3.12)) *and* if a small amount of heat produces a large temperature rise per unit volume (small ρc in the denominator).

It takes a time $\sim y^2/\kappa$ for a temperature rise to diffuse a distance y into a cold mass.

3.4 Convection

3.4.1 *Free and forced convection*

Convection is the transfer of heat to or from a moving fluid. Since the movement continually brings unheated fluid to the source or sink of heat, convection will always produce more rapid heat transfer than conduction through the stationary fluid.

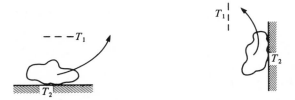

Fig. 3.3 Fluid movement by free convection, away from the hotter surface ($T_2 > T_2$)

In *free convection* (sometimes called *natural convection*) the movement is caused by the heat flow itself. Consider the volumes of fluid in contact with the hot surfaces of Fig. 3.3. Initially the fluid absorbs energy by conduction from the hot surface, and consequently the fluid density decreases by volume expansion. The heated portion then rises through the unheated fluid to a position of lower temperature, thereby carrying heat across the temperature gradient.

In *forced convection* the fluid is moved across a surface by an external agency such as a pump or wind. This movement is independent of the heat transfer. Obviously there are some conditions where convection is partly forced and partly free.

3.4.2 *Nusselt number* $\mathcal{N}$

The analysis of convection proceeds from a gross simplification of the processes. We imagine the fluid near the surface to be stationary. We then consider the heat flowing across an idealized boundary layer of stationary fluid of thickness δ and cross-sectional area A (Fig. 3.4). The temperatures across the fictitious boundary are T_f, the fluid temperature away from the surface, and T_s, the surface temperature. This being so, the heat transfer by conduction across the stationary fluid is given by

$$q = \frac{P}{A} = \frac{k(T_s - T_f)}{\delta} \tag{3.13}$$

where k is the thermal conductivity of the fluid.

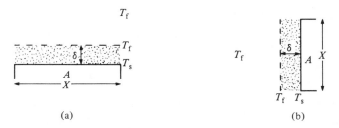

Fig. 3.4 Idealized thermal boundary layer in free convection. (a) Hot surface horizontal. (b) Hot surface vertical

As described here, δ is fictitious and cannot be measured. We can however measure X, a 'characteristic dimension' specified rather arbitrarily for each particular surface (see Fig. 3.4 and Appendix C).

From (3.13),

$$q = \frac{P}{A} = \frac{k(T_s - T_f)}{\delta} = \frac{X}{\delta} \frac{k(T_s - T_f)}{X} = \mathcal{N} \frac{k(T_s - T_f)}{X} \tag{3.14}$$

$\mathcal{N}$ is the *Nusselt number* for the particular circumstance. It is a dimensionless scaling factor, useful for all bodies of the same shape in equivalent conditions of fluid flow. Tables of values of $\mathcal{N}$ are available for specified conditions, with the appropriate characteristic dimension identified (Appendix C).

From Section 3.2 it follows that:

thermal resistance of convection $\qquad R_v = X/\mathcal{N}kA \tag{3.15}$

convective thermal resistivity of unit area $\quad r_v = R_vA = X/\mathcal{N}k \tag{3.16}$

convective heat transfer coefficient $\qquad h_v = 1/r_v = \mathcal{N}k/X \tag{3.17}$

The amount of heat transferred by convection will depend on three factors:

(1) The properties of the fluid
(2) The speed of the flow
(3) The shape and size of the surface.

The Nusselt number $\mathcal{N}$ is a dimensionless measure of the heat transfer. Therefore it can depend only on dimensionless measures of the three factors listed. In choosing these measures, it is convenient to separate the cases of forced and free convection.

3.4.3 *Forced convection*

For a given shape of surface, a nondimensional measure of the speed of the flow is the *Reynolds number*:

$$\mathcal{R} = uX/\nu \tag{3.18}$$

We saw in Section 2.5 that $\mathcal{R}$ determines the pattern of the flow, and in particular whether it is laminar or turbulent. In flow over a flat plate (Fig. 3.5) turbulence occurs when $\mathcal{R} \geq 3 \times 10^5$, with subsequent increase in the heat transfer because of the perpendicular motions involved.

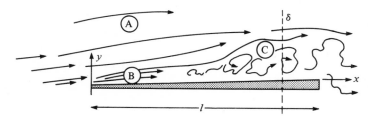

Fig. 3.5 Fluid flow over a hot plate. General view of pathlines, showing regions (A) well away from the surface (B) laminar flow near the leading edge (C) turbulent flow in the downstream region

The flow of heat into or from a fluid depends on the thermal diffusivity κ of the fluid. It will also depend on the kinematic viscosity ν (which may be considered to be the diffusivity of momentum), since ν affects the Reynolds number and thus the character of the flow. These are the only two properties of the fluid that influence the Nusselt number in forced convection, since the separate effects of k, ρ, and c are combined in κ.

A nondimensional measure of the properties of the fluid is the *Prandtl number*:

$$\mathcal{P} = \nu/\kappa \tag{3.19}$$

If $\mathcal{P}$ is large, changes in momentum diffuse more quickly through the fluid than do changes in temperature. Appendix B shows that many common fluids have $\mathcal{P} \sim 1$.

Thus, for each shape of surface, the heat transfer by forced convection can be expressed in the form

$$\mathcal{N} = \mathcal{N}(\mathcal{R}, \mathcal{P}) \tag{3.20}$$

That is, for each shape, the Nusselt number $\mathcal{N}$ is a function only of the Reynolds number and the Prandtl number. (These formulas are sometimes expressed instead in terms of other closely related dimensionless parameters, e.g. the Stanton number $\mathcal{N}/\mathcal{R}\mathcal{P}$ and the Péclet number $\mathcal{R}/\mathcal{P}$. Neither of these are used in this book.) The precise value of $\mathcal{N}$ has to be determined by experiment.

Formulas summarizing these experimental data are given in Appendix C, and an example of their use is given in Section 3.4.5. It is important to realize that these formulas are mostly only accurate to $\pm 10\%$, partly because they are approximations to the experimental conditions, and partly because the experimental data themselves usually contain both random and systematic errors.

3.4.4 *Free convection*

In free convection, the fluid flow speed depends on the heat transfer, rather than vice versa. We replace (3.20) by

$$\mathcal{N} = \mathcal{N}(\mathcal{A}, \mathcal{P}) \tag{3.21}$$

where the *Rayleigh number*

$$\mathcal{A} = \frac{g\beta X^3 \Delta T}{\kappa \nu} \tag{3.22}$$

is essentially a dimensionless measure of the driving temperature difference ΔT. g is the acceleration due to gravity, β is the coefficient of thermal expansion and the other symbols are as before.

The Rayleigh number expresses the balance of processes shown in Fig. 3.6. Heated fluid is driven upwards by a buoyancy force proportional to $g\beta\Delta T$, and retarded by a viscous force proportional to ν. It loses excess temperature (and therefore buoyancy) at a rate proportional to κ. Therefore the vigor of convection increases with $g\beta\Delta T/\kappa\nu$, i.e. with $\mathcal{A}$. The factor X^3 is inserted to make this ratio dimensionless. It is found experimentally that free convection is nonexistent if $\mathcal{A} \lesssim 10^3$ and is turbulent if $\mathcal{A} \gtrsim 10^5$.

This argument shows that the Nusselt number in free convection depends mainly on the Rayleigh number. The dimensional argument used in Section 3.4.3 suggests that $\mathcal{N}$ may also depend on the Prandtl number, as indicated in (3.21). Formulas to calculate these Nusselt numbers are given in Appendix C for various geometries. These cannot be expected to give better than 10% accuracy. (In some engineering texts, these formulas are expressed in terms of the Grashof number $\mathcal{G} = \mathcal{A}/\mathcal{P}$. Since $\mathcal{G}$ has less direct physical significance than $\mathcal{A}$, it will not be used in this text.)

Note that the Nusselt number (and thus the thermal resistance) in free convection depends on ΔT, through the dependence on $\mathscr{A}$. This is because a bigger temperature difference drives a stronger flow, which transfers heat more efficiently. By contrast, in forced convection the Nusselt number and thermal resistance are virtually independent of ΔT.

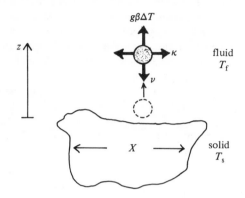

Fig. 3.6 Schematic diagram of a blob of fluid moving upward in free convection. It is subject to an upward buoyancy force, a retarding viscous force, and a sideways temperature loss

3.4.5 *Calculation of convective heat transfer*

Because of the complexity of fluid flows, there is no purely analytical way of calculating convective heat transfer. Instead we have to use the results of experiments on geometrically similar objects. By expressing these results in nondimensional form, they can be applied to different sizes of objects and for different fluids. A brief collection of working formulas is given in Appendix C; much more extensive collections are given in textbooks of heat transfer.

Thus a systematic procedure for calculating convective heat transfer is as follows:

(1) Draw a diagram of the heated object.
(2) Section the diagram into standard geometries (i.e. parts corresponding to the illustrations in Appendix C).
(3) For each such section:

   (a) Identify the characteristic dimensions (X).
   (b) Calculate the Reynolds number $\mathscr{R}$ and/or the Rayleigh number $\mathscr{A}$ for each section of the object.
   (c) Choose the formula for $\mathscr{N}$ from tables appropriate to that range of $\mathscr{R}$ or $\mathscr{A}$. (The different formulas usually correspond to laminar or turbulent flow.)
   (d) Calculate the Nusselt number $\mathscr{N}$ and hence the heat flow $P = qA$.

(4) Add the heat flows from each section to obtain the total heat flow from the object.
(5) If data give the Grashof number $\mathscr{G}$, then $\mathscr{A} = \mathscr{G}\mathscr{P}$.

Example 3.2   Free convection between parallel plates
Two flat plates each $1\,m \times 1\,m$ are separated by $3\,cm$ of air. The lower is at $70\,°C$ and the upper at $45\,°C$. The edges are sealed together by insulating material acting as walls to prevent air movement beyond the plates. Calculate the convective thermal resistance of unit area and the heat flux between the top and the bottom plate.

Solution
Fig. 3.7 corresponds to the standard geometry given for (C.7) in Appendix C. Since the edges are sealed, no outside air can enter between the plates and only free convection occurs. Using (3.22) and Table B.1 in Appendix B, for mean temperature 57°C ($= 330\,K$)

$$\mathscr{A} = \frac{g\beta X^3 \Delta T}{\kappa \nu} = \frac{g\beta}{\kappa \nu} X^3 \Delta T$$

$$= \frac{(9.8\,m\,s^{-1})\,(1/330\,K)}{(2.6 \times 10^{-5}\,m^2\,s^{-1})(1.8 \times 10^{-5}\,m^2\,s^{-1})}(0.03\,m)^3\,(25\,K) = 4.1 \times 10^4$$

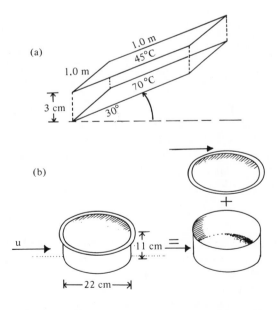

Fig. 3.7 Diagrams for worked examples on convection. (a) Parallel plates. Example 3.2 (b) Cooking pot with lid, Example 3.3

Using (C.7) a reasonable value for $\mathcal{N}$ can be obtained, although $\mathcal{A}$ is slightly less than 10^5:

$$\mathcal{N} = 0.062\,\mathcal{A}^{0.33} = 2.06$$

From (3.16),

$$r_v = \frac{X}{\mathcal{N}k} = \frac{0.03\,\text{m}}{(2.06)(0.028\,\text{m}^{-1}\text{K}^{-1})} = 0.52\,\text{K}\,\text{W}^{-1}\text{m}^2$$

From (3.5) the heat flux is

$$P = \frac{A\Delta T}{r} = \frac{(1\,\text{m}^2)(25\,\text{K})}{0.52\,\text{K}\,\text{W}^{-1}\text{m}^2} = 48\,\text{W}$$

Note the following:

(1) The factor $g\beta/\kappa\nu\ (=\mathcal{A}/X^3\Delta T)$ is tabulated in Appendix B.
(2) The fluid properties are evaluated at the mean temperature (57°C in this case).
(3) It is *essential* to use consistent units (e.g. SI) in evaluating dimensionless parameters like $\mathcal{A}$.

Example 3.3   Convective cooling of a cooking pot
A metal cooking pot with a shiny outside surface, of the dimensions shown in Fig. 3.7(b), is filled with food and water and sits on top of a stove. What is the minimum energy required to maintain it at boiling temperature for one hour (1) if it is sheltered from the wind (2) if it is exposed to a breeze of $3\,\text{m}\,\text{s}^{-1}$?

Solution
We assume that the lid is tight, so that there is no heat loss by evaporation. We also neglect heat loss by radiation: this is justified in Problem 3.4. Since the conductive resistance of the pot wall is negligible, the problem is then reduced to calculating the convective heat loss from the top and the sides of a cylinder with a surface temperature of 100°C. We shall consider the ambient (air) temperature to be 20°C. Therefore heat transfer properties of the air are evaluated at the mean temperature $\bar{T} = 60°\text{C}$.

(1) *Free convection alone.* For the top (using Table B.1 and (3.22)),

$$\mathcal{A} = (5.8 \times 10^7\,\text{m}^{-3}\text{K}^{-1})(0.22\,\text{m})^3(80\,\text{K}) = 4.9 \times 10^7$$

and from (C.2))

$$\mathcal{N} = 0.14\,\mathcal{A}^{0.33} = 48.4$$

and

$$P_{top} = Ak\mathcal{N}\Delta T/X$$

$$= (\pi/4)(0.22\,\text{m})^2(0.027\,\text{W}\,\text{m}^{-1}\text{K}^{-1})(48.4)(80\,\text{K})/(0.22\,\text{m}) = 18\,\text{W}$$

For the sides, $X = 0.11\,\text{m}$:

$$\mathcal{A}_{side} = \mathcal{A}_{top}(0.11\,\text{m}/0.22\,\text{m})^3 = 6.1 \times 10^6$$

and (from (C.5))

$$\mathcal{N} = 0.56\mathcal{A}^{0.25} = 27.8$$

so

$$P_{side} = \pi(0.22\,\text{m})(0.11\,\text{m})(0.027\,\text{W}\,\text{m}^{-1}\text{K}^{-1})(27.8)(80\,\text{K})/(0.11\,\text{m}) = 41\,\text{W}$$

Hence

$$P_{free} = P_{top} + P_{side} = 59\,\text{W}$$

and the energy required to make up this heat loss for 1 hour is $(0.059\,\text{kW})(3600\,\text{s}) = 0.21\,\text{MJ}$.

(2) *Forced plus free convection.* Here we calculate the forced convective power losses separately, and add them to those already calculated for free convection, in order to obtain an estimate of the total convective heat loss P_{total}.

For the top,

$$\mathcal{R} = (3\,\text{m}\,\text{s}^{-1})(0.22\,\text{m})/(1.9 \times 10^{-5}\,\text{m}^2\text{s}^{-1}) = 3.5 \times 10^4$$

which suggests the use of (C.8):

$$\mathcal{N} = 0.664\mathcal{R}^{0.5}\mathcal{P}^{0.33} = 110$$

So

$$P_{top} = Ak\mathcal{N}\Delta T/X = 42\,\text{W}$$

For the sides, as for the top,

$$\mathcal{R} = 3.5 \times 10^4$$

which suggests the use of (C.11):

$$\mathcal{N} = 0.26\mathcal{R}^{0.6}\mathcal{P}^{0.3} = 124$$

and

$$P_{\text{side}} = \pi(0.22\,\text{m})(0.11\,\text{m})(0.027\,\text{W}\,\text{m}^{-1}\text{K}^{-1})(124)(80\,\text{K})/(0.22\,\text{m}) = 93\,\text{W}$$

Hence

$$P_{\text{forced}} = 93 + 42 = 135\,\text{W}$$

The total estimate is

$$P_{\text{total}} = P_{\text{forced}} + P_{\text{free}} = 194\,\text{W} = 0.7\,\text{MJ}\,\text{h}^{-1}$$

The overall accuracy of these calculations may be no better than ±50%, although the individual formulas are better than this. This is because forced and free convection may both be significant, but their separate contributions do not simply add because the flow induced by free convection may oppose or reinforce the pre-existing flow. Similarly the flows around the 'separate' sections of the object interact with each other.

In Example 3.3 there is an additional confusion about whether the flow is laminar or turbulent. For example, on the top of the pot $\mathscr{A} > 10^5$, suggesting turbulence, but using the external flow speed to calculate $\mathscr{R}$ (as earlier) gives $\mathscr{R} < 10^5$, suggesting laminar flow. In practice such a flow would be turbulent, since it is difficult to smooth out streamlines which have become tangled by turbulence. The only safe way to accurately evaluate a convective heat transfer, allowing for all these interactions, is by experiment! Some formulas (such as (C.15)) based on such specialized experiments are available, but have a correspondingly narrow range of applicability.

3.5 Radiative heat transfer

3.5.1 *Introduction*

Surfaces emit energy by electromagnetic radiation according to fundamental laws of physics. Absorption of radiation is a closely related process. The literature and terminology concerning radiative heat transfer are confusing. Symbols and names for the same quantities vary, and the same symbol and name may be given for totally different quantities. Here we shall attempt to follow the recommendations of the International Solar Energy Society (1978).

In this chapter we consider radiative heat transfer in general. In Chapter 4 we shall consider solar radiation in particular, and in Chapters 5 and 6 solar energy heating devices.

3.5.2 *Radiant flux density (RFD)*

Radiation is energy transported by electromagnetic propagation through space or transparent media. Its properties relate to its wavelength λ and frequency $\nu = c/\lambda$, where c is the speed of light. The named regions of the spectrum are shown in Fig. 3.8. The flux of energy per unit area is the *radiant flux density* (abbreviation RFD, unit Wm^{-2}, symbol ϕ). The variation of RFD with wavelength is described by the *spectral RFD* (symbol ϕ_λ, unit $(Wm^{-2})m^{-1}$ or more usually $Wm^{-2}\mu m^{-1}$), which is simply the derivative $d\phi/d\lambda$. Thus $\phi_\lambda\Delta\lambda$ gives the power per unit area in a (narrow) wavelength range $\Delta\lambda$, and integration of ϕ_λ gives the total RFD $\phi = \int \phi_\lambda d\lambda$.

Some writers use special terms to distinguish between radiation coming *onto* a surface from that coming *away* from the surface. The former is usually called *irradiance*; the latter has a variety of names, none of which we shall use.

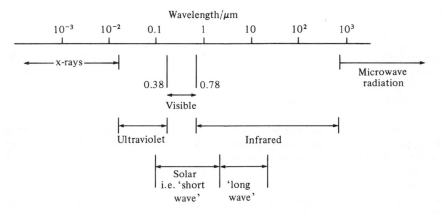

Fig. 3.8 Some of the named portions of the electromagnetic spectrum (the spectrum extends both 'longwards' and 'shortwards' from that shown)

It is obvious that radiation has directional properties, and that these need to be specified. Understanding is always helped by:

(1) Drawing pictures of the radiant fluxes and the methods of measurement
(2) Clarifying the units of the parameters.

Consider a small test instrument for measuring radiation parameters in an ideal manner. This could consist of a small, totally absorbing, black plane (Fig. 3.9) that can be adapted to (a) absorb on both sides, (b) absorb on one side only, (c) absorb from one direction only and (d) absorb from one three-dimensional solid angle only.

The energy ΔE absorbed in time Δt could be measured from the temperature rise of the plane of area ΔA knowing its thermal capacity. From Fig. 3.9(a) the radiant flux density from *all* directions would be $\phi = \Delta E/2\Delta A\Delta t$. In Fig. 3.9(b) the radiation is incident from the hemisphere above one side of the test plane (which may be labelled + or −), so

$$\phi = \frac{\Delta E}{\Delta A\Delta t} \tag{3.23}$$

In Fig. 3.9(c) a vector quantity is now measured where the direction of the radiation flux can be defined to be perpendicular to the receiving plane. In Fig. 3.9(d) the radiation flux is measured within a solid angle $\Delta\omega$ from a central direction perpendicular to the plane of measurement, and the unit of measurement will be $W\,m^{-2}\,sr^{-1}$.

The wavelength properties of the received radiation need not be specified for any of these measurements since the absorbing surface is assumed to be totally black. However if a dispersing device is placed in front of the instrument which

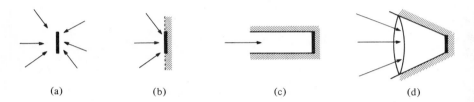

(a)   (b)   (c)   (d)

Fig. 3.9 Measurements of various radiation parameters using a small totally absorbing plane. (a) Absorbs all directions. (b) Absorbs from hemisphere above one side only. (c) Absorbs from one direction only. (d) Absorbs from one solid angle only

passes only a small range of wavelength from $\lambda - \Delta\lambda/2$ to $\lambda + \Delta\lambda/2$, then the spectral radiant flux density may be measured as

$$\phi_\lambda = \frac{\Delta E}{\Delta A\,\Delta t\,\Delta\lambda} \quad (\text{unit}\,W\,m^{-2}\,m^{-1}) \tag{3.24}$$

This quantity can also be given directional properties as with ϕ.

Difficulty may sometimes arise, especially regarding certain measuring instruments. There are two systems of units relating to the measurement of radiation quantities – photometric and radiometric units (Kaye and Laby, 1975). Photometric units have been established to quantify responses as recorded by the human eye, and relate to the SI unit of the candela. Radiometric units quantify total energy effects irrespective of visual response, and relate to the basic energy units of the joule and watt. For our purposes, only radiometric units need be used.

3.5.3 *Absorption, reflection and transmission of radiation*

Radiation incident on matter may be reflected, absorbed or transmitted (Fig. 3.10). These interactions will depend on the type of material, the surface properties, the wavelength of the radiation, and the angle of incidence θ. Normal incidence ($\theta = 0$) may be inferred if not otherwise mentioned, but at grazing incidence ($90° > \theta \geq 70°$) there are significant changes in the properties.

At wavelength λ, within wavelength interval $\Delta\lambda$, the *monochromatic absorptance* α_λ is the fraction absorbed of the incident flux density $\phi_\lambda\Delta\lambda$. Note that α_λ is a property of the surface alone. It depends, for example, on the energy levels of the atoms in the surface. It does *not* depend on the incident radiation: α_λ merely specifies what proportion of radiation at a particular wavelength would be absorbed if that wavelength was present in the incident radiation. Whether that wavelength is actually present is irrelevant. The subscript on α_λ, unlike that on ϕ_λ, does *not* indicate differentiation.

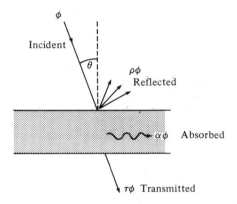

Fig. 3.10 Reflection, absorption and transmission of radiation (ϕ is the incident radiation flux density)

Similarly, we define the *monochromatic reflectance* ρ_λ and the *monochromatic transmittance* τ_λ.

Conservation of energy implies that

$$\alpha_\lambda + \rho_\lambda + \tau_\lambda = 1 \tag{3.25}$$

and that $0 \leq \alpha_\lambda, \rho_\lambda, \tau_\lambda \leq 1$. All of these properties are almost independent of the angle of incidence θ, unless θ is near grazing incidence ($\theta = 90°$). In practice, the radiation incident on a surface contains a wide spectrum of wavelengths, and not just one small interval. We define the *absorptance* α to be the absorbed proportion of the total incident radiant flux density:

$$\alpha = \phi_{abs}/\phi_{in} \tag{3.26}$$

It follows that

$$\alpha = \frac{\int_{\lambda=0}^{\infty} \alpha_\lambda \phi_{\lambda,in} \, d\lambda}{\int_{\lambda=0}^{\infty} \phi_{\lambda,in} \, d\lambda} \tag{3.27}$$

Equation (3.27) makes clear that the total absorptance α, unlike α_λ, depends on the spectrum of the incident radiation. For example, a body which looks blue in daylight looks black under sodium lamps because it absorbs orange light but reflects blue.

The total reflectance $\rho = \phi_{refl}/\phi_{in}$ and the total transmittance $\tau = \phi_{trans}/\phi_{in}$ are similarly defined, and again

$$\alpha + \rho + \tau = 1 \tag{3.28}$$

Example 3.4   Calculation of absorbed radiation

A certain surface has α_λ varying with wavelength as shown in Fig. 3.11(a) (this is a typical variation for a 'selective surface'). Calculate the power absorbed by $1\,m^2$ of this surface from each of the following incident spectral distributions of radiant flux density:

(1) ϕ_λ given by curve I of Fig. 3.11(b) (this approximates a source at 6000 K).
(2) ϕ_λ given by curve II (approximating a source at 1000 K).
(3) ϕ_λ given by curve III (approximating a source at 500 K).

Solution

(1) Over the entire range of λ, $\alpha_\lambda = 0.8$. Therefore from (3.27) $\alpha = 0.8$ also, and the absorbed power is

$$P = \alpha(1\,m^2) \int \phi_{\lambda, in} \, d\lambda$$

$$= (0.8) \, (1\,m^2)[(\tfrac{1}{2})(2000\,W\,m^{-2}\,\mu m^{-1})(2\,\mu m)] = 1600\,W$$

(The integral is the area under curve I.)

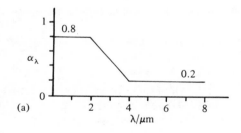

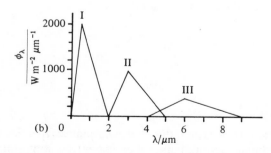

Fig. 3.11 Data for Example 3.4. The maxima of curves I, II, III in (b) are at (0.5, 2000), (3.0, 1000) and (6.0, 400) respectively

(2) Here we have to explicitly calculate the integral $\int \alpha_\lambda \phi_{\lambda,\text{in}} d\lambda$ of (3.27). Tabulate as follows (the interval of λ is chosen to match the accuracy of the data; here the spectra are obviously linearized, so an interval $\Delta\lambda \sim 1\,\mu m$ is adequate):

$\dfrac{\lambda}{\mu m}$	$\dfrac{\Delta\lambda}{\mu m}$	α_λ	$\dfrac{\phi_\lambda}{W\,m^{-2}\mu m^{-1}}$	$\dfrac{\alpha_\lambda \phi_\lambda \Delta\lambda}{W\,m^{-2}}$
2.5	1	0.62	500	310
3.5	1	0.33	750	250
4.5	1	0.2	200	40
				Total 600

Therefore the power absorbed is approximately 600 W.

(3) In a manner similar to part 1 of this solution, $\alpha_\lambda = 0.2$ over the relevant wavelength interval. Thus the power absorbed is

$$P = (0.2)(1\,m^2)[(\tfrac{1}{2})(400\,W\,m^{-2}\mu m^{-1})(5\,\mu m)] = 200\,W.$$

3.5.4 *Black bodies, emittance and Kirchhoff's laws*

In radiation theory, an idealized surface which absorbs all radiation incident on it is called a *black body*. (The term is used because a surface with the colour we call 'black' absorbs all visible radiation.) By definition, a black body has $\alpha_\lambda = 1$ for all λ, and therefore also has total absorptance $\alpha = 1$. Thus no real body can absorb more radiation than a similarly dimensioned black body placed in the same incident radiation.

Kirchhoff showed that, in addition, no real body can *emit* more radiation than a similar black body at the same temperature.

We define the *emittance* ϵ of a surface to be the ratio of the RFD emitted by the surface to the RFD emitted by a black body at the same temperature. The *monochromatic emittance* ϵ_λ of a surface is similarly defined as the corresponding ratio of RFD in the wavelength range $(\lambda - \tfrac{1}{2}d\lambda, \lambda + \tfrac{1}{2}d\lambda)$. It follows from this argument that

$$0 \le \epsilon, \epsilon_\lambda \le 1 \tag{3.29}$$

Note that the emittance ϵ of a real body may vary with temperature.

Kirchhoff extended his theoretical argument to reach an important general conclusion about surface properties, namely

$$\alpha_\lambda = \epsilon_\lambda \tag{3.30}$$

for each surface, provided the two parameters are measured *at the same wavelength and temperature*. This result, called *Kirchhoff's law*, still holds even if the bodies are (1) not in a constant temperature enclosure or (2) not in thermal equilibrium with their environment.

It is extremely important to realize that the concept of a constant temperature enclosure for all the bodies concerned is quite wrong for solar energy devices, since the sun is one of these bodies. Often the flux of incident radiation is that of the sun at 5800 K with peak intensity at $\lambda \sim 0.5\,\mu m$, and the receiving surface is at about 350 K emitting at about $\lambda \sim 10\,\mu m$. The dominant monochromatic absorptance is therefore $\alpha_{\lambda = 0.5\,\mu m}$ and the dominant monochromatic emittance is $\epsilon_{\lambda = 10\,\mu m}$. These two coefficients need not be equal.

Kirchhoff's law does *not* imply that the outgoing flux of radiation from a body at a particular wavelength is equal to the incoming flux at that wavelength. Only the coefficients α_λ and ϵ_λ are equal, not the fluxes $A\phi_{\lambda,\text{in}}$ and $A\phi_{\lambda,\text{out}}$.

3.5.5 *Radiation emitted by a body*

The monochromatic radiant flux density emitted by a black body of *absolute temperature T* is

$$\phi_{B\lambda} = \frac{C_1}{\lambda^5[\exp(C_2/\lambda T) - 1]} \tag{3.31}$$

where $C_1 = 3.74 \times 10^{-16}\,\text{W m}^2$ and $C_2 = 0.0144\,\text{m K}$ are fundamental constants. This formula, *Planck's radiation law*, can also be derived from quantum theory, thereby expressing C_1, C_2 in terms of Planck's constant, the speed of light and Boltzmann's constant. Fig. 3.12 shows how $\phi_{B\lambda}$ varies with wavelength λ and temperature T. It can be seen that the wavelength λ_m where the radiation is strongest (i.e. where $\phi_{B\lambda}$ is a maximum) increases as T decreases. In fact, by differentiating (3.31) and setting $d\phi_{B\lambda}/d\lambda = 0$, we find that

$$\lambda_m T = 2898\,\mu m\,K \tag{3.32}$$

(This relation is called Wien's displacement law.) It can be seen that for $T \gtrsim 700\,K$ significant radiation is emitted in the visible region and the surface

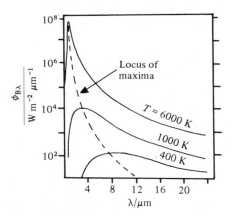

Fig. 3.12 Spectral distribution of black body radiation. After Duffie and Beckman (1980)

does not appear black. Fig. 3.12 shows that hot bodies radiate much more power than cold bodies. Indeed (3.31) implies that the total RFD emitted by a black body is

$$\phi_B = \int_0^\infty \phi_{B\lambda}\, d\lambda = \sigma T^4 \tag{3.33}$$

where $\sigma = 5.67 \times 10^{-8}\,\mathrm{W\,m^{-2}\,K^{-4}}$ is the *Stefan-Boltzmann constant*, another fundamental constant.

It follows that the heat flow from a *real* body of emittance ϵ, area A and *absolute* (surface) temperature T is

$$P_r = \epsilon\sigma A T^4 \tag{3.34}$$

Note that this formula gives the radiation emitted *by* the body. The *net* radiative flux away from the body may be much less (e.g. (3.39)). More convenient for calculation than (3.31) is the dimensionless function

$$D = \int_0^\lambda \phi_{B\lambda}\, d\lambda / \sigma T^4 \tag{3.35}$$

which turns out to be a function of the single variable $v = \lambda T$. This function is tabulated in Table 3.1.

Example 3.5
Find the radiant flux density emitted in the wavelength range 0.4 to 0.7 μm by a black body at 2000 °C.

Solution
Note that in all radiation formulas T is the absolute temperature. Thus $T = 2000 + 273 = 2273\,\mathrm{K}$, and

for $\lambda_1 = 0.4\,\mu\mathrm{m}$, $v_1 = \lambda_1 T = 910\,\mu\mathrm{m\,K}$
for $\lambda_2 = 0.7\,\mu\mathrm{m}$, $v_2 = \lambda_2 T = 1590\,\mu\mathrm{m\,K}$

Referring to (3.35) and Table 3.1, we see that the required quantity is

$$\int_{\lambda_1}^{\lambda_2} \phi_{B\lambda}\, d\lambda = \int_0^{\lambda_2} \phi_{B\lambda}\, d\lambda - \int_0^{\lambda_1} \phi_{B\lambda}\, d\lambda$$

$$= \sigma T^4 [D(v_2) - D(v_1)]$$
$$= (5.67 \times 10^{-8}\,\mathrm{W\,m^{-2}\,K^{-4}})(2273\,\mathrm{K})^4 [0.0190 - 0.0001]$$
$$= 28.6\,\mathrm{kW\,m^{-2}}$$

Table 3.1   Fraction of black body radiation between zero and λT (adapted from Sargent, 1972)
$D(\lambda T) = (\int_0^\lambda \phi_{B\lambda}\, d\lambda)/\sigma T^4$

λT	$D(\lambda T)$	λT	$D(\lambda T)$	λT	$D(\lambda T)$
μm K		μm K		μm K	
500	0.0000	4400	0.5488	8300	0.8676
600	0.0000	4500	0.5843	8400	0.8711
700	0.0000	4600	0.5793	8500	0.8745
800	0.0000	4700	0.5937	8600	0.8778
900	0.0001	4800	0.6075	8700	0.8810
1000	0.0003	4900	0.6209	8800	0.8841
1100	0.0009	5000	0.6337	8900	0.8871
1200	0.0021	5100	0.6461	9000	0.8899
1300	0.0043	5200	0.6579	9100	0.8927
1400	0.0077	5300	0.6693	9200	0.8954
1500	0.0128	5400	0.6803	9300	0.8980
1600	0.0197	5500	0.6909	9400	0.9005
1700	0.0285	5600	0.7010	9500	0.9030
1800	0.0393	5700	0.7107	9600	0.9054
1900	0.0521	5800	0.7201	9700	0.9076
2000	0.0667	5900	0.7291	9800	0.9099
2100	0.0830	6000	0.7378	9900	0.9120
2200	0.1009	6100	0.7461	10000	0.9141
2300	0.1200	6200	0.7541	11000	0.9318
2400	0.1402	6300	0.7618	12000	0.9450
2500	0.1613	6400	0.7692	13000	0.9550
2600	0.1831	6500	0.7763	14000	0.9628
2700	0.2053	6600	0.7831	15000	0.9689
2800	0.2279	6700	0.7897	16000	0.9737
2900	0.2506	6800	0.7961	17000	0.9776
3000	0.2732	6900	0.8022	18000	0.9807
3100	0.2958	7000	0.8080	19000	0.9833
3200	0.3181	7100	0.8137	20000	0.9855
3300	0.3401	7200	0.8191	30000	0.9952
3400	0.3617	7300	0.8244	40000	0.9978
3500	0.3829	7400	0.8295	50000	0.9988
3600	0.4036	7500	0.8343	60000	0.9993
3700	0.4238	7600	0.8390	70000	0.9995
3800	0.4434	7700	0.8436	80000	0.9996
3900	0.4624	7800	0.8479	90000	0.9997
4000	0.4829	7900	0.8521	100000	0.9998
4100	0.4987	8000	0.8562	∞	1.0000
4200	0.5160	8100	0.8601		
4300	0.5327	8200	0.8639		

3.5.6 *Radiative exchange between black surfaces*

All bodies, including the sky, emit radiation. We do not need to calculate how much radiation each body emits individually, but rather what is the *net* gain (or loss) of radiant energy by each body.

Fig. 3.13 shows two surfaces 1 and 2, each exchanging radiation. The net rate of exchange depends on the surface properties and on the geometry. In particular we must know the proportion of the radiation emitted by 1 actually reaching 2, and vice versa.

Consider the simplest case with both surfaces diffuse and black, and with no absorbing medium between them. (A *diffuse* surface is one which emits equally in all directions; its radiation is not concentrated into a beam. Most opaque surfaces, other than mirrors, are diffuse.) The shape factor F_{ij} is the proportion of radiation emitted by surface i reaching surface j. It depends only on the geometry and not on the properties of the surfaces. Let ϕ_B be the radiant flux

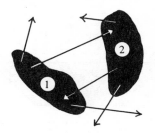

Fig. 3.13 Exchange of radiation between two (black) surfaces

density emitted by a black body surface into the hemisphere above it. The radiant power reaching 2 from 1 is

$$P'_{12} = A_1 \phi_{B1} F_{12} \tag{3.36}$$

Similarly the radiant power reaching 1 from 2 is

$$P'_{21} = A_2 \phi_{B2} F_{21} \tag{3.37}$$

If the two surfaces are in thermal equilibrium, $P'_{12} = P'_{21}$ and $T_1 = T_2$: so by (3.33)

$$\phi_{B1} = \sigma T_1^4 = \sigma T_2^4 = \phi_{B2}$$

Therefore

$$A_1 F_{12} = A_2 F_{21} \tag{3.38}$$

This is a geometrical relationship independent of the surface properties and temperature.

If the surfaces are *not* at the same temperature, then the net radiative heat flow from 1 to 2, using (3.38), is

$$
\begin{aligned}
P_{12} &= P'_{12} - P'_{21} \\
&= \phi_{B1} A_1 F_{12} - \phi_{B2} A_2 F_{21} \\
&= \sigma T_1^4 A_1 F_{12} - \sigma T_2^4 A_2 F_{21} \\
&= \sigma (T_1^4 - T_2^4) A_1 F_{12} \tag{3.39}
\end{aligned}
$$

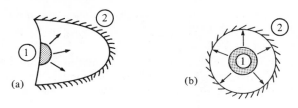

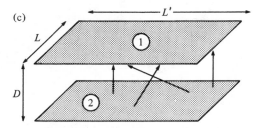

Fig. 3.14 Geometries with shape factor $F_{12} = 1$.
(a) convex or flat surface (1) completely surrounded by surface (2). (b) One long cylinder
(1) inside another (2). (c) Closely spaced large parallel plates (L/D, $L'/D >> 1$).

Or, if it is easier to calculate F_{21},

$$P_{12} = \sigma(T_1^4 - T_2^4) A_2 F_{21} \tag{3.40}$$

In general, the calculation of F_{ij} requires a complicated integration, and results
are tabulated in handbooks, e.g. Wong (1977). Solar collector configurations
frequently approximate to Fig. 3.14, where the shape factor becomes unity.

3.5.7 *Radiative exchange between gray surfaces*

A gray body has a diffuse opaque surface with $\epsilon = \alpha = 1 - \rho =$ constant,
independent of surface temperature, wavelength and angle of incidence. This is
a reasonable approximation for most opaque surfaces in common solar energy
applications where maximum temperatures are $\sim 200\,°C$ and wavelengths are
between $0.3\,\mu m$ and $15\,\mu m$.

The radiation exchange between any number of gray bodies may be analyzed
allowing for absorption, re-emission and reflection. The resulting system of
equations can be solved to yield the heat flow from each body if the temperatures
are known, or vice versa. If there are only two bodies, the heat flow from body 1
to body 2 can be expressed in the form

$$P_{12} = \sigma A_1 F'_{12}(T_1^4 - T_2^4) \tag{3.41}$$

where the *exchange factor* F'_{12} depends on the geometric shape factor F_{12}, the area ratio (A_1/A_2) and the surface properties ϵ_1, ϵ_2. Comparison with (3.39) shows that for black bodies $F'_{12} = F_{12}$.

Exchange factors for the most commonly encountered geometries are listed in Appendix C. More exhaustive lists are given in specialized texts (Wong, 1977; McAdams, 1954; Rohsenow and Hartnett, 1973).

3.5.8 *Thermal resistance formulation*
Equation (3.41) can be factorized into the form

$$P_{12} = A_1 F'_{12} \sigma (T_1^2 + T_2^2)(T_1 + T_2)(T_1 - T_2) \tag{3.42}$$

Comparing this with (3.1) we see that the resistance to radiative heat flow from body 1 is

$$R_r = [A_1 F'_{12} \sigma (T_1^2 + T_2^2)(T_1 + T_2)]^{-1} \tag{3.43}$$

In general, R_r depends strongly on temperature. However, T_1 and T_2 in (3.43) are *absolute* temperatures, so that it is often true that $(T_1 - T_2) << T_1, T_2$. In this case (3.43) can be simplified to

$$R_r \approx 1/(4\sigma A_1 F'_{12} \bar{T}^3) \tag{3.44}$$

where $\bar{T} = \frac{1}{2}(T_1 + T_2)$ is the mean temperature.

Example 3.6 Typical values of R_r, P_r
Two parallel plates of area $1\,\text{m}^2$ have emittances of 0.9 and 0.2 respectively. If $T_1 = 350\,\text{K}$ and $T_2 = 300\,\text{K}$ then, using (3.44) and (C.18), Appendix C,

$$R_r = \frac{(1/0.9) + (1/0.2) - 1}{4(1\,\text{m}^2)(5.67 \times 10^{-8}\,\text{W}\,\text{m}^{-2}\,\text{K}^{-4})(325\,\text{K})^3} = 0.66\,\text{K}\,\text{W}^{-1}$$

This is comparable to the typical convective resistances of Example 3.2. The corresponding heat flow is

$$P_r = 50\,\text{K}/(0.66\,\text{K}\,\text{W}^{-1}) = 75\,\text{W}$$

3.6 Properties of 'transparent' materials

An ideal transparent material has transmittance $\tau = 1$, reflectance $\rho = 0$ and absorptance $\alpha = 0$. However, in practice 'transparent' materials (such as glass) have $\tau \sim 0.9$ at angles of incidence $\leq 70°$, and rapidly reducing τ and increasing ρ as angles of incidence approach $90°$.

According to Maxwell's equations of electromagnetism, the reflectance of a material depends on its refractive index and on the angle of incidence. For most common glasses at angles of incidence less than 40° (the important range in practice) $\rho \approx 0.08$ for visible light. Thus with no absorption the transmittance would be

$$\tau_r = 1 - \rho \approx 0.92 \tag{3.45}$$

However, some radiation is absorbed as it passes through a partially transparent medium. The proportion reaching a depth x below the surface decreases with x according to the Bouger-Lambert law. The transmitted proportion at x is

$$\tau_a = e^{-Kx} \tag{3.46}$$

where the *extinction coefficient K* varies from about $0.04 \, \text{cm}^{-1}$ (for good quality 'water white' glass) to about $0.30 \, \text{cm}^{-1}$ for common window glass (with iron impurity, having greenish edges). Iron-free glass has a smaller extinction coefficient than normal window glass, and so is better for solar energy applications.

Using the terms τ_r and τ_a, the transmittance becomes

$$\tau = \tau_r \tau_a \tag{3.47}$$

Fig. 3.15(a) shows the variation with wavelength and thickness of the overall monochromatic transmittance $\tau_\lambda = \tau_{\lambda r} \tau_{\lambda a}$, for a typical glass. Note the very low transmittance in the thermal infrared region ($\lambda > 3 \, \mu$m). Glass is a good absorber in this waveband. Fig. 3.15(b) shows that polythene, on the other hand, is

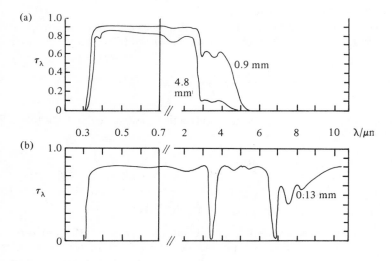

Fig. 3.15 Monochromatic transmittance of: (a) glass (0.15% Fe$_2$O$_3$) of thickness 4.8 mm and 0.9 mm (b) polythene. Note the scale change at $\lambda = 0.7 \, \mu$m. Data from Dietz (1954) and Meinel and Meinel (1976)

transparent in both the visible and infrared. Plastics such as Mylar, with greater molecular complication, have transmittance characteristics lying between those of glass and polythene.

3.7 Heat transfer by mass transport

Free and forced convection (Section 3.4) is heat transfer by the movement of fluid mass. Analysis proceeds by considering thermal interactions between a (solid) surface and the moving fluid. However there are frequent practical applications where energy is transported by a moving fluid or solid without considering heat transfer across a surface – for example, when heat is transported from a solar collector to a storage tank. These systems of heat transfer by mass transport are analyzed by considering the fluid alone.

3.7.1 *Single phase heat transfer*
Consider the fluid flow through a heated pipe shown in Fig. 3.16. According to (2.6), the net heat flow out of the control volume (i.e. out of the pipe) is

$$P_m = \dot{m}c(T_3 - T_1) \tag{3.48}$$

where $\dot{m}$ is the mass flow rate through the pipe ($kg\,s^{-1}$) and T_1, T_3 are the temperatures of the *fluid* on entry and exit respectively. If both T_1 and T_3 are

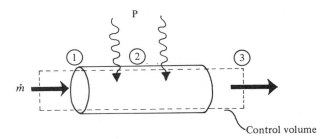

Fig. 3.16 Mass flow through a heated pipe. Heat is taken out by the fluid at a rate $P_m = \dot{m}(T_3 - T_1)$ regardless of how the heat enters the fluid at (2)

measured experimentally, P_m may be calculated without knowing the details of the transfer process at the pipe wall. The thermal resistance for this process is defined as

$$R_m = (T_3 - T_1)/P_m = 1/\dot{m}c \tag{3.49}$$

Note that the heat flow may be determined by external factors controlling the rate of mass flow $\dot{m}$. Thus temperature is not a driving function for the heat transfer, unlike conduction, radiation and free convection. Therefore (3.49) should be used only with great caution.

3.7.2 *Phase change*

A most effective means of heat transfer is as latent heat of vaporization/condensation. For example, 2.4 MJ of heat vaporizes 1 kg of water, which is much greater than the 0.42 MJ to heat 1 kg through 100 °C. Heat taken from the heat source (as in Fig. 3.17) is carried to wherever the vapor condenses (the 'heat sink'). The associated heat flow is

$$P_m = \dot{m}\Lambda \tag{3.50}$$

where $\dot{m}$ is the rate at which fluid is being evaporated (or condensed), and Λ is the latent heat of vaporization. This expression is most useful when $\dot{m}$ is known (e.g. from experiment).

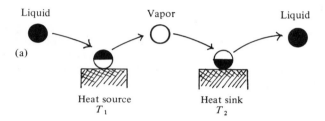

Fig. 3.17 Heat transfer by phase change. Liquid absorbs heat, changes to vapor, then condenses, so releasing heat

Theoretical prediction of evaporation rates is very difficult, because of the multitude of factors involved, such as the density, viscosity, specific heat and thermal conductivity of both the liquid and the vapor; the latent heat, the pressure and the temperature difference; and the size, shape and nucleation properties of the surface. Some guidance and specific empirical formulas are given in the specialized textbooks cited at the end of the chapter.

Since evaporation and condensation are both nearly isothermal processes, the heat flow by this mass transport is not determined directly by the source temperature T_1 and the sink temperature T_2. The associated thermal resistance can however be defined as

$$R_m = (T_1 - T_2)/\dot{m}\Lambda \tag{3.51}$$

3.8 Multimode transfer and circuit analysis

3.8.1 *Resistances only*

Section 3.2 showed in general terms how thermal resistances could be combined in series or parallel (or indeed in more complicated networks). The resistances that are combined do not have to refer to the same mode: conduction, convection, radiation and mass transfer can be treated together by this formalism. In setting up the circuit analog, it is important to consider all significant heat paths and resistances. Many examples will be found in later chapters, especially Chapter 5.

3.8.2 *Thermal capacitance*

The circuit analogy can be developed further. Thermal energy can be stored in bodies, similar to electrical energy stored in capacitors.

For example, consider a tank of hot water standing in a constant temperature environment at T_0 (Fig. 3.18a)). The water (of mass m and specific heat capacity c) is at some temperature T_1 above the ambient temperature T_0. Heat flows from the water to the environment according to the equation

$$-mc \frac{\mathrm{d}}{\mathrm{d}t}(T_1 - T_0) = \frac{(T_1 - T_0)}{R_{10}} \tag{3.52}$$

where the minus sign indicates that T_1 decreases when $(T_1 - T_0)$ is positive, and R_{10} is the combined thermal resistance of heat loss by convection, radiation and conduction (Fig. 3.18(b)). Similarly in the electrical circuit of Fig. 3.18(c), electrical current flows from one side of the capacitor (at voltage V_1) to the other (at voltage V_2) according to the equation

$$\frac{\mathrm{d}q}{\mathrm{d}t} = -C_\mathrm{e} \frac{\mathrm{d}}{\mathrm{d}t}(V_1 - V_2) = \frac{V_1 - V_2}{R'_{12}} \tag{3.53}$$

where $q = C_\mathrm{e}(V_1 - V_2)$ is the charge held by the capacitor.

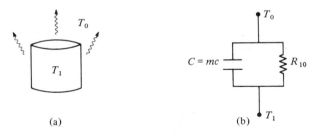

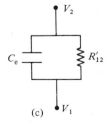

Fig. 3.18 A hot object loses heat to its surroundings. (a) Physical situation. (b) Thermal circuit analog. (c) Electrical circuit analog.

We see that these equations are exactly analogous. That is, if we write $v = T_1 - T_0$, $C = mc$ in (3.52), and $v = V_1 - V_2$, $C = C_e$ in (3.53) we obtain exactly the same equation. Since the mathematical boundary conditions also correspond (in each case we would have an initial value for v), the solutions are exactly the same.

Thus if we define the *thermal capacitance* of a body to be its heat capacity $C = mc$ (unit $J K^{-1}$) we have the complete analogy shown in Table 3.2.

In drawing analog circuits for thermal systems, some care is needed to ensure that the capacitances connect across the correct temperatures (cf. voltages). A useful check is to refer to the differential equations (e.g. (3.52), (3.53)) to ensure that they do in fact correspond with the circuit.

Table 3.2   Analogous electrical and thermal quantities

Thermal			Electrical		
Quantity	Symbol	Unit	Quantity	Symbol	Unit
Temperature	T	kelvin, K	Potential	V	volt, V
Heat flow	P	watt, W	Current	I	ampere, A
Resistance	R	$K W^{-1}$	Resistance	R	ohm $\Omega = V A^{-1}$
Thermal capacitance	C	$J K^{-1}$	Capacitance	C	farad F $= A s V^{-1}$

Problems

3.1 Show explicitly that for thermal resistances in series, as in Fig. 3.2(b), $R_{13} = R_{12} + R_{23}$.
   Hint: what is the relation between the heat flows in the various resistances?

3.2 Verify from the definitions (3.17) and (3.22) that $\mathscr{N}$ and $\mathscr{A}$ are indeed dimensionless.

3.3 A layer of fluid is confined between two horizontal plates, as for (C.7), Appendix C. The hot bottom plate is at height $z = 0$ and temperature $T_1 + \Delta T$, and the cold top plate is at height $z = d$ and temperature T_1. Use the measured profiles of Fig. 3.19 to measure the thermal thickness δ at each Rayleigh number shown. Taking $X = d$, find the Nusselt number $\mathscr{N}$ in each case.
   Note: this is one of the few cases where the thermal thickness δ of (3.13) can be directly measured. It is the thickness of the layer (near the plate) where the temperature varies almost linearly with z. Note that (C.7) does not apply here because $\mathscr{A} \ll 10^5$ and the flow is laminar.

3.4 Calculate the heat lost per hour by radiation from the pot of Example 3.3, and check that it is indeed less than the heat lost by convection.

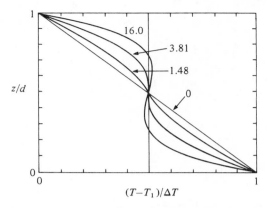

Fig. 3.19 Observed mean temperature profiles in a convecting fluid layer between plates of separation d. The curves are labeled by the value of $\mathcal{A}/\mathcal{A}_{\text{crit}}$, where $\mathcal{A}_{\text{crit}} = 1700$ is the Rayleigh number below which no convection occurs. See (C.7) From Gille (1967)

3.5 Newton's law of cooling ($c.1666$) states that the rate of loss of heat from a body is proportional to the temperature difference between the body and its surroundings. According to this statement, the rate of loss of heat is independent of other variables, such as wind speed.

Calculate the heat flow lost by convection from a flat plate 1 m square at 50°C to air at 20°C, at wind speeds of (a) 0 (b) $5\,\text{m s}^{-1}$ (c) $10\,\text{m s}^{-1}$. Use (C.15).

What is the percentage error incurred by using a single mean value of heat transfer coefficient h, as suggested by Newton's law, to cover the three cases?

3.6 For each of the wind speeds cited, calculate the heat flow from the plate of Problem 3.5 by combining separate calculations for forced and free convection. Use (C.2) and (C.8).

Compare the results and the ease of calculation with those of Problem 3.5.

3.7 *Heat loss through windows*
A room has two glass windows each 1.5 m high, 0.80 m wide and 5.0 mm thick (Fig. 3.20). The temperature of the air and wall surface inside the room is 20°C. The temperature of the outside air is 0°C. There is no wind. Calculate the heat loss through the glass (a) assuming (falsely) that the only resistance to heat flow is from conduction through the material of the glass, and (b) allowing (correctly) for the thermal resistance of the air boundary layers against the glass, as in Fig. 3.20.
Hint: assume as a first approximation that $T_2 \approx T_3 \approx \frac{1}{2}(T_1 + T_4)$. Justify this assumption afterwards.

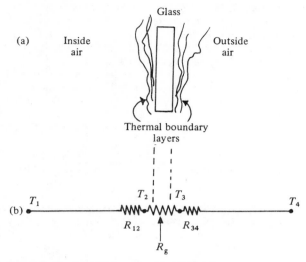

Fig. 3.20 Heat loss through a window; see Problem 3.7

Solutions

3.3 $\mathscr{A}/\mathscr{A}_c$:   0   1.48   3.8   16
$\mathscr{N}$(expt):   1   1.5   2.1   3.3

3.4 $26\,\text{kJ}\,\text{h}^{-1}$, using $\epsilon = 0.10$, $T_1 = 373\,\text{K}$ in (C.17).

3.5 (a) 5.7, (b) 25, (c) $44\,\text{W}\,\text{m}^{-2}\text{K}^{-1}$, using (C.15). Error $\sim 50\%$.

3.6 $h_{\text{total}} = h_{\text{forced}} + h_{\text{free}} = 4.8, 13.7, 43.2\,\text{W}\,\text{m}^{-2}\text{K}^{-1}$, using (C.2), (C.8).

3.7 (a) $10\,\text{k}\,\text{W}$ (b) $r_v = 0.032\,\text{m}^2\,\text{K}\,\text{W}^{-1}$ on both sides;
$R_{14} = 0.029\,\text{K}\,\text{W}^{-1}$; $P_{14} = 0.7\,\text{kW}$.

Bibliography

General textbooks on heat transfer

There are many good texts written for engineering students and practicing engineers. Without exception these will include heat transfer under large temperature differences using complex situations. Solar energy systems seldom require such complex analysis since temperature differences are relatively small and simple conditions exist. Therefore do not be daunted by the fearsome format of some of these books, of which a small but representative sample follows.

Ede, A. J. (1967) *An Introduction to Heat Transfer Principles and Calculations*, Pergamon Press, Oxford.
Clear and simple account for the nonspecialist.

Kay, J. M. and Nedderman, R. M. (1974) *An Introduction to Fluid Mechanics and Heat Transfer*, 3rd edn, Cambridge University Press.
A terse account for engineering students.

Kreith, F. R. (1973) *Principles of Heat Transfer*, 3rd edn, Intext Press, New York.
Detailed but clear textbook for would-be specialists.

McAdams, W. C. (1954) *Heat Transmission*, 3rd edn, McGraw-Hill, New York.
A handbook for specialists, covering many special cases. British units.

Mikheyev, M. (1964) *Fundamentals of Heat Transfer*, MIR, Moscow, English translation.
Similar in scope to Ede.

Rohsenow, W. R. and Hartnett, J. P. (eds) (1973) *Handbook of Heat Transfer*, McGraw-Hill, New York.
Comprehensive and detailed handbook for practitioners, but in FPS units. Includes an article by V. Paschkis describing physical construction of analog circuits.

Wong, H. Y. (1977) *Handbook of Essential Formulae and Data on Heat Transfer for Engineers*, Longmans, London.
Easy to use collection, priced for student use. Strongly recommended for practical application and quick reference. Not of fearsome format!

Heat transfer for solar energy applications

The following books include useful chapters of heat transfer formulas that are particularly useful in solar applications. These chapters mostly assume that the basics are known already.

Duffie, J. A. and Beckman, W. A. (1980) *Solar Engineering of Thermal Processes*, John Wiley and Sons, New York.
Very reliable and widely used. SI units.

Lunde, P. J. (1979) *Solar Thermal Engineering*, John Wiley and Sons, New York.
FPS and SI units.

Meinel, A. B. and Meinel, M. P. (1976) *Applied Solar Energy*, Addison-Wesley, Massachusetts.
Chapter 10 is specifically on heat transfer.

Sayigh, A. (ed.) (1977) *Solar Energy Engineering*, Academic Press, London.
Chapter 5 by J. Sabbagh is a brief but specialized formulary.

Specific references

De Vries, H. F. W. and Francken, J. C. (1980) 'Simulation of a solar energy system by means of electrical resistance', *Solar Energy*, **25**, 279–81.
Uses an actual physical network.

Dietz, A. G. H. (1954) 'Diathermanous materials and properties of surfaces', in Hamilton, R. W. (ed.) *Space Heating with Thermal Energy*, MIT Press, Boston.

Gille, J. (1967) 'Interferometric measurement of temperature gradient reversal in a layer of convecting air'. *J. Fluid Mech.*, **30**, 371–84.

International Solar Energy Society (1978) 'Units and symbols in solar energy', *Solar Energy*, **21**, 65–68.

Kaye, H. and Laby, G. (1975) *Tables of Physical and Chemical Constants*, 14th edn, Longmans, London.
Includes an excellent brief account of radiation units.

Sargent, S. L. (1972) 'A compact table of black body radiation functions', *Bull. Am. Meteorol. Soc.*, **53**, 360.

Turner, J. S. (1973) *Buoyancy Effects in Fluids*, Cambridge University Press.
Especially Chapter 7.

4 *Solar radiation*

4.1 Introduction

Solar radiation arrives on the earth at a maximum flux density of about $1\,\mathrm{kW\,m^{-2}}$ in a wavelength band between 0.3 and $2.5\,\mu$m. This is called *short wave radiation* and includes the visible spectrum. For habited areas received fluxes vary widely from about 3 to $30\,\mathrm{MJ\,m^{-2}\,day^{-1}}$, depending on place, time and weather. The quality of the radiation is characterized by the photon energy of around $2\,\mathrm{eV}$ as determined by the 6000 K surface temperature of the sun. This is an energy flux of very high thermodynamic quality, from an accessible source at a temperature very much greater than those of conventional engineering sources. The flux can be used thermally with devices relating to conventional engineering (e.g. steam turbines), and more importantly with methods developed from photochemical and photophysical interactions.

Radiant energy fluxes relating to the earth's atmospheric and surface temperature are also of the order of $1\,\mathrm{kW\,m^{-2}}$, but occur in a wavelength band between about 5 and $25\,\mu$m, called *long wave radiation*, peaking at about $10\,\mu$m.

The short and long wave radiation regions can be treated as quite distinct from each other.

The main aim of this chapter is to calculate the solar radiation likely to be available as input to a solar device at a specific location, orientation and time. First, we discuss how much radiation is available outside the earth's atmosphere (Section 4.2). The proportion of this that reaches our device depends on geometric factors such as latitude (Sections 4.4, 4.5) and on atmospheric factors such as absorption by water vapor (Section 4.6). Two final sections deal briefly with the measurement of solar radiation, and with the more difficult problem of how to use other meteorological data to estimate a solar measurement.

The most basic information is contained in Figs 4.7 and 4.15. Readers whose main interest is solar energy devices are advised to acquaint themselves with these summaries of the solar input.

4.2 Extraterrestrial solar radiation

Nuclear fusion reactions in the active core of the sun produce inner temperatures of about $10^7\,$K and an inner radiation flux of uneven spectral distribution. This internal radiation is absorbed in the outer passive layers which are heated to about 5800 K and so become a source of radiation with a relatively continuous form of spectral distribution.

Fig. 4.1 shows the spectral distribution of the solar irradiance at the earth's mean distance, uninfluenced by any atmosphere. Note how this distribution is like that from a black body at 5800 K in shape, peak wavelength and total power emitted. (Compare Fig. 3.12.) The area beneath this curve is the *solar constant* $G_0^* = 1353 \, \text{W m}^{-2}$. This is the radiant flux density incident on a plane directly

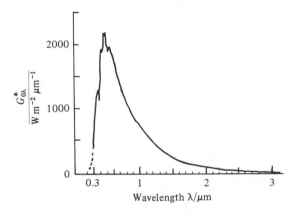

Fig. 4.1 Spectral distribution of extraterrestrial solar irradiance, $G_{0\lambda}^*$ (NASA standard curve, NASA, 1971)

facing the sun's rays, above the atmosphere at a distance of $1.496 \times 10^8 \, \text{km}$ from the sun (i.e. at the earth's mean distance from the sun). The RFD actually received by the top of the atmosphere differs from the solar constant by less than $\pm 1.5\%$ because of fluctuations in the sun's radiant output, and by $\pm 4\%$ through the year because of the predictable changes in the sun–earth distance arising from the earth's slightly elliptic path.

The solar spectrum can be divided into three main regions:

(1) Ultraviolet region ($\lambda < 0.4 \, \mu\text{m}$): 9% of the irradiance
(2) Visible region ($0.4 \, \mu\text{m} < \lambda < 0.7 \, \mu\text{m}$): 45% of the irradiance
(3) Infrared region ($\lambda > 0.7 \, \mu\text{m}$): 46% of the irradiance

The contribution to the solar radiation flux from wavelengths greater than $2.5 \, \mu\text{m}$ is negligible, and all three regions are classed as solar short wave radiation.

4.3 Components of radiation

Solar radiation incident on the atmosphere from the direction of the sun is the solar extraterrestrial beam radiation. Beneath the atmosphere at the earth's surface, the radiation will be observable from the direction of the sun's disk in the *direct beam*, and also from other directions as *diffuse radiation*. Fig. 4.2 is a sketch of how this happens.

The practical distinction between the two components is that only the beam component can be focused. Even on a clear day there is some diffuse radiation. The ratio between the beam irradiance and the total irradiance thus varies from about 0.9 on a clear day to zero on a completely overcast day.

It is important to distinguish the various components of solar radiation and to distinguish the plane on which the irradiance is being measured. We use subscripts as illustrated in Fig. 4.3: b for beam, d for diffuse, t for total, h for

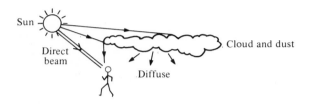

Fig. 4.2 Origin of direct beam and diffuse radiation

horizontal plane and c for the plane of a collector. The asterisk * denotes the plane perpendicular to the beam. Subscript 0 denotes values above the atmosphere. Subscripts c and t are assumed if no subscripts are given, so that G (no subscript) $\equiv G_{tc}$. Fig. 4.3 shows that

$$G_{bc} = G_b^* \cos\theta \tag{4.1}$$

where θ is the angle between the beam and the normal to the collector surface. In particular,

$$G_{bh} = G_b^* \cos\theta_z \tag{4.2}$$

where θ_z is the angle between the beam and the vertical.

The total irradiance on any plane is the sum of the beam and diffuse components:

$$G_t = G_b + G_d \tag{4.3}$$

4.4 Geometry of the earth and sun

4.4.1 *Definitions*

You will find it helpful to manipulate a sphere on which you mark the points and planes indicated in the following diagrams.

Fig. 4.4 shows the earth. It rotates in 24 hours about its own axis, which defines the points of the north and south poles N and S. The axis of the poles is normal to the earth's *equatorial plane*. In Fig. 4.4, C is the center of the earth. The point P

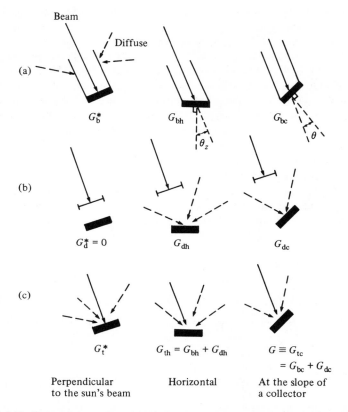

Fig. 4.3 Techniques to measure various components of solar radiation. The detector is assumed to be a black surface of unit area with a filter to exclude long wave radiation. (a) Diffuse blocked. (b) Beam blocked. (c) Total

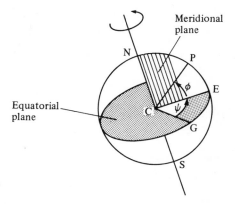

Fig. 4.4 Definition sketch for latitude ϕ and longitude ψ (see text for detail)

on the earth's surface is determined by its *latitude* ϕ and *longitude* ψ. ϕ is positive for points north of the equator, negative south of the equator. ψ is measured positive eastwards from Greenwich, England. The vertical north–south plane through P is the local *meridional plane*. E and G in Fig. 4.4 are the points on the equator having the same longitude as P and Greenwich respectively.

Once every 24 hours the meridional plane CEP includes the sun. This is noon on *solar time* for all points having that longitude. Clocks do not necessarily read 12.00 at solar noon, mainly because *civil time*, on which they are set, is defined so that large parts of a country, covering up to 15° of longitude, share the same civil time, i.e. those places are all in the same *time zone*. Resetting the clocks for 'summer time' means that solar time and civil time may differ by over one hour. Moreover, the ellipticity of the earth's orbit around the sun means that it is not exactly 24 hours between successive solar noons, although the average interval is 24.0000 hours. The correction for this is called the 'equation of time'; it never exceeds 15 minutes. The *hour angle* ω at P is simply the angle through which the earth has rotated since solar noon. Since the earth rotates $(360°/24\,h) = 15°\,h^{-1}$, the hour angle is given by

$$\omega = (15°\,h^{-1})(t_{solar} - 12\,h)$$

$$= (15°\,h^{-1})(t_{zone} - 12\,h) + \omega_{eq} + (\psi - \psi_{zone}) \tag{4.4}$$

where t_{solar} and t_{zone} are respectively the local solar and civil times (measured in hours), ψ_{zone} is the longitude where the sun is overhead when t_{zone} is noon (i.e. where solar time and civil time coincide). The small correction term ω_{eq} is the equation of time, and can be neglected for most purposes (see Duffie and Beckman, 1980). According to (4.4), ω is positive in the evening and negative in the morning.

The earth revolves around the sun once per year. The direction of the earth's axis remains fixed in space, at an angle $\delta_0 = 23.5°$ away from the normal to the plane of revolution (Fig. 4.5). The angle between the sun's direction and the equatorial plane is called the *declination* δ, and provides a convenient measure of seasonal changes. Suppose the line from the center of the earth to the sun cuts the earth's surface at P in Fig. 4.4. Then δ is precisely the angle ϕ of Fig. 4.4. That is, the declination is the latitude of the point where the sun is overhead at solar noon. Then, as can be seen in Fig. 4.6, δ varies smoothly from $+\delta_0 = +23.5°$ at midsummer in the northern hemisphere, to $-\delta_0 = -23.5°$ at northern midwinter. Analytically,

$$\delta = \delta_0 \sin[360°(284 + n)/365] \tag{4.5}$$

where n is the day in the year ($n = 1$ on 1 January).

4.4.2 *Latitude, season and daily insolation*

The *daily insolation H* is the total energy per unit area received in one day from the sun:

$$H = \int G\,dt \tag{4.6}$$

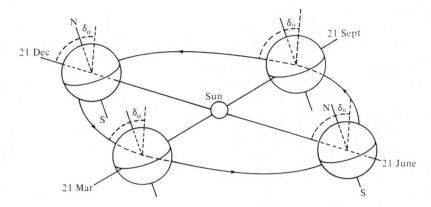

Fig. 4.5 The earth revolving around the sun, as viewed from a point obliquely above the orbit (not to scale!). The heavy line on the earth is the equator

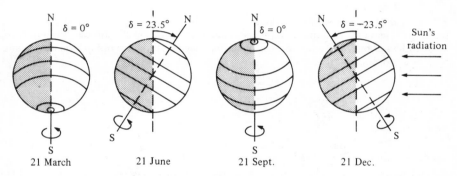

Fig. 4.6 The earth, as seen from a point further along its orbit, at various times of the year. Circles of latitude 0°, ± 23.5°, ± 66.5° are shown. Note how the declination δ varies. (After Dickinson and Cheremisinoff, 1982)

Fig. 4.7 illustrates how the daily insolation varies with latitude and season. The seasonal variation at high latitudes is very great. The quantity plotted is the clear sky solar radiation on a horizontal plane. Its seasonal variation arises from three main factors:

(1) *Variation in the length of the day*   Problem 4.5 shows that the number of hours between sunrise and sunset is

$$N = (2/15)\cos^{-1}(-\tan\phi\tan\delta) \tag{4.7}$$

At latitude 48°, for example, N varies from 16h in midsummer to 8h in midwinter.

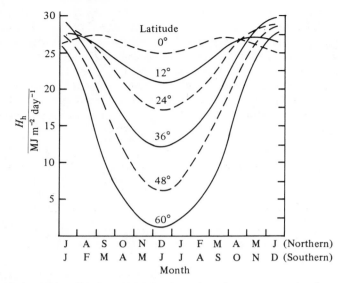

Fig. 4.7 Variation with season and latitude of H_h, the solar energy received on a horizontal plane on a clear day. In summer, H_h is about $25\,\mathrm{MJ\,m^{-2}\,day^{-1}}$ at all latitudes. In winter, H_h is much less at high latitudes because of shorter day length, more oblique incidence and greater atmospheric attenuation

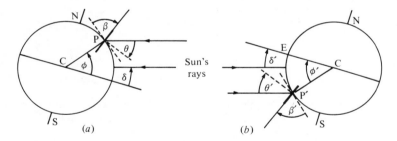

Fig. 4.8 Cross-sections through the earth at solar noon, showing the relation between latitude ϕ, declination δ, and slope β of a collector at P. θ is the angle of incidence on the north/south-facing collector. (a) Northern hemisphere in summer: $\phi, \delta, \beta > 0$. (b) Corresponding case in southern hemisphere ($\phi' = -\phi, \delta' = -\delta, \beta' = \beta, \theta' = \theta$)

In the polar regions ($|\phi| > 66.5°$), $|\tan\phi\tan\delta|$ may exceed 1. In this case $N = 24\,\mathrm{h}$ (in summer) or $N = 0$ (in winter) (see Fig. 4.6).

(2) *Orientation of receiving surface*   Fig. 4.8 shows that the horizontal plane at a location P is oriented much more towards the solar beam in summer than in winter. Therefore even if G_b^* in (4.2) remains the same, the factor $\cos\theta_z$ reduces G_{bh} in winter, and proportionately reduces H_h. Thus the curves of Fig. 4.7 are approximately proportional to $\cos\theta_z = \cos(\phi - \delta)$ (Fig. 4.8). For the insolation on surfaces of different slopes, see Fig. 4.16.

(3) *Variation in atmospheric absorption* The clear sky radiation plotted in Fig. 4.7 is less than the extraterrestrial radiation because of atmospheric attenuation. This attenuation increases with θ_z, so that G_b^* is lower in winter, thereby increasing the seasonal variation beyond that due to the geometric effects (1) and (2) (see Section 4.6).

Although the clear sky radiation is a rather notional quantity, and actual weather conditions vary widely from those assumed in its calculation, Fig. 4.7 still gives a useful guide to the *average* insolation as a function of latitude and season.

4.5 Geometry of collector and the solar beam

4.5.1 *Definitions*

For the tilted surface (collector) of Fig. 4.9 we define the following angles:

(1) *Slope β* The angle between the plane surface in question and the horizontal ($0 < \beta < 90°$ for a surface facing towards the equator; $90° < \beta < 180°$ for a surface facing away from the equator.)
(2) *Surface azimuth angle γ* The deviation from the local meridian of the projection on a horizontal plane of the normal to the surface. ($\gamma = 0$ for a surface facing due south, >0 for a surface facing west of south, and <0 eastward. For a horizontal surface, we take $\gamma = 0$.)
(3) *Angle of incidence θ* The angle between the beam radiation to the surface and the normal to that surface.

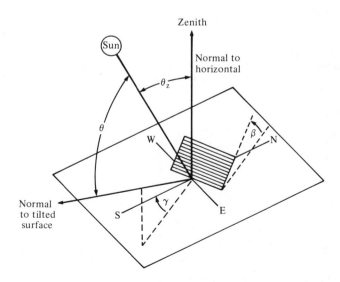

Fig. 4.9 Zenith angle θ_z, slope β and solar azimuth angle γ for a tilted surface. (Note: for the case drawn $\gamma < 0$.) After Duffie and Beckman (1980)

The sign conventions used here for β, γ, θ and for δ, ω (defined in Section 4.4.1) are those of Duffie and Beckman (1980).

4.5.2 *Angle between beam and collector*

With this sign convention, the general relation between the various angles can be shown to be

$$\cos\theta = (A - B)\sin\delta + [C\sin\omega + (D + E)\cos\omega]\cos\delta \qquad (4.8)$$

where

$A = \sin\phi\cos\beta$

$B = \cos\phi\sin\beta\cos\gamma$

$C = \sin\beta\sin\gamma$

$D = \cos\phi\cos\beta$

$E = \sin\phi\sin\beta\cos\gamma$

Example 4.1 Calculation of angle of incidence
Calculate the angle of incidence of beam radiation on a surface located at Glasgow (56°N, 4°W) at 10 a.m. on 1 February, if the surface is oriented 20° east of south, and tilted at 40° to the horizontal.

Solution
February 1 is day 32 of the year ($n = 32$), so from (4.5)

$$\delta = 23.5°\sin[360°(284 + 32)/365] = -17.5°$$

Civil time in Glasgow winter is Greenwich Mean Time, which is solar time ($\pm 15\,\text{min}$) at longitude $\psi_{\text{zone}} = 0$. Hence $t_{\text{solar}} \approx 10\,\text{h}$, so (4.4) gives $\omega = -30°$.
We also have $\phi = +56°$, $\gamma = -20°$ and $\beta = +40°$, so that in (4.8)

$A = \sin 56° \cos 40°$ $= 0.635$

$B = \cos 56° \sin 40° \cos(-20°) = 0.338$

$C = \sin 40° \sin(-20°)$ $= -0.220$

$D = \cos 56° \cos 40°$ $= 0.428$

$E = \sin 56° \sin 40° \cos(-20°) = 0.500$

and so

$$\cos\theta = (0.635 - 0.338)\sin(-17.5°)$$
$$+ [-0.220\sin(-30°) + (0.428 + 0.500)\cos(-30°)]\cos(-17.5°)$$
$$= 0.783$$

Thus

$$\theta = 38.5°$$

For several special geometries, the complicated formula (4.8) becomes greatly simplified. For example, Fig. 4.8 suggests that a collector oriented towards the equator will directly face the solar beam at noon if its slope β is equal to the latitude ϕ. In this case ($\gamma = 0$, $\beta = \phi$), (4.8) reduces to

$$\cos\theta = \cos\omega\cos\delta \qquad (4.9)$$

For a horizontal plane, $\beta = 0$ and (4.8) reduces to

$$\cos\theta_z = \sin\phi\sin\delta + \cos\phi\cos\omega\cos\delta \qquad (4.10)$$

Two cautions should be noted about (4.8), and other formulas like it that may be encountered.

(1) θ may exceed 90° (i.e. $\cos\theta$ negative) in early morning or late evening, when the sun is near the observer's horizon. In this case, the irradiance will be on the back of a fixed collector. Such effects must be recognized when analyzing particular systems.
(2) Formulas are normally derived for the case when all angles are positive, and in particular $\phi > 0$. Some northern writers pay insufficient attention to sign, with the result that their formulas do not apply in the southern hemisphere. Southern readers will be wise to check all such formulas, e.g. by constructing complementary diagrams such as Figs. 4.8(a), (b) in which $\theta' = \theta$ and checking that the formula in question agrees with this.

4.5.3 *Optimum orientation of a collector*
A concentrating collector (Section 6.8) should always point towards the direction of the solar beam (i.e. $\theta = 0$). The optimum direction of a fixed flat plate collector is perhaps not obvious, however. The insolation H_c received is the sum of the beam and diffuse components:

$$H_c = \int (G_b^* \cos\theta + G_d)\,dt \qquad (4.11)$$

A suitable collector orientation for most purposes is facing the equator (e.g. due north in the southern hemisphere) with a slope equal to the latitude, as in (4.9).

Other considerations will modify this for particular cases, e.g. the orientation of existing buildings and whether more heat is regularly required (or available) in mornings or afternoons. However, since $\cos\theta \approx 1$ for $\theta < 30°$, variations of $\pm 30°$ in azimuth or slope should have little effect on the total energy collected. Over the course of a year the angle of solar noon varies considerably, however, and it may be sensible to adjust the 'fixed' collector slope.

4.5.4 *Hourly variation of irradiance*
Some examples of the hourly variation of G_h are given in Fig. 4.10(a) for clear days and Fig. 4.10(b) for a cloudy day. On clear days (following Monteith, 1973), the form of Fig. 4.10(a) is

$$G_h \approx G_h^{max} \sin(\pi t')/N \qquad (4.12)$$

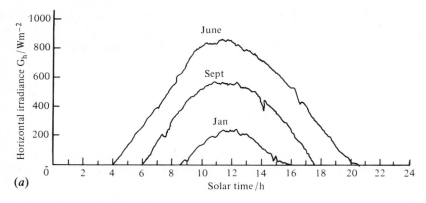

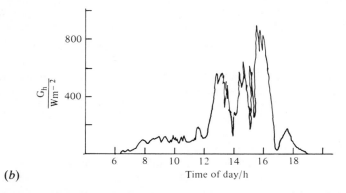

Fig. 4.10 (a) Irradiance on a horizontal surface, measured on three different almost *clear* days at Rothamsted (52°N, 0°W). Note how both the maximum value of G_h and the length of the day are much less in winter than summer. After Monteith (1973). (b) Typical variation of irradiance on a horizontal surface for a day of variable cloud. Note the low values during the overcast morning, and the large irregular variations in the afternoon due to scattered cloud

where t' is the time after sunrise and N is the duration of daylight for the particular clear day (see (4.7) and Fig. 4.10(a)). Integrating (4.12) over the daylight period for a clear day,

$$H_h \approx (2N/\pi)\, G_h^{max} \tag{4.13}$$

Thus for example at latitude $\pm 50°$ in midsummer, if $G_h^{max} \approx 900\,\mathrm{W\,m^{-2}}$ and $N \approx 16\,\mathrm{h}$, then $H_h \approx 33\,\mathrm{MJ\,m^{-2}\,day^{-1}}$ In the midwinter at the same latitude, $G_h^{max} \approx 200\,\mathrm{W\,m^{-2}}$ and $N \approx 8\,\mathrm{h}$, so $H_h \approx 3.7\,\mathrm{MJ\,m^{-2}\,day^{-1}}$. In the tropics $G_h^{(max)} \approx 950\,\mathrm{W\,m^{-2}}$, but the daylight period does not vary greatly from $12\,\mathrm{h}$ throughout the year. Thus $H_h \approx 26\,\mathrm{MJ\,m^{-2}\,day^{-1}}$.

These calculations make no allowances for cloud or dust, and so average measured values of H_h are always less than those mentioned. In most regions average values of H_h are typically 50–70% of the clear sky value. Only desert areas will have higher averages.

4.6 Effects of the earth's atmosphere

4.6.1 *Air mass ratio*

The distance travelled by the direct beam through the atmosphere depends on the angle of incidence to the atmosphere (the zenith angle) and the height above sea level of the observer (Fig. 4.11). We consider a clear sky with no cloud, dust or air pollution. Since the top of the atmosphere is not well defined, of more importance than the distance travelled is the mass of atmospheric gases and vapors encountered. For the direct beam at normal incidence passing through the atmosphere at normal pressure, a standard mass of atmosphere will be

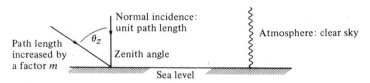

Fig. 4.11 Air mass ratio $m = \sec\theta_z$

encountered. If the beam is at zenith angle θ_z, the increased path length compared with the normal path is called the *air mass ratio* (or air mass), symbol m. The abbreviation AM is used for air mass ratio. AM0 refers to zero atmosphere, i.e. radiation in outer space; AM1 refers to $m = 1$, i.e. sun overhead; AM2 refers to $m = 2$; and so on.

From Fig. 4.11, since no account is usually taken of the curvature of the earth,

$$m = \sec\theta_z \tag{4.14}$$

The differing air mass ratio encountered because of change in atmospheric pressure or change in height of the observer is considered separately.

4.6.2 *Atmospheric absorption and related processes*

As the solar short wave radiation passes through the earth's atmosphere a complicated set of interactions occurs. The interactions include *absorption*, the conversion of radiant energy to heat and the subsequent re-emission as long wave radiation; *scattering*, the wavelength dependent change in direction, so that usually no extra absorption occurs and the radiation continues at the same frequency; and *reflection*, which is independent of wavelength. These processes are outlined in Fig. 4.12.

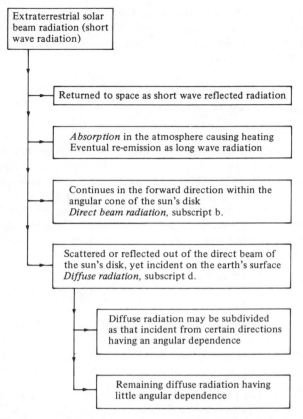

Fig. 4.12 Effects occurring as extraterrestrial solar radiation is incident upon the atmosphere

The effects and interactions that occur may be summarized as follows:

(1) *Reflection*   On average about 30% of the extraterrestrial solar intensity is reflected back into space. Most of the reflection occurs from clouds, with a small proportion from the earth's surface (especially snow and ice). The continuing short wave solar radiation has flux density $\sim(1-\rho_0)\times 1.3\,\mathrm{kW\,m^{-2}} \approx 1\,\mathrm{kW\,m^{-2}}$. This reflectance is called the albedo.

(2) *Greenhouse effect and long wave radiation*   If the radius of the earth is R, and the extraterrestrial solar irradiance (the solar constant) is G_0, then the received power is $\pi R^2(1-\rho_0)\,G_0$. This is equal to the power radiated from the earth system, of emittance $\epsilon = 1$ and mean temperature T_e as observed from space. At thermal equilibrium, since geothermal and tidal energy effects are negligible,

$$\pi R^2(1-\rho_0)\,G_0 = 4\pi R^2 \sigma T_e^4 \tag{4.15}$$

and hence $T_e \approx 250\,\mathrm{K} = -23°\mathrm{C}$.

In space, the long wave radiation from the earth has approximately the spectral distribution of a black body at 250 K. The peak spectral distribution at this temperature occurs at $10\,\mu$m, and the distribution does not overlap with the solar distribution (Fig. 4.13).

It is obvious from Fig. 4.13 that a definite distinction can be made between the spectral distribution of the sun's radiation (short wave) and that from the thermal sources of the earth and the earth's atmosphere (long wave). The infrared long wave fluxes at the earth's surface are themselves complex and large. The atmosphere radiates down to this surface as well as up and out into space. When measuring radiation or when determining the energy balance of an area of ground or a device, it is extremely important to be aware of the invisible infrared fluxes in the environment which often reach intensities of $1\,\mathrm{kW\,m^{-2}}$.

The effective black body temperature of the earth's system is equivalent to that of the outer atmosphere and not of the earth's surface. The earth's average surface temperature of about 14 °C is about 40 °C above the temperature of the outer atmosphere which acts as an infrared 'blanket'. This increase in temperature is called the *greenhouse effect*, since the glass of a horticultural glasshouse (a greenhouse) likewise prevents the transmission of infrared radiation from inside to out, but does allow the short wave solar radiation to be transmitted.

Since air is nearly transparent, a body on the earth's surface exchanges radiation not with the air immediately surrounding it, but with the air higher up in the atmosphere which is cooler. Considering this in terms of Fig. 3.14, the sky behaves as an enclosure at a temperature T_s, the *sky temperature*, which is lower than the ambient temperature T_a. A common estimate is

$$T_s \approx T_a - 6\,°\mathrm{C} \tag{4.16}$$

although in desert regions $(T_a - T_s)$ may be as large as 25 °C.

(3) *Absorption in the atmosphere* The solar short wave and the atmospheric long wave spectral distributions may be divided into regions to explain the important absorption processes. Fig. 4.14 will be useful to help the explanation.

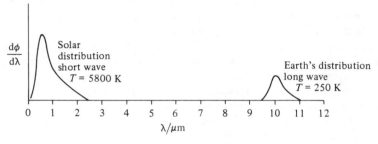

Fig. 4.13 The short and long wave spectral distributions

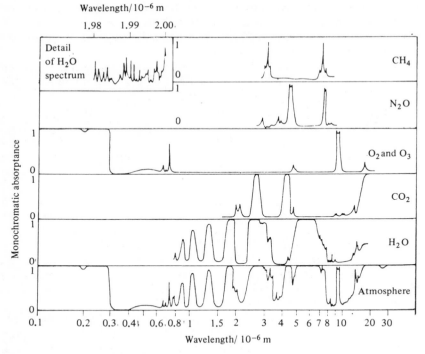

Fig. 4.14 Monochromatic absorptance versus wavelength of the atmosphere. The contributions (not to relative scale) of some main constituents are also shown. From Fleagle and Businger (1963)

(a) *Short wave ultraviolet region,* $\lambda < 0.3\,\mu m$ Solar radiation is completely removed at sea level by absorption in O_2, O_3, O and N_2 gases and ions.

(b) *Near ultraviolet region,* $0.3\,\mu m < \lambda < 0.4\,\mu m$ Only a little radiation is transmitted, but enough to cause sunburn.

(c) *Visible region,* $0.4\,\mu m < \lambda < 0.7\,\mu m$ The pure atmosphere is almost totally transparent to visible radiation, and becomes an open 'window' for solar energy to reach the earth. About half of the solar irradiance is in this spectral region. Note however that aerosol particulate matter and pollutant gases can cause significant absorption effects.

(d) *Near infrared (short wave) region,* $0.7\,\mu m < \lambda < 2.5\,\mu m$ Nearly 50% of the extraterrestrial solar radiation is in this region. Up to about 20% of this may be absorbed, mostly by water vapor and also by carbon dioxide in the atmosphere (Figs 4.14 and 4.15). The CO_2 concentration in the atmosphere is relatively constant at 0.03% by volume, but water vapor concentrations can vary greatly up to about 4% by volume. Thus fluctuations of absorption by water vapor could be of significance in practical applications; however, cloud associated with increased water vapor is likely to be of far greater significance.

(e) *Far infrared region,* $\lambda > 12\,\mu m$ The atmosphere is almost completely opaque in this part of the spectrum.

Fig. 4.15 shows the cumulative effect on the solar spectrum of these absorptions. The lower curve is the spectrum of the sun, seen through air mass ratio $m = 1$. This represents the radiation received near midday in the tropics (with the sun vertically above the observer). The spectrum actually received depends on dustiness and humidity, even in the absence of cloud (see Thekaekara, 1977 for details).

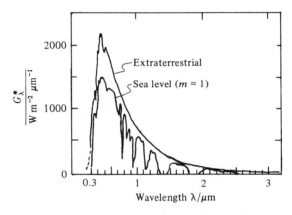

Fig. 4.15 Spectral distributions of solar radiation received above the atmosphere (upper curve) and at sea level (lower curve). About half the irradiance occurs in the visible region (0.4–0.7 μm). There is a gradual decrease of G_λ^* as λ increases into the infrared, with dips in the sea level spectrum due to absorption by H_2O and CO_2

4.7 Measurements of solar radiation

4.7.1 *Instruments*

Table 4.1 lists the commonest instruments used for measuring solar radiation. They are mostly variations on two basic types: a *pyroheliometer*, which measures the beam irradiance G_b^*, and a pyranometer or *solarimeter*, which measures total irradiance G_{tc} (Fig. 4.3).

Only the reference standard pyroheliometer gives an absolute reading. In this instrument, the solar beam falls on an absorbing surface of area A, whose temperature rise is measured and compared with the temperature rise in an identical (shaded) absorber heated electrically. In principle, then,

$$\alpha A G_b^* = P_{elec} \tag{4.17}$$

In practice there are many complications, e.g. the independent measurement of α.

Table 4.1 Classification of solar radiometers

Type	Measures	Stability $(\%y^{-1})$	Absolute accuracy (%)	Typical output for $1\,kW\,m^{-2}$	Auxiliary equipment needed	Approx. price (US$)	Notes
Reference standard pyroheliometer	Direct radiation (absolute, i.e. against electrical heating)	0.2	2	—	Varies with use	5000	Used mainly by standards bureaux to calibrate other instruments
WMO grade 1 solarimeter	Global radiation G_t	1	3	10 mV	Voltage integrator	2000 (total system)	Thermopile sensor
Bimetallic strip	Global radiation G_t	5	15	5 cm deflection	None	300	Comes with built-in chart recorder, driven by clockwork. Purely mechanical, but delicate
Solar cells	Global radiation G_t	2	15	10 mA	(Milli-ammeter) current integrator	100 (total system)	Nonuniform spectral response. Compact. Easily mounted on a collector
Actinometer (WMO grade 2)	Direct radiation G_b	2	4	10 mV	Voltage integrator	2000	Comes with tracking mechanism. Thermopile
Campbell-Stokes	Sunshine hours	10	20	Burnt chart	Special cardboard strips	100	Already in wide use
Human eye	% cloud	10	20	Visual scale	Training	—	One site only
Satellite photo	% cloud	10	20	Photo	Radio receiver + special plotter	20 000	Covers whole region. Can be used for general forecasting

4.8 Estimation of solar radiation

4.8.1 *Need for estimation*

Before installing a collector of solar energy, it is necessary to determine how much solar energy there will be to collect. Knowing this, and the projected pattern of energy usage from the device, it is possible to calculate the size of collector.

Ideally, the data required would be several years of measurements of irradiance on the proposed collector plane. This is very rarely available, so the required (statistical) measures have to be estimated from meteorological data available (1) at the site or (2) (more likely) at some 'nearby' site, which appears to have similar irradiance.

4.8.2 *Statistical variation*

In addition to the regular variations depicted in Figs 4.7 and 4.10(a) there are also substantial irregular variations. Of these the most significant for engineering purposes are perhaps the day-to-day fluctuations (e.g. Fig. 4.10(b)) as they affect the amount of energy storage that a solar energy system will require.

Thus even a complete record of past irradiance can be used to predict *future* irradiance only in a statistical sense. Therefore design methods usually rely on approximate averages, such as monthly means of daily insolation. To estimate these cruder data from other measurements is easier than to predict the whole pattern of irradiance.

4.8.3 *Sunshine hours as a measure of insolation*

All major meteorological stations measure daily the hours of bright sunshine, n. Records of this quantity are available for several decades. It is usually measured by a Campbell-Stokes recorder (Table 4.1), which comprises a specially marked card placed behind a magnifying glass. When the sun is 'bright' a hole is burnt in the card. The observer measures n from the total burnt length on each day's card.

Many attempts have been made to correlate insolation with sunshine hours, usually by an expression of the form

$$H = H_0'[a + b(n/N)] \tag{4.18}$$

where (for the day in question) H_0' is the clear day horizontal radiation (Fig. 4.7) and N is the length of the day ((4.7)).

Unfortunately, it has been found that the regression coefficients a and b vary from site to site. Moreover, the correlation coefficient is usually only about 0.7, i.e. the data are widely scattered from those predicted from the equation.

Sunshine hour data give a useful guide to the *variations* in irradiance. For example, it is safe to say that a day with $n < 1$ will contribute no appreciable energy to any solar energy system. The requirement for energy storage can therefore be assessed from the daily data. The records can also be used to assess whether, for instance, mornings are statistically sunnier than afternoons at the site.

Many other climatological correlations with insolation have been proposed, using such variables as latitude, ambient temperature, humidity and cloud cover. Most have a limited accuracy and range of applicability.

4.8.4 *Proportion of beam radiation*

As noted in Section 4.3, the proportion of incoming radiation that is focusable (beam component) depends on the cloudiness and dustiness of the atmosphere.

These factors can be measured by the *clearness index* K_T, which is the ratio of radiation received on a horizontal surface in a period (usually one day) to the radiation that would have been received on a parallel extraterrestrial surface in the same period:

$$K_T = H_h/H_{oh} \tag{4.19}$$

The clearest days have air mass ratio $m = 1$ and therefore $K_T \approx 0.8$. For these days the diffuse fraction is about 0.2; it rises to 1.0 on completely overcast days ($K_T = 0$). On a sunny day with significant aerosol or with a little cloud, the diffuse fraction can be as high as 0.5.

The proportion of beam radiation can be found by subtraction:

$$H_{bh}/H_{th} = 1 - H_{dh}/H_{th} \tag{4.20}$$

These values of H_{bh}/H_{th} suggest that it is difficult to operate focusing systems successfully in any but the most cloud-free locations. However notice that such systems track the sun, and therefore collect not the horizontal beam component H_{bh} but the larger normal beam component H_b^*.

4.8.5 *Effect of inclination*

If all solar radiation came in the solar beam, it would be straightforward to convert irradiances measured on one plane (plane 1) to those on another plane (plane 2). This is particularly important for transforming data from the horizontal plane. Equation (4.8) gives the angle of incidence of the beam to each plane. Then, for the beam component,

$$G_{1b}/\cos\theta_1 = G_{2b}/\cos\theta_2 = G_b^* \tag{4.21}$$

The calculation for the diffuse component, however, cannot be so precise.

Duffie and Beckman (1980) discuss many refinements of these methods. Although the uncertainty is more than 10%, the results are still instructive. For example, Fig. 4.16 shows the variation in estimated daily radiation on various slopes as a function of time of year, at a latitude of 45°N, and with clearness index $K_T = 0.5$. Note that at this latitutde, the average insolation on a vertical sun-facing surface varies remarkably little with season, and in winter exceeds $10 \, MJ \, m^{-2} \, day^{-1}$. This is double the insolation on a horizontal surface in winter, and is certainly high enough to provide a useful input to passive and some active heating systems (Section 6.4).

Problems

4.1 (a) Consider the sun and earth to be equivalent to two spheres in space. From the data given, determine the solar constant outside the atmosphere ($W \, m^{-2}$).

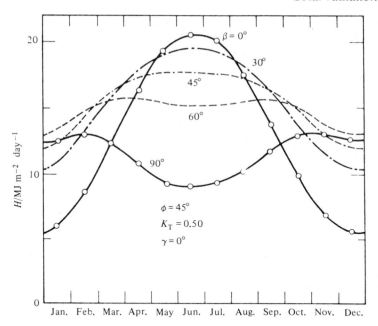

Fig. 4.16 Variation in estimated average daily insolation H on a surface at various slopes as a function of time of year. For latitude 45 °N, with $K_T = 0.50$, $\gamma = 0°$ and ground reflectance 0.20. From Duffie and Beckman (1980, p. 98)

(b) Suppose the earth has an average absorptance α and an average emittance ϵ. Determine the ratio α/ϵ for the earth to have equilibrium temperature of (i) 10 °C and (ii) 25 °C.

Data:

Sun diameter	$= 1.389 \times 10^9 \, \text{m}$
Earth diameter	$= 1.278 \times 10^7 \, \text{m}$
Sun–earth distance	$= 1.498 \times 10^{11} \, \text{m}$
Equivalent black body temperature	$= 5760 \, \text{K}$

4.2 Assume that the sign conventions for ω (hour angle) in Section 4.4.1, and for β (slope) and γ (surface azimuth) in Section 4.5.1 are correct for the northern hemisphere. By considering diagrams of appropriate special cases (e.g. Fig. 4.8) verify that the conventions are correct also for the southern hemisphere (e.g. a north-facing collector in the southern hemisphere has $\beta > 0$, $\gamma = 180°$).

4.3 At Suva ($\phi = -18°$) at 9 a.m. on 20 May, the irradiance measured on a horizontal plane was $G_h = 1.0 \, \text{MJ} \, \text{h}^{-1} \text{m}^{-2}$.

(a) Determine the angle θ_z between the beam radiation and the vertical, and hence find the irradiance $G^* = (G_b + G_d)^*$ measured in the beam direction.

(b) Determine the angle θ_c between the beam and a collector of slope 30° facing due north. Hence find the irradiance G_c on the collector. (Assume that all the radiation came from the beam direction.)

(c) What difference does it make to assume instead that the diffuse radiation is uniform around the sky?

4.4 Show that the radiative heat loss from a surface at temperature T to the sky (effectively at temperature T_s) may be written as

$$P_r = A_1 \epsilon \sigma (T_1^4 - T_s^4)$$
$$= A_1 h_r (T_1 - T_a)$$

where

$$h_r = \epsilon \sigma (T_1^2 + T_s^2)(T_1 + T_s)(T_1 - T_s)/(T_1 - T_a)$$

4.5 (a) From (4.10) find the hour angle at sunrise (when the zenith angle $\theta_z = 90°$). Hence show that the number of hours between sunrise and sunset is given by (4.7).

(b) Calculate the length of the day at midsummer and midwinter at latitudes of (i) 12° (ii) 60°.

4.6 (a) If the orbit of the earth were circular, then the irradiance on a horizontal plane outside the atmosphere would be

$$G'_{oh} = G_0^* \cos \theta_z \tag{4.22}$$

where G_0^* is the solar constant.

If ω_s is the hour angle at sunset (in degrees), show that the integrated daily extraterrestrial radiation on a horizontal surface is

$$H'_{oh} = (t_s G_0^*/\pi)[\cos \phi \cos \delta \sin \omega_s + (2\pi\omega_s/360) \sin \phi \sin \delta] \tag{4.23}$$

where t_s is the length of the day.

Note: because of the slight ellipticity of the earth's orbit, the extraterrestrial radiation is not H'_{oh} but

$$H_{oh} = [1 + e \cos(360n/365)]H'_{oh} \tag{4.24}$$

where $e = 0.033$ is the eccentricity of the orbit and n is the day number (e.g. $n = 1$ for 1 January).

(b) Use (4.24) to calculate H_{oh} for $\phi = 48°$ in midsummer and midwinter. Compare your answers with the clear sky radiation given in Fig. 4.7.

4.7 Derive (4.9), i.e.

$$\cos \theta = \cos \omega \cos \delta$$

from first principles. (This formula gives the angle θ between the beam and the normal to a surface having azimuth $\gamma = 0$, slope $\beta = |\text{latitude}|$.

Hint: Construct an (x, y, z) coordinate system centered on the earth's center with the N pole on Oz and the sun in the plane $y = 0$, and find the direction cosines of the various directions.

Note: The derivation of the full formula (4.8) is similar but complicated. See Coffari (1977) for details.

Solutions

4.1 (a) $G_0 = \sigma T_s^4 (4\pi R_s^2)/4\pi l^2 = 1340\,\text{W}\,\text{m}^{-2}$.

(b) In equilibrium,

$$\alpha G_0 \pi R_E^2 = \epsilon \sigma T_E^4 4\pi R_E^2$$

So $\alpha/\epsilon = $ (i) 1.08, (ii) 1.33 for T_E (i) 283 K (ii) 298 K.

4.3 (a) Use (4.8) with $\phi = -18°$, $\delta = 20.0°$, $\beta = 0$, γ arbitrary but irrelevant (γ terms all multiplied by $\sin\beta$), $\omega = 45°$. Hence $\theta_1 = \cos^{-1}(0.526) = 58°$, with $G_d = 0$, $G_b = G_b^* = G_h/\cos\theta_1 = 0.9\,\text{MJ}\,\text{h}^{-1}\text{m}^{-2}$.

(b) As for (a), with $\beta = +30°$, $\gamma = -180°$; $\theta_z = 40°$

(c) Assume $G_{bh} = G_{dh} = \frac{1}{2}G_h = 0.5\,\text{MJ}\,\text{m}^{-2}\text{h}^{-1}$. Then $G_b^* = G_{bh}/\cos\theta_1 = 0.9$, and so

$$G^* = G_b^* + G_d^* = 0.9 + 0.5 = 1.4$$

(assuming $G_d^* = G_{dh} = G_{dc}$) and

$$G_c = G_b^* \cos\theta_c + G_{dc} = 1.2\,\text{MJ}\,\text{m}^{-2}\text{h}^{-1}.$$

4.5 (b) (i) 12.7 h, 11.3 h (ii) 18.5 h, 5.5 h.

4.6 (b) Summer $H_0 = 42.9\,\text{MJ}\,\text{m}^{-2}$, winter $H_0 = 8.2\,\text{MJ}\,\text{m}^{-2}$; $K_T = 0.7$.

Bibliography

General

Dickinson, W.C. and Cheremisinoff, P.N. (eds) (1982) *Solar Energy Technology Handbook*, Butterworths, London.
    Clear diagrams of geometry.

Duffie, J. A. and Beckman, W. A. (1980) *Solar Engineering of Thermal Processes*, Wiley, New York.

Meinel, A. B. and Meinel, M. P. (1977) *Applied Solar Energy*, Addison-Wesley, London.
    Chapters 2, 3 on solar flux availability are among the best in the book.

Monteith, J. (1973) *Principles of Environmental Physics*, Edward Arnold, London.
    Includes a concise description of the radiation environment near the ground.

Sayigh, A. A. M. (ed.) (1977) *Solar Energy Engineering*, Academic Press, London.
See especially the following chapters:

Thekaekara, M. P. 'Solar irradiance, total and spectral': first hand and first rate account.
Sayigh, A. A. M. 'Solar energy prediction from climatological data'.
Wood, B. D. 'Solar energy measuring equipment'.
Coffari, E. 'The sun and the celestial vault'.

Particular

Fleagle, R. C. and Businger, J. A. (1963) *An Introduction to Atmospheric Physics*, Academic Press, London.
NASA (1971) *Solar Electromagnetic Radiation*, SP-8005, Washington DC.
The present standard for irradiance (reprinted in Sayigh (1977) and elsewhere).
Revfeim, K. J. A. (1981) 'Estimating solar radiation income from "bright" sunshine records', *Q.J. Roy. Met. Soc.*, **107**, 427–35.

5 *Solar water heating*

5.1 Introduction

An obvious use of solar energy is for heating air and water. Dwellings in cold climates need heated air for comfort, and hot water for washing and other domestic purposes. Large volumes of hot water are used for process heat in industry. For example, some 20% of Australia's energy consumption is used for heating fluids to 'low' temperatures ($<100\,^\circ$C). Because of this, the manufacture of solar water heaters has become a thriving industry in several countries, especially Australia, Israel, the USA and Japan.

The principles and analysis that apply to solar water heaters apply also to many other systems which harness the sun's energy as heat, e.g. air heaters, solar stills for distilling water, crop driers, and solar 'power towers'. These other applications will be dealt with in Chapter 6.

In this chapter we discuss only water heating, starting with essentials and then discussing successively the various refinements depicted in Fig. 5.1. These refinements either increase the proportion of radiation absorbed by the heater or decrease the heat lost from the system. Table 5.1 shows that although each successive refinement increases the performance of the system, it also increases the cost.

The main part of a solar heating system is the collector, where solar radiation is absorbed and energy is transferred to the fluid. Collectors considered in this chapter are classed as *flat plate collectors*, in contrast to the focusing collectors discussed in Section 6.8. Flat plate collectors absorb both beam and diffuse radiation, and therefore still function when beam radiation is cut off by cloud. This advantage, together with their favorable cost (Table 5.1), means that flat plate collectors are preferred for heating to temperatures less than about $100\,^\circ$C.

The simpler collectors (Fig. 5.1(a)–(e)) hold all the water that is to be heated. The more refined collectors, Fig. 5.1(f)–(i), heat only a little water at a time. The heated water is then usually accumulated in a separate storage tank. As discussed in Section 5.5, this enables us to reduce the heat losses from the *system* as a whole.

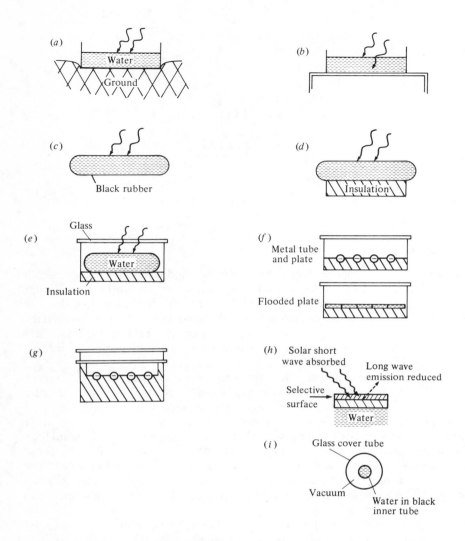

Fig. 5.1 A sequence of solar collectors, in order of increasing efficiency and cost. For detailed discussion see the text. (a) Open trough on ground. Heat flows easily to ground. (b) Open trough off ground. Clear water not a good absorber; loses heat by evaporation. (c) Black bag. Used in Japan for evening bath; high heat loss, especially to wind; no overnight storage. (d) Black bag insulated underneath. Heat losses confined to top surface, therefore only half those of (c). (e) Sheltered black bag. Cheap, but materials degrade. (f) Metal tube and plate, and flooded plate. Standard commercial collector; fluid moves through the collector, e.g. to a separate storage tank; flooded plate more efficient than tube and plate. (g) Double glazed flat plate. Better insulated version of (f); can operate up to ~100°C; iron-free glass less absorbing than window glass. (h) Selective surface. $\alpha_{short} \gg \epsilon_{long}$, radiative losses reduced. (i) Evacuated collector. No convection losses to the cover

Table 5.1 Summary of the typical performance for different types of collectors

Surface	Glazing	Figure	r_{pa} $\mathrm{m^2\,K\,W^{-1}}$	$T_p^{(m)}$ °C	Price $\mathrm{\$\,m^{-2}}$
Black	None	5.1(c)	0.031	40	20
Black	Single sheet	5.1(e), 5.1(f)	0.13	95	50–200
Black	Two sheets	5.1(g)	0.22	140	300
Selective	Single sheet	5.1(f), 5.1(h)	0.40	240	300
Selective	Two sheets	5.1(g), 5.1(h)	0.45	270	400
Selective	Evacuated tube	5.1(i)	0.48	160	500

Notes
(1) r_{pa} is the resistance to heat losses through the top of the collector for $T_p = 90\,°C$, $T_a = 20\,°C$, $u = 5\,\mathrm{m\,s^{-1}}$.
(2) $T_p^{(m)}$ is the (stagnation) temperature for which an irradiance of $750\,\mathrm{W\,m^{-2}}$ just balances the heat lost through r_{pa}. The actual working temperature is substantially less than this (see text).
(3) Prices are in US dollars as at 1983, and are very approximate ($\pm$ factor of 2). They do, however, give some relative indication.
(4) Calculations of r_{pa} and $T_p^{(m)}$ are in Examples 5.1, 5.2 and 5.5 and in Problems 5.3, 5.4, 5.5.

5.2 Calculation of heat balance: general remarks

All solar collectors include an absorbing surface called the *plate*. In Fig. 5.2 the radiant flux striking the plate is $\tau_{cov}A_p G$, where G is the irradiance on the collector, A_p is the exposed area of the plate and τ_{cov} is the transmittance of any transparent cover that may be used to protect the plate from the wind (e.g. Fig. 5.1(e)). The heat transfer terms are all defined in Chapters 3 or 4.

Only a fraction α_p of this flux is actually absorbed. Since the plate is hotter than its surroundings, it loses heat at a rate $(T_p - T_a)/R_L$, where R_L is the resistance to heat loss from the plate (temperature T_p) to the outside environment (temperature T_a). The *net* heat flow into the plate is

$$P_{net} = \tau_{cov}\alpha_p A_p G - [(T_p - T_a)/R_L]$$
$$= \eta_{sp}A_p G \tag{5.1}$$

where η_{sp} is the capture efficiency (<1). This is the Hottel-Whillier equation.

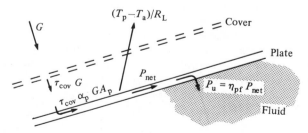

Fig. 5.2 Heat transfer from solar radiation to a fluid in a collector.

In general, only a fraction η_{pf} of P_{net} is transferred to the fluid at temperature T_f. In a well-designed collector the temperature difference between the plate and the fluid is small, and the *transfer efficiency* η_{pf} is only slightly less than 1. Thus the useful output power from the collector is

$$P_u = \eta_{pf} P_{net} \tag{5.2a}$$

$$= mc \, dT_f/dt \text{ if a static mass } m \text{ of fluid is being heated} \tag{5.2b}$$

$$= \dot{m}c(T_2 - T_1) \text{ if a mass } \dot{m} \text{ flows through the collector in unit time} \tag{5.2c}$$

In the second case, (5.2c), T_1 is the temperature of the fluid as it enters the collector and T_2 as it leaves the collector.

These equations are most commonly used to determine the output P_u for a given irradiance G. The parameters A, τ, α of the collector are usually specified, leaving R_L to be calculated using the methods of Chapter 3. Although T_p depends on P_u, one can usually make a reasonable first estimate and refine it later if required. This is illustrated in the following sections.

5.3 Unsheltered heaters

5.3.1 *Bare trough on the ground*

This is the simplest possible water 'heater' (Fig. 5.1(a)). A trough of water resting on the ground is exposed to the sun. An outdoor swimming pool is an example. On a hot sunny day the water is warmed, but the temperature rise is limited by the ease with which heat is conducted to the ground.

5.3.2 *Bare trough off the ground*

Supporting the trough off the ground decouples that heat sink (Fig. 5.1(b)), but the heating is still severely limited by the low absorptance of water ($\alpha = 1 - \tau \ll 1$). Moreover, much of the heat that is absorbed goes into increased evaporation, thus lessening the temperature increase.

5.3.3 *The black bag*

Here the water is enclosed in a shallow matt black bag, usually placed on a roof (Fig. 5.1(c)). So no heat is lost by evaporation. The black outer surface absorbs radiation much better than transparent water (typically $\alpha = 0.9$). Some of this absorbed heat is then passed to the water inside by conduction.

This type of heater is cheap, easy to make, and gives moderately hot water ($\sim 45°$). It is widely used in Japan to provide hot water for the evening bath.

Loss of heat by forced convection from wind severely limits the performance. Another problem is that many cheap black waterproof materials (e.g. old inner tyres, polythene bags) degrade in sunshine, and cease to be waterproof within a few months. However it may be cheaper to replace the container periodically than to use a more sophisticated design. Note that the bag material need not be black as supplied, but it is almost always worth making it black with matt paint or dye.

Example 5.1 The heat balance of an unsheltered black bag
A rectangular black rubber bag $1\,\text{m} \times 1\,\text{m} \times 0.1\,\text{m}$ with walls 5 mm thick is filled with 100 liters of water, supported on a thin, nonconductive, horizontal grid well above the ground, and exposed to a solar irradiance $G = 750\,\text{W m}^{-2}$ (Fig. 5.3). The ambient temperature T_a is 20 °C, and the wind speed is $5\,\text{m s}^{-1}$. Calculate the resistance to heat losses from the bag. Hence estimate the maximum average temperature of the water, and also the time taken to reach that temperature.

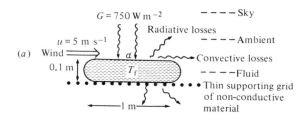

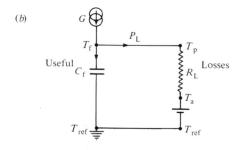

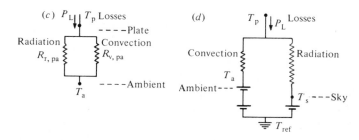

Fig. 5.3 Black bag solar water heater. (a) Physical diagram in section. (b) Simplified circuit analog. (c) R_L shown as parallel radiation and convection resistances from the plate to the same ambient temperature T_a. (d) R_L shown as parallel components losing heat to sinks at different, and possibly changing, temperatures

Solution

The heat going into or out of the water is conducted through the rubber skin, which is the 'plate' of this system (see Fig. 5.3(a)). Therefore the maximum temperature of the water cannot exceed that of the rubber. Indeed in equilibrium, when the water is hot, they are equal. Since the thermal capacity of the thin skin is much less than that of the water, and the conductive resistance of the skin is negligible, we shall treat the bag and contents as one composite object. This has temperature $T_p = T_f$, absorptance $\alpha = 0.9$, and thermal capacity $C_f = mc$. (In practice, the water at the top of the bag will be hotter than T_f, and that at the bottom colder; we neglect this for simplicity.) With this approximation, the resistance $R_{pf} = 0$ and $\eta_{pf} = 1$ in (5.2a).

From (5.1) and (5.2b), and with $\tau_{cov} = 1$ since there is no cover,

$$mc\frac{dT_f}{dt} = \alpha AG - (T_f - T_a)/R_L \tag{5.3}$$

In the circuit diagrams of Fig. 5.3 G acts as a current source in the analogy with electrical circuits. We shall set up the analysis as in Chapter 3, but in this example we shall also allow for the environmental temperature to change.

The capacitance C_f is shown connected between T_f and a reference T_{ref} in Fig. 5.3(b). T_{ref} is an arbitrary but fixed temperature which is independent of time. This corresponds to the fact that dT_f/dt on the left hand side of (5.3) can be replaced by $d(T_f - T_{ref})/dt$ if $dT_{ref}/dt = 0$. A convenient choice is $T_{ref} = 0°C$. Only if the ambient temperature is independent of time can we set $T_{ref} = T_a$ and still preserve the analogy between the circuit and the heat balance equation (5.3). The battery in the right arm of the analog circuit supplies the (possibly variable) 'voltage' $T_a - T_{ref}$.

The resistance R_L between the plate and the environment includes losses from both the top and the bottom of the collector. For this system, the top and bottom are similarly exposed to the environment, so that from a total exposed area $A_L = 2\,m^2$ there is a single outward heat flow by convection and radiation in parallel (Fig. 5.3(c)).

In many situations the heat sink temperatures for convection and for radiation are not equal. In general, convective loss is to the ambient air temperature, and radiative loss is to the sky and/or the environment. In Fig. 5.3(d) we establish the full circuit diagram for the heat loss component R_L. This circuit allows for the different, and possibly changing, heat sink temperatures. In this example, however, T_{sky} and T_a will be treated as constant.

The resistance to convective heat loss is

$$R_{v,pa} = 1/(h_v A_L) \tag{5.4}$$

where h_v is given by (C.15) as

$$h_v = a + bu = 24.7\,W\,m^{-2}\,K^{-1} \tag{5.5}$$

for the values given. The radiative heat flow to the sky is given by (C.17) as

$$P_{r,ps} = \epsilon_p \sigma A_L (T_p^4 - T_s^4) \tag{5.6}$$

where the effective temperature of the sky $T_s = T_a - 6\,\text{K}$ (see (4.16)).

It is convenient to write the heat flow (5.6) in the form

$$P_{r,ps} = h_{r,pa} A_L (T_p - T_a) \tag{5.7}$$

which will be identically equal to (5.6) if we take

$$\frac{1}{AR_{r,pa}} = h_{r,pa} = \frac{\epsilon_p \sigma (T_p^2 + T_s^2)(T_p + T_s)(T_p - T_s)}{T_p - T_a} \tag{5.8}$$

In this way, we can represent the losses as in Fig. 5.3(c), where the loss resistance $R_L = (1/R_{v,pa} + 1/R_{r,pa})$ is connected between the plate and ambient, as (5.3) would suggest. It can be verified that $h_{r,pa}$ depends only weakly on T_p. Numerically, taking $T_p = 40\,°C$ as a likely value, we find $h_{r,pa} = 7.2\,\text{W}\,\text{m}^{-2}\,\text{K}^{-1}$, $r_{pa} = 0.031\,\text{m}^2\,\text{K}\,\text{W}^{-1}$ and $R_L = 0.015\,\text{K}\,\text{W}^{-1}$.

The maximum temperature obtainable occurs when the input balances the losses and (5.3) reduces to (5.1) with $P_{net} = 0$:

$$(T_f - T_a)/R_L = \tau \alpha A_p G$$

Hence $T_f = 31\,°C$ for this uninsulated bag having $A_p = 1\,\text{m}^2$.

We estimate the time taken to reach this temperature by using (5.3) to find the rate at which T_f is increasing at the halfway temperature $T_f = 25\,°C$. Using the value of R_L calculated, we find $(dT_f/dt)_{25°C} = 8.1 \times 10^{-4}\,\text{K}\,\text{s}^{-1}$. The time for the temperature to increase by $11\,°C$ is then approximately

$$\Delta t = \Delta T/(dT_f/dt) = 1.3 \times 10^4\,\text{s} = 3.7\,\text{h}$$

In practice, the irradiance G varies through the day, so that the calculations give only the order of magnitude of ΔT and Δt. To obtain a more accurate answer we would have to evaluate (5.3) on an hour-by-hour basis, and also allow for the stratification of water.

5.3.4 *Black sack with bottom insulation*
The heat losses of the system of Fig. 5.1(c) can be almost halved simply by insulating the bottom of the collector (Fig. 5.1(d)). Almost any material that traps air in *small* holes ($\lesssim 1\,\text{mm}$) is useful as an insulator, e.g. fiberglass, expanded polystyrene or wood shavings. The thermal conductivity of all these materials is comparable with that of still air ($k \sim 0.03\,\text{W}\,\text{m}^{-1}\,\text{K}^{-1}$ (see Table B.3). The holes must not be too large or the air will carry heat by convection, and they must be dry since water is a much better conductor than air (see Appendix B).

Problem 5.2 shows that only a few centimeters of insulation are required to raise the bottom resistance to ten times the top resistance. This is almost always cheap enough to be cost effective.

5.4　Sheltered heaters

5.4.1　*Sheltered black sack*

Placing the bag of Fig. 5.1(d) in an open sunny location usually exposes it to cooling breezes. The bag can be sheltered from the wind by placing it in a covered box with a transparent lid (Fig. 5.1(e)).

Glass is the best material for this purpose. Clear polythene sheet is cheaper initially, but has to be replaced very frequently, as it disintegrates in sunlight. Also, Fig. 3.15 shows that glass has a significantly smaller transmittance for infrared radiation than polythene, so that glass offers more resistance to radiative heat loss from the plate. Special plastic covers are available for solar collectors that have similar properties to glass, but are tougher.

This system, with a total capital cost of about US$200, is often worth while in favorable conditions, although only in applications (e.g. domestic) where the supply of hot water is not to be relied on and not required in the early morning. The bag may have to be filled by hand.

Example 5.2　Heat balance of a sheltered collector
The black bag of Example 5.1 is placed inside a box with a glass lid 3 cm above it and 10 cm insulation below. For the same external conditions, again calculate the resistance to heat losses from the bag, the theoretical maximum average temperature of the water, and the time taken to attain it.

Solution
Fig. 5.4(a) shows the physical system and Fig. 5.4(b) its circuit analog. As before, we shall treat the bag and contents as a composite system having absorptance $\alpha = 0.9$ and thermal capacity $C_f = mc$. The temperature at which heat is lost from the system is the outside temperature of the bag T_p. To a first approximation $T_p = T_f$, the mean temperature of the water.

The plate loses heat by conduction through the base, which has an outside temperature $T_b \approx T_a$, so

$$P_b = (T_p - T_b)/R_b \approx (T_p - T_a)/R_b \tag{5.9}$$

Using (3.9) with approximate data,

$$P_b \approx \frac{(T_p - T_a)kA}{x} \approx \frac{(70 - 20)\,\text{K}\,(0.03\,\text{W m}^{-1}\text{K}^{-1})(1\,\text{m}^2)}{0.1\,\text{m}}$$

$$\approx 15\,\text{W}$$

which is negligible.

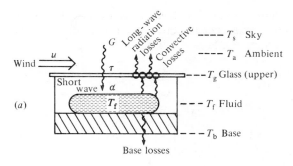

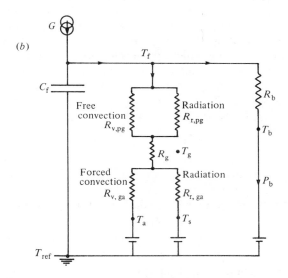

Fig. 5.4 (a) Sheltered black bag collector. (b) Circuit analog of (a)

We should incorporate the effective bottom resistance R_b into the loss resistance R_L of (5.1). In practice, as noted earlier, we can usually make R_b great enough so P_b is negligible. Thus the heat balance of the water is given by (5.1) in the form

$$mc\frac{dT_f}{dt} = \tau\alpha AG - (T_f - T_a)/R_{pa} \qquad (5.10)$$

The outward heat transfer occurs in the three stages indicated in Fig. 5.4(b):

(1) Free convection by the air in the gap carries heat to the glass. In parallel with this the plate radiates heat at wavelengths $\sim 10\,\mu m$. At these wavelengths

glass is not transparent but strongly absorbing (see Fig. 3.15). Therefore this radiation is not exchanged directly with the sky but is absorbed by the glass.
(2) The heat reaching the glass by these two mechanisms is then conducted to the outer surface of the glass.
(3) From here it is transferred to the surroundings by free and/or forced convection, and radiation.

Thus, the overall resistance between the top of the plate and the surroundings is

$$R_{pa} = \left\{ \frac{1}{R_{v,pg}} + \frac{1}{R_{r,pg}} \right\}^{-1} + R_g + \left\{ \frac{1}{R_{v,ga}} + \frac{1}{R_{r,ga}} \right\}^{-1} \tag{5.11}$$

In Fig. 5.4(b) the resistance R_g is negligible since the glass is thin ($\sim 5\,mm$) and a moderately good conductor ($k \approx 1\,W\,m^{-1}\,K^{-1}$). Therefore the temperature difference across the glass is also negligible. The convective and radiative resistances vary only slowly with the temperatures in the circuit, so the calculation can proceed with initial estimates for these temperatures:

$$T_p = 70\,°C$$

$$T_g = \tfrac{1}{2}(T_p + T_a) = 45\,°C$$

For our $1\,m^2$ collector, the convective resistance $R_{v,pg}$ follows directly from Example 3.2:

$$R_{v,pg} = 0.52\,K\,W^{-1}$$

Taking $\epsilon_p = \epsilon_g = 0.9$ for long wave radiation, the resistance to radiative heat transfer is, by a calculation similar to that in Example 3.6,

$$R_{r,pg} = 0.16\,K\,W^{-1}$$

Thus the total plate-to-glass resistance is given by

$$R_{pg} = [(1/R_{v,pg}) + (1/R_{r,pg})]^{-1} \tag{5.12}$$

$$= 0.12\,K\,W^{-1}$$

The resistance between the outside of the glass and the surroundings is just that already calculated for the unsheltered bag, namely $R_{ga} = 0.031\,K\,W^{-1}$. Putting these values into (5.11), we obtain $R_{pa} = 0.15\,K\,W^{-1}$. Then (5.10) with $dT_f/dt = 0$, $\tau = \alpha = 0.9$ and $G = 750\,W\,m^{-2}$ implies $T_p^{(m)} = 95\,°C$. Estimating $(dT/dt)_{60°C}$ as in Example 5.1 gives a time of 31 hours to reach maximum.

The calculation can be iterated with a better estimate of T_p, but the accuracy of the calculation hardly warrants this.

The calculated value of the maximum obtainable temperature is over-optimistic because we have neglected the periodicity of the solar radiation, which does not provide sufficient time for the maximum temperature to be reached. Nevertheless we can correctly conclude from Example 5.2 that

(1) The presence of a glass cover approximately quadruples the thermal resistance between the hot water and the outside air.
(2) A simple sheltered collector can yield water temperatures in excess of 50 °C.

5.4.2 *Metal plate collectors with moving fluid*

In the plate and tube collector (Fig. 5.1(f)) the water is confined in parallel tubes which are attached to a black metal plate. It is essential to have low thermal resistance between the plate and the tube, and across the plate between the tubes.

Typically the tube diameter is ~2 cm, the tube spacing ~20 cm and the plate thickness ~0.3 cm. The plate and tubes are sheltered from the wind in a box with a glass top. This collector has essentially the same circuit analog as the sheltered black bag (Fig. 5.4(b)) and therefore similar resistances. Flooded plate collectors are potentially more efficient than tube collectors because of increased thermal contact area. The heated fluid may be used immediately, or it may be stored and/or recirculated, as in Fig. 5.6.

5.4.3 *Efficiency of a flat plate collector*

A collector of area A_p exposed to irradiance G (measured in the plane of the collector) gives a useful output

$$P_u = A_p q_u = \eta_c A_p G \tag{5.13}$$

According to (5.1) and (5.2(a)) the collector efficiency η_c can be divided into two stages, the capture efficiency η_{sp} and the transfer efficiency η_{pf}:

$$\eta_c = \eta_{sp}\eta_{pf} \tag{5.14}$$

Equation (5.1) also shows that as the plate gets hotter its losses increase until, at the 'equilibrium' temperature $T_p^{(m)}$ η_{sp} decreases to zero. In a well-designed

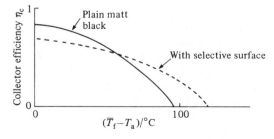

Fig. 5.5 Typical efficiency curve of a flat plate collector. $\overline{T}_f$ is the mean temperature of the working fluid

collector the temperature difference between the plate and the fluid is small and η_{pf} is high (see Problem 5.8). Typically $\eta_{pf} = 0.85$ and is almost independent of the operating conditions. The capture efficiency η_{sp} would vary linearly with temperature if R_L were constant, but the radiative resistance decreases appreciably as T_p increases. Therefore a plot of η_c against operating temperature has a slight curvature, as shown in Fig. 5.5.

The performance of a flat plate collector, and in particular its efficiency at high temperatures, can be substantially improved by

(1) Reducing the convective transfer between the plate and the outer glass cover, by inserting an extra glass cover (see Fig. 5.1(g) and Problem 5.5); and/or
(2) Reducing the radiative loss from the plate by making its surface not simply black but selective, i.e. strongly absorbing but weakly emitting (see Section 5.6).

The resulting gains in performance are summarized in Table 5.1.

5.5 Systems with separate storage

5.5.1 *Forced circulation*
The collectors of Fig. 5.1(f) can heat only a small volume of water which must be passed to an insulated tank for storage (Fig. 5.6). For domestic systems, tanks

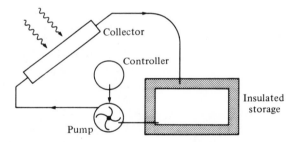

Fig. 5.6 Collector coupled to a separate storage tank by a pump

with a volume of about 100 to 200 liters can store a day's supply of hot water. For forced circulation only a small pump is needed. It is common to set the pumping rate so that the water temperature rises by about 4 °C in each passage through the collector. Single speed pumps are usually used, as they are the cheapest. Since the temperature rise depends on the solar irradiance G and the inlet temperature T_1, the '4 °C' rise will be achieved only for one set of conditions.

Example 5.3 Temperature rise through a collector
A flat plate collector measuring $2\,m \times 0.8\,m$ has a loss resistance $r_L = 0.13\,m^2\,K\,W^{-1}$ and a plate transfer efficiency $\eta_{pf} = 0.85$. The glass cover

has transmittance $\tau = 0.9$ and the absorptance of the plate is $\alpha = 0.9$. Water enters at a temperature $T_1 = 40°C$. The ambient temperature $T_a = 20°C$ and the irradiance in the plane of the collector is $G = 750\,W\,m^{-2}$.

(a) Calculate the flow rate needed to produce a temperature rise of $4°C$.
(b) Suppose the pump continues to pump at night, when $G = 0$. What will be the temperature *fall* in each passage through the collector? (Assume that $T_1 = 40°C$, $T_a = 20°C$ still.)

Solution
(a) From (5.1) and (5.2), the useful power per unit area is

$$q_u = (\rho c Q/A)(T_2 - T_1) = \eta_{pf}[\tau \alpha G - (T_p - T_a)/r_L] \tag{5.15}$$

Assuming $T_p = 42°C$ (the mean temperature of the fluid), this yields

$$Q = 3.5 \times 10^{-5}\,m^3\,s^{-1} = 130\,l\,h^{-1}$$

(b) From (5.15) with $G = 0$, $T_p = 38°C$ and the calculated value of Q,

$$T_2 - T_1 = -1.3°C$$

If the collector of Example 5.3 was part of a hot water system with a volume of 130 liters, circulating once per hour, then the water temperature at night would fall by $1.3°C\,h^{-1}$ because the pumping was continued. Therefore the pump has to be switched according to the prevailing conditions, and protection is needed in cold climates to prevent the water in the collector freezing.

An advantage of forced circulation is that an existing water heater system can easily be converted to solar input by adding collectors and a pump. The system is also likely to be more efficient, and the storage tank need not be higher than the collectors. A disadvantage however is that the system is probably dependent on mains electricity for the pump, which may be expensive or unreliable.

5.5.2 *Thermosyphon circulation*
The water circulation in a *thermosyphon* system (Fig. 5.7) is driven by the density difference between hot and cold water. Consider the simple system shown in Fig. 5.8(a), a closed vertical loop of pipe filled with fluid.

At the section aa′,

$$\int_{a(left)}^{b} \rho g\,dz - \int_{a(right)}^{b} \rho g\,dz > 0 \tag{5.16}$$

The left column of fluid is exerting a greater pressure at aa′ than the right column, thus setting the whole loop of fluid in motion.

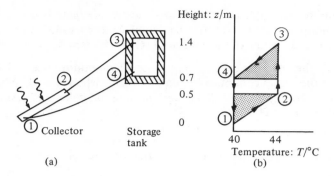

(a)   (b)

Fig. 5.7 Collector and storage tank with thermosyphon circulation. (a) Physical diagram. (b) Temperature distribution (see Example 5.4)

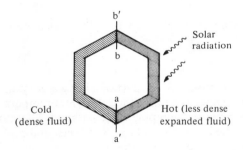

Fig. 5.8 Principle of thermosyphon flow

The driving pressure, which is precisely the left hand side of (5.16), can be expressed more generally as

$$p_{th} = \oint \rho g \, dz \tag{5.17}$$

where the circle denotes that the integral is taken around a *closed* loop. Note that dz in (5.17) is the vertical increment, and not the increment of length along the pipe. Equation (5.17) can be rewritten as

$$p_{th} = \rho_0 g H_{th} \tag{5.18}$$

where the *thermosyphon head*

$$H_{th} = \oint (\rho/\rho_0 - 1) \, dz \tag{5.19}$$

represents the energy gain per unit weight of the fluid and ρ_0 is any convenient reference density. This energy gain of the fluid can be lost by other processes, and in particular by pipe friction represented by the friction head H_f of Section 2.6.

The expansion coefficient β is usually constant,

$$\beta = -(1/\rho)\,d\rho/dT$$

Then (5.19) reduces to

$$H_{th} = -\beta I_T = -\beta \oint (T - T_0)\,dz \tag{5.20}$$

where T_0 is a reference temperature. Flow is in the direction for which I_T is positive.

Example 5.4 Calculation of thermosyphon flow
In the heating system of Fig. 5.7, water enters the collector at temperature $T_1 = 40\,°C$, is heated by $4\,°C$, and goes into the top of the tank without loss of heat at $T_3 = T_2 = 44\,°C$. If the system holds 100 liters of water, calculate the time for all the water to circulate once round the system. Assume the tank has time to achieve stable stratification.

Solution
The circulation and insulation insure that the coldest water at the bottom of the tank is at the same temperature as the inlet to the collector (i.e. $T_4 = T_1$). The integral $\oint (T - T_0)\,dz$ around the contour 1234 is just the area inside the curve (Fig. 5.7(b)). This area is the sum of the shaded triangles plus the middle rectangle, i.e.

$$I_T = \tfrac{1}{2}(0.5\,m)(4\,°C) + (0.2\,m)(4\,°C) + \tfrac{1}{2}(0.7\,m)(4\,°C) = +3.2\,mK$$

and is positive since the portion 123 (z increasing) lies to the right of the portion 341 (z decreasing). Therefore flow goes in the direction 1234. Taking a mean value $\beta = 3.5 \times 10^{-4}\,K^{-1}$ in (5.20) gives $H_{th} = -0.0010\,m$. This value will be sufficiently accurate for most purposes, but a more accurate value could be derived by plotting a contour of $\rho(z)$, using Table B.2 for $\rho(T)$, and evaluating (5.19) directly.

To calculate the flow speed, we equate the thermosyphon head to the friction head (2.12) opposing it. Most of the friction will be in the thinnest pipes, namely the riser tubes in the collector. Suppose there are four tubes, each of length $L = 2\,m$ and diameter $D = 12\,mm$. Then in each tube, using the symbols of Chapter 2,

$$H_{th} = 2fLu^2/Dg$$

where u is the flow speed in the tube and $f = 16\,\nu/uD$ for laminar flow.

Hence

$$u = \frac{gD^2 H_{th}}{32 L \nu}$$

$$= \frac{(1.0 \times 10^{-3} \mathrm{m})(12 \times 10^{-3} \mathrm{m})^2(9.8 \,\mathrm{m\,s^{-2}})}{(32)(2 \,\mathrm{m})(0.7 \times 10^{-6} \mathrm{m^2 s^{-1}})}$$

$$= 0.031 \,\mathrm{m\,s^{-1}}$$

Checking for consistency, we find the Reynolds number $uD/\nu = 540$, so that the flow is laminar as assumed.

The volume flow rate through the four tubes is

$$Q = 4(u\pi D^2/4) = 1.4 \times 10^{-5} \mathrm{m^3 s^{-1}}$$

Thus, if the system holds 100 liters of water, the whole volume circulates in a time of

$$(100)(10^{-3}\mathrm{m}^3)\left(\frac{1}{1.4 \times 10^{-5}\mathrm{m^3 s^{-1}}}\right)\left(\frac{1\,\mathrm{h}}{3.6 \times 10^3 \mathrm{s}}\right)$$

$$= 2.0 \,\mathrm{h}$$

5.6 Selective surfaces

5.6.1 *Ideal*

A solar collector absorbs radiation at wavelengths around $0.5\,\mu$m (from a source at $6000\,\mathrm{K}$) and emits radiation at wavelengths around $10\,\mu$m (from a source at $\sim 350\,\mathrm{K}$). Therefore an ideal surface for a collector would maximize its energy gain and minimize its energy loss, by having a high monochromatic absorptance α_λ at $\lambda \sim 0.5\,\mu$m and low monochromatic emittance ϵ_λ at $\lambda \sim 10\,\mu$m, as indicated schematically in Fig. 5.9. Such a surface has $\alpha_{short} >> \epsilon_{long}$, in contrast to (3.30). With a selective surface α and ϵ are weighted means of α_λ and ϵ_λ respectively, over *different* wavelength ranges (cf. (3.27)).

5.6.2 *Metal–semiconductor stack*

Some semiconductors have $\alpha_\lambda/\epsilon_\lambda$ characteristics which resemble those of an ideal selective surface. A semiconductor absorbs only those photons with energies greater than E_g, the energy needed to promote an electron from the valence to the conduction band (see Chapter 7). The critical energy E_g corresponds to a wavelength of $1.1\,\mu$m for silicon and $2\,\mu$m for Cu_2O; shorter wavelengths are strongly absorbed (Fig. 5.9). However, the low mechanical strength, low thermal conductivity and high cost of semiconductors make them unsuitable for the entire collector material.

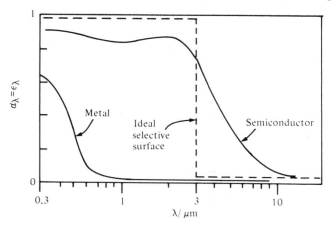

Fig. 5.9 Spectral characteristics of various surfaces. The metal shown is Cu, the semiconductor is Cu_2O

Metals, on the other hand, are usually strong, good conductors and relatively cheap. They are also unfortunately good reflectors (i.e. poor absorbers) in the visible and infrared.

When light (or other electromagnetic radiation) is incident on a metal, the free electrons near the surface vibrate rapidly in response to the varying electromagnetic field. Consequently the electrons constitute a varying current which radiates electromagnetic waves, as in a radio aerial. It appears to an outside observer that the incident radiation has been reflected. The power of the reflected wave is only just below that of the incident wave (Born and Wolf, 1975), so for $\lambda \geq 1\,\mu m$, $\rho_\lambda \approx 0.97$ (i.e. $\alpha_\lambda = \epsilon_\lambda \approx 0.03$) (see Fig. 5.9).

Some metals exhibit an increase in absorptance below a fairly short wavelength λ_p. For copper $\lambda_p \approx 0.5\,\mu m$ (see Fig. 5.9), so copper absorbs blue light more than red and appears reddish in colour. The wavelength λ_p corresponds to the 'plasma frequency' $f_p = c/\lambda_p$, which is the natural frequency of oscillation of an electron displaced about a positive ion. Net energy has to be fed to the electrons to make them oscillate faster than this frequency, so α_λ increases to about 0.5 for frequencies above f_p (i.e. wavelengths below λ_p).

By placing a thin layer of semiconductor over a metal, we can combine the desirable characteristics of both. Fig. 5.10 shows how the incoming short wave radiation is absorbed by the semiconductor. The absorbed heat is then passed by conduction to the underlying metal. Since the thermal conductivity of a semiconductor is low, the semiconductor layer should be thin to ensure efficient transfer to the metal. But it should not be too thin; otherwise some of the radiation would reach the metal and be reflected back.

Fortunately the absorption length of a semiconductor at $\lambda = 0.6\,\mu m$ is typically only $\sim 1\,\mu m$, i.e. 63% of the incoming radiation is absorbed in the top $1\,\mu m$, and 95% in the top $3\,\mu m$ (see Section 3.6), so the absorptance for solar radiation is high. The emitted radiation is at wavelengths $\sim 10\,\mu m$, for which the

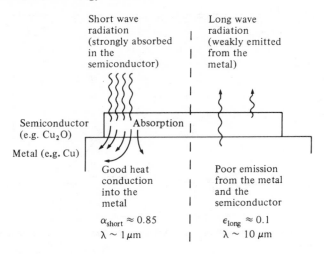

Short wave radiation (strongly absorbed in the semiconductor)

Long wave radiation (weakly emitted from the metal)

Semiconductor (e.g. Cu_2O)

Metal (e.g. Cu)

Absorption

Good heat conduction into the metal

Poor emission from the metal and the semiconductor

$\alpha_{short} \approx 0.85$
$\lambda \sim 1\,\mu m$

$\epsilon_{long} \approx 0.1$
$\lambda \sim 10\,\mu m$

Fig. 5.10 Heat flow in one type of selective surface. Here a semiconductor (which strongly absorbs solar short wave radiation) is deposited on a metal (which is a weak emitter of thermal long wave radiation)

emittance of both the metal and the semiconductor is low ($\epsilon \approx 0.1$, as in Fig. 5.10).

The result is a composite surface which has much lower radiative losses than a simple black-painted surface (which is black to both visible and infrared radiation, and therefore has $\alpha = \epsilon \approx 0.9$). The absorptance is not quite so high as that of a pure black surface, because α_λ of the selective surface decreases for $\lambda \geq 1\,\mu m$ (see Fig. 5.9), and 30% of the solar radiation is at wavelengths greater than $1\,\mu m$ (see Fig. 4.1).

The low emittance of the selective surface becomes more of an advantage as the working temperature increases, since the radiative losses increase as ϵT^4. For example, at a plate temperature of 40°C with $\epsilon > 0.9$, radiative losses are typically only 20% of the total (e.g. calculate these in Example 5.1); however, at a plate temperature of 400°C they would be 50% if $\epsilon = 0.9$ but only 10% if $\epsilon = 0.1$. (But see caution after (6.24) for $T > 1000$°C.)

One method for preparing an actual selective surface involves dipping a sheet of copper into an alkaline solution, so that a film of Cu_2O (which is a semiconductor) is formed on it. Fig. 5.9 shows the absorptance characteristics of such a surface, as prepared commercially.

Many other selective surfaces have been developed in laboratories, but few of these have yet come into commercial practice. This is because they are too expensive, or too fragile or cannot be reproduced, or because their characteristics deteriorate with time ('aging'). Many of these surfaces work by mechanisms different from that of the Cu/Cu_2O described, e.g. destructive interference of electromagnetic waves in thin films, multiple reflections in a roughened metal surface, and dispersed oxides within a metal (Meinel and Meinel, 1977; Seraphin, 1979).

All of these surfaces cost more to prepare than simply applying matt black paint. This, together with their poorer absorption at low temperatures, makes them uneconomic at present for collectors working below 60°C. But many of the advanced uses of solar heat now under development (e.g. the power towers described in Section 6.9.2) work at temperatures of several hundred degrees Celsius, and will require selective surfaces capable of withstanding years of fluctuating high temperatures while retaining $\alpha_{short}/\epsilon_{long}$ as high as possible (e.g. ~30).

5.7 Evacuated collectors

Using a selective absorbing surface substantially reduces the radiative losses from a collector. To obtain yet higher temperatures (e.g. to *deliver* heat at temperatures around or above 100°C, for which there is substantial industrial demand) it is necessary to reduce the convective losses as well. One way is to use extra layers of glass above a flat plate collector ('double glazing': see Fig. 5.1(g) and Problem 5.3). A method better but technically more difficult is to evacuate the space between the plate and its glass cover.

The basic component of an evacuated collector is a double tube, such as that shown in Fig. 5.11(a). The outer tube is made of glass because it is transparent to solar radiation but not to thermal radiation. The inner tube is also usually made of glass, which holds a vacuum better than most other materials. The outgassing rate from baked Pyrex glass is such that the pressure can be held below $0.1\,N\,m^{-2}$ for 300 years, which is about 10^{12} times longer than for a copper tube. The inner tube has a circular cross-section. This helps the fairly weak glass withstand the tension forces produced in it by the pressure difference between the fluid inside and the vacuum outside. Typically the tubes have outer diameter $D = 2\,cm$ and inner diameter $d = 1\,cm$. By suitably connecting an array of these tubes, collectors may receive both direct and diffuse solar radiation. Many variations on the basic geometry have been tried, but that of Fig. 5.11(a) is the simplest to analyze.

Example 5.5 Heat balance of an evacuated collector
Calculate the loss resistance of the evacuated collector of Fig. 5.11(a) and estimate its stagnation temperature. Take $D = 2\,cm$, $d = 1\,cm$ and $G = 750\,W\,m^{-2}$, $T_a = 20°C$, $u = 5\,m\,s^{-1}$.

Solution
The symbols and methods of Chapter 3 are used.

The circuit analog is shown in Fig. 5.11(b). It has no convective pathway between the plate (inner tube) and glass (outer tube). The only convection is from the outer glass to the environment.

Consider a unit length of tube, and assume $T_p = 100°C = 373\,K$, $T_g = 40°C = 313\,K$. (This value of $T_g < \frac{1}{2}(T_p + T_a)$, because the idea of the design is to

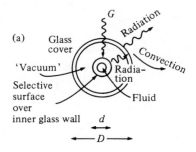

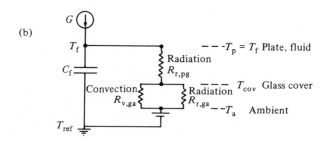

Fig. 5.11 (a) Evacuated collector. (b) Circuit analog of (a)

make $r_{pg} > r_{ga}$.) Treating the two tubes as close parallel surfaces, and taking $\epsilon_p = 0.1$ (a fairly conservative value) and $\epsilon_g = 1$, then by (3.6), (3.43) and (C.18) we obtain

$$1/r_{pg} = \sigma\epsilon_p(T_p^2 + T_g^2)(T_p + T_g) = 0.92\,\mathrm{W\,m^{-2}\,K^{-1}}$$

Taking the characteristic area A_{pg} to be that of a cylinder of length 1 m and mean diameter 1.5 cm, we find

$$R_{pg} = r_{pg}/A_{pg} = 23\,\mathrm{K\,W^{-1}}$$

For the outside surface of area $A_g = 2\pi\,(2\,\mathrm{cm})(1\,\mathrm{m}) = 0.125\,\mathrm{m^2}$, the loss coefficients are approximately, by (C.15),

$$h_{v,ga} = (5.7 + au)\,\mathrm{W\,m^{-2}\,K^{-1}}$$

where a $= 3.8\,\mathrm{m^{-1}\,s}$
By (3.6), (3.8) and (3.44)

$$h_{r,ga} = 4\sigma\left(\frac{T_g + T_a}{2}\right)^3 = 6.3\,\mathrm{W\,m^{-2}\,K^{-1}}$$

Hence

$$R_{ga} = (h_{v,ga} + h_{r,ga})^{-1}/A_g = 0.25\,\mathrm{K\,W^{-1}}$$

and

$$R_{pa} = R_{ga} + R_{pg} = 23.2\,\mathrm{K\,W^{-1}}$$

Note how the radiation resistance R_{pg} dominates, with no convection to 'short circuit' it. It does not matter that the mixed convection formula (C.15) does not strictly apply to this geometry, since it will underestimate the resistance from a curved surface.

Since each 1 m of tube occupies the same collector area as a flat plate of area $0.02\,\mathrm{m^2}$, we could say that the equivalent resistance of unit area of this collector is $r_{pg} = 0.48\,\mathrm{m^2\,K\,W^{-1}}$, although this figure does not have the same significance as for true flat plates.

To calculate the heat balance on a single tube, we note that the input is to the projected area of the inner tube, whereas the losses are from the entire outside of the larger outer tube. At equilibrium input equals output,

$$\tau\alpha_p Gd = (T_p - T_a)/R_{pa}$$

giving, for the maximum (stagnation) temperature

$$T_p^{(m)} = 156\,^\circ\mathrm{C} \text{ for } \tau = 0.9,\ \alpha_p = 0.85$$

This temperature is less than that listed for the double glazed flat plate in Table 5.1, but $T_p^{(m)}$ – and more importantly the outlet temperature T_2 when there is flow in the tubes – can be raised by increasing the input to each tube, e.g. by placing a mirror behind it. Even a flat white sheet helps, especially since it feeds some of the radiation losses back to the outer tube.

Problems

5.1 In a sheltered flat plate collector, the heat transfer between the plate and the outside air above it can be represented by the network of Fig. 5.12, where T_p, T_g and T_a are the mean temperatures of plate, glass and air respectively.

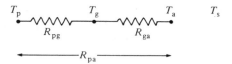

Fig. 5.12 Thermal resistances for Problem 5.1

(a) Show that

$$T_g = T_a + (R_{ga}/R_{pa})(T_p - T_a)$$

Verify that, for $T_p = 70°C$ and the resistances calculated in Example 5.2, this implies $T_g = 32°C$.

(b) Recalculate the resistances involved, using this second approximation for T_g instead of the first approximation of $\frac{1}{2}(T_p + T_a) = 45°C$ used in the example, and verify that the effect on the overall resistance r_{pa} is small.

(c) Use the resistances calculated in (b) in the formula in (a) to calculate a third approximation for T_g. Is a further iteration justified?

5.2 The collector of Example 5.2 had a resistivity to losses from the top of $r_{pa} = 0.13\,m^2\,K\,W^{-1}$. Suppose the *bottom* of the plate is insulated from the ambient (still) air by glass wool insulation with $k = 0.034\,W\,m^{-1}\,K^{-1}$. What thickness of insulation is required to insure that the resistance to heat loss at the bottom is (a) equal to (b) 10 times the resistance of the top?

5.3 A certain flat plate collector has *two* glass covers. Draw a resistance diagram showing how heat is lost from the plate to the surroundings, and calculate the resistance (for unit area) r_{pa} for losses through the covers. (Assume the standard conditions of Example 5.2.) Why will this collector need thicker *bottom* insulation than a single glazed collector?

5.4 Calculate the top resistance r_{pa} of a flat plate collector with a single glass cover and a *selective* surface. (Assume the standard conditions of Example 5.2.) See Fig. 5.1(h).

5.5 Calculate the top resistance of a flat plate collector with double glazing and a selective surface. (Again under the standard conditions.) See Fig. 5.1(g).

5.6 Bottled beer is pasteurized by passing hot water (at 70°C) over it for 10 minutes. It is found that to do this properly 50 liters of hot water has to be passed over each bottle. The water is recycled, so that its minimum temperature is 40°C.

(a) A brewery proposes to use solar energy to heat this water. What form of collector would be most suitable for this purpose? Given that the brewery produces 65000 bottles in an 8h working day, and that the irradiance at the brewery is $20\,MJ\,m^{-2}\,day^{-1}$ (on a horizontal surface), calculate the minimum collector area required, assuming no losses.

(b) Refine your estimate of the required collector area by allowing for the usual losses from a single glazed flat plate collector. (Make suitable estimates for G, T_a, u.)

(c) For this application, would it be worth while using collectors with (i) double glazing (ii) selective surface?

Justify your case as quantitatively as you can.
Hint: Use the results summarized in Table 5.1.

5.7 Some of the radiation reaching the plate of a glazed flat plate collector is reflected from the plate to the glass and back to the plate, where a fraction α of that is absorbed, as shown in Fig. 5.13.

   (a) Allowing for multiple reflections, show that the product $\tau\alpha$ in (5.1) and (5.10) should be replaced by

$$(\tau\alpha)_{\text{eff}} = \frac{\tau\alpha}{1-(1-\alpha)\rho_{\text{d}}}$$

   where ρ_{d} is the reflectance of the cover system for diffuse light.

   (b) The reflectance of a glass sheet increases noticeably for angles of incidence greater than about 45° (why?). The reflectance ρ_{d} can be estimated as the value for incidence of 60°; typically $\rho_{\text{d}} \approx 0.7$. For $\tau = \alpha = 0.9$, calculate the ratio $(\tau\alpha)_{\text{eff}}/\tau\alpha$, and comment on its effect on the heat balance of the plate.

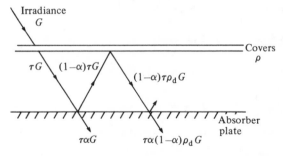

Fig. 5.13 Multiple reflections between collector cover(s) and plate (for Problem 5.7)

5.8 *Fin efficiency*
   Figure 5.14 shows a tube and plate collector. An element of the plate $dx\,dy$ absorbs some of the heat reaching it from the sun, loses some to the surroundings, and passes the rest by conduction along the plate (in the x direction) to the bond region above the tube. Suppose the plate has conductivity k and thickness δ, and that the section of plate above the tube is at constant temperature T_{b}.

   (a) Show that in equilibrium the energy balance on the element of the plate can be written

$$k\delta\frac{d^2 T}{dx^2} = (T - T_{\text{a}} - \tau G r_{\text{pa}})/r_{\text{pa}}$$

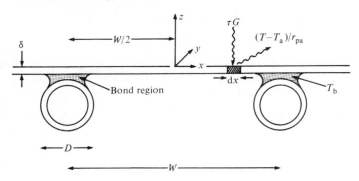

Fig. 5.14 Cross-section of a tube and plate collector (for Problem 5.8)

(b) Justify the boundary conditions

$$\frac{dT}{dx} = 0 \text{ at } x = 0$$

$$T = T_b \text{ at } x = (W - D)/2$$

(c) Show that the solution of (a), (b) is

$$\frac{T - T_a - \tau Gr_{pa}}{T_b - T_a - \tau Gr_{pa}} = \frac{\cosh mx}{\cosh m(W - D)/2}$$

where $m^2 = 1/(k\delta r_{pa})$, and that the *heat* flowing into the bond region from the side is

$$(W - D)F[\tau C - (T_b - T_a)/r_{pa}]$$

where the fin efficiency is given by

$$F = \frac{\tanh m(W - D)/2}{m(W - D)/2}$$

(d) Evaluate F for $k = 385 \, \text{W m}^{-1} \text{K}^{-1}$, $\delta = 1 \, \text{mm}$, $W = 100 \, \text{mm}$, $D = 10 \, \text{mm}$.

5.9 What happens to a thermosyphon system at night? Show that if the tank is wholly above the collector then the system can stabilize with $H_{th} = 0$, but that a system with the tank lower (in parts) will have a reverse circulation. *Hint*: construct temperature–height diagrams like Fig. 5.7(b).

Solutions

5.1 (b) $h_{v,pg} = 2.4\,\text{W}\,\text{m}^{-2}\text{K}^{-1}$

$h_{r,pg} = 6.2\,\text{W}\,\text{m}^{-2}\text{K}^{-1}$

hence

$r_{pg} = 0.12\,\text{m}^2\text{K}\,\text{W}^{-1}$

$h_{v,ga} = 24.7\,\text{W}\,\text{m}^{-2}\text{K}^{-1}$

$h_{r,ga} = 7.9\,\text{W}\,\text{m}^{-2}\text{K}^{-1}$

hence

$r_{ga} = 0.030\,\text{m}^2\text{K}\,\text{W}^{-1}$

$r_{pa} = 0.15\,\text{m}^2\text{K}\,\text{W}^{-1}$ (cf. 0.13 for $T_g = 45\,°\text{C}$).

(c) 30°C. No.

5.2 (a) 4 mm (b) 44 mm.

5.3 In effect, the plate to glass circuit of Fig. 5.4(b) occurs twice.
$r_{pa} = 0.22\,\text{m}^2\text{K}\,\text{W}^{-1}$.

5.4 Same circuit as Fig. 5.4(b) but with $\epsilon_p = 0.1$, $r_{pa} = 0.40\,\text{m}^2\text{K}\,\text{W}^{-1}$.

5.5 Circuit as for Problem 5.3 but with $\epsilon_p = 0.1$. $r_{pa} = 0.45\,\text{m}^2\text{K}\,\text{W}^{-1}$.

5.6 Use (5.2) in the form

$$mc\,\Delta T/\Delta t = A\tau_{cov}\alpha G - (T_p - T_a)/r_{pa}$$

with $m = 65\,000$ bottles $\times$ 50 kg/bottle, $\Delta T = 30\,°\text{C}$, $\Delta t = 8\,\text{h}$, $G = (20\,\text{MJ}$ $\text{m}^{-2})/(8\,\text{h}) = 700\,\text{W}\,\text{m}^{-2}$ (average), to give

	τ_{cov}	α	$\dfrac{r_{pa}}{\text{m}^2\,\text{K}\,\text{W}^{-1}}$	$\dfrac{A}{10^4\,\text{m}^2}$	Collector price $\$\,\text{m}^{-2}$	Collector price M\$
(a)	1.0	1.0	∞	2.0	–	–
(b)	0.9	0.9	0.13	4.3	100	4.3
(c) (i)	$(0.9)^2$	0.9	0.22	3.8	130	4.9
(ii)	$(0.9)^2$	0.9	0.40	3.3	160	52

5.7 (a) Note that

$$\tau_\alpha + \tau_\alpha(1-\alpha)\rho_d + \tau_\alpha(1-\alpha)^2\rho_d^2 + \ldots = \tau_a/[1-(1-\alpha)\rho_d]$$

(b) 1.08

5.8 (a) Conduction term is $k\delta\,dy[(dT/dx)_x - (dT/dx)_{x+dx}]$

(c) 0.98

Bibliography

General

Charters, W. W. S. (1977) 'Solar energy utilization – liquid flat-plate collectors', in Sayigh, A. (ed.) *Solar Energy Engineering*, Academic Press, London.
Brief but clear account.

Duffie, J. A. and Beckman, W. A. (1980) *Solar Engineering of Thermal Processes*, John Wiley and Sons, New York.
The standard work on this subject, including not just the collectors but the *systems* of which they form part.

Specialized references

Born, M. and Wolf, W. (1975) *Principles of Optics*, Pergamon Press, Oxford.
Electromagnetic theory of absorption etc. Heavy going!

Close, D. J. (1982) 'The performance of solar water heaters with natural circulation', *Solar Energy*, **5**, 33–40.
Theory and experiments on thermosyphon systems. Later articles in the same journal elaborate on this.

Meinel, A. B. and Meinel, M. P. (1977) *Applied Solar Energy*, Addison-Wesley, London.
Chapter 9 describes many types of selective surface.

Seraphin, B. O. (ed.) (1979) *Solar Energy Conversion: Solid State Physics Aspects*, Springer-Verlag.
Includes two substantial articles on selective surfaces.

6 Other uses for solar heat

6.1 Introduction

There are many more uses for solar heat than just heating water. However, the theory of heat transfer and storage for all uses has much in common with Chapter 5, and so in this chapter we shall introduce main concepts only.

Solar heat can be used to heat air, either for drying crops (Section 6.3) or for heating buildings (Section 6.4). Both these applications are of major economic importance. Much of the present world grain harvest is lost to fungal attack, which could be prevented by proper drying. Keeping buildings warm in winter accounts for up to half of the energy requirements of cold countries (see Fig. 16.2). Even a partial contribution to this load, by designing or redesigning buildings to make use of solar heating, could save billions of dollars' worth of fuel per year.

Crop drying requires the transfer not only of heat but also of water vapor. This is especially so in the solar distillation systems discussed in Section 6.6, which use solar heat to distil fresh (drinking) water from brackish impure water.

Heat engines convert heat into work (which may in turn be converted to electricity), and can be powered by solar radiation. Indeed, since the efficiency of heat engines rises with their working temperature, there are theoretical advantages in using solar radiation which arrives at a thermodynamic temperature of 6000 K, as discussed in Section 6.8. High temperatures – though not as high as 6000 K – can be obtained by concentrating the incident radiation on to a small area. Devices for doing this are treated in Section 6.8, and their application to electricity production and other systems in Section 6.9.

6.2 Air heaters

Hot air is required for two main purposes: drying crops (Section 6.3) and warming people (Section 6.4). Solar air heaters are similar to the solar water heaters of Chapter 5 in that the fluid is warmed by contact with a radiation absorbing surface. In particular, the effects on their performance of orientation and heat loss by wind etc. are very similar for both types.

Two typical designs are shown in Fig. 6.1. Equation (2.6) gives the useful heat flow into the air:

$$P_u = \rho c Q (T_2 - T_1) \qquad (6.1)$$

The density of air is 0.001 that of water, and so for the same energy input air can be given a much greater volumetric flow rate Q. Since the thermal conductivity of air is much lower than that of water for similar circumstances, the heat transfer from the plate to the fluid is much poorer. Therefore air heaters of the type shown in Fig. 6.1(a) are often built with roughened or grooved plates, to increase the surface area and turbulence available for heat transfer to the air. An alternative strategy is to increase the contact area by using porous or grid collectors (Fig. 6.1(b)). The analysis of internal heat transfer is complicated, because the same molecules carry the useful heat and the convective heat loss, i.e. the flow 'within' the plate and from the plate to the cover are coupled, as indicated in Fig. 6.2. The usual first approximation is to ignore this coupling and to use (5.1) as for water heaters; this is done in the following sections. Note that

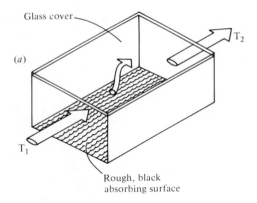

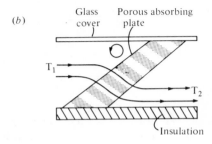

Fig. 6.1 Two designs of air heater

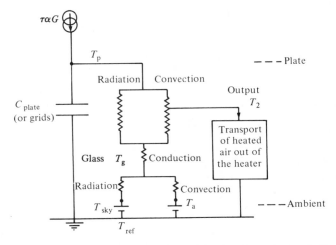

Fig. 6.2 Heat circuit for the air heater of Fig. 6.1 (a). Note how air circulation within the heater makes the exit temperature T_2 less than the plate temperature T_p. Symbols of Chapter 5

air heaters are cheap because they do not have to contain a heavy fluid, can be built of light, local materials, and do not require frost protection.

6.3 Crop driers

Most agricultural crops which are intended to be stored before use have to be dried first. Otherwise insects and fungi, which thrive in moist conditions, render them unusable. Examples include wheat, rice, coffee, copra (coconut flesh) and timber. We shall consider grain drying, but the other cases are similar. All crop drying involves transfer of water from the crop to the air around it, so we must first determine how much water the air can accept as water vapor.

6.3.1 *Water vapor and air*

The *absolute humidity* (or 'vapor concentration') χ is the mass of water vapor in $1\,m^3$ of air. At a particular temperature T, if we try to increase χ beyond saturation (e.g. by blowing in steam), liquid water condenses. The saturation humidity χ_s depends strongly on temperature (Table B.2(b)). A plot of χ_s (or some related measure of humidity) against T is called a *psychometric chart* (Fig. 6.3). The ratio χ/χ_s is called the *relative* humidity, and ranges from 0% (completely dry air) to 100% (saturated air). Many other measures of humidity are also used (Monteith, 1973).

Consider air with the composition of point B in Fig. 6.3. If it is cooled without change in moisture content, its representative point moves horizontally to A. Alternatively the air can be cooled by using it to evaporate water. If this happens in a closed system with no other heat transfer (i.e. the air cools adiabatically), the

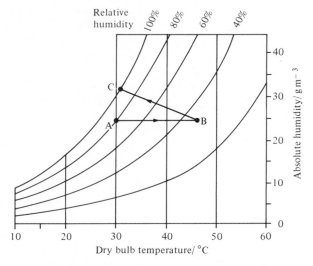

Fig. 6.3 Psychometric chart (for standard pressure $101.3\,kN\,m^{-2}$)

humidity of the air rises, and its representative point moves diagonally upwards (BC).

6.3.2 *Water content of crop*

The percentage *moisture content* (dry basis) w of a sample of grain is defined by

$$w = (m - m_0)/m_0 \tag{6.2}$$

where m is the total mass of the sample 'as is' and m_0 is the mass of the dry matter in the sample (m_0 can be determined for wood by drying the sample in an oven at $105\,°C$ for 24 hours). We shall always use this definition of moisture content ('dry weight' basis), which is standard in forestry. In other areas of agriculture, moisture content on a 'wet weight' basis may be used:

$$w' = (m - m_0)/m$$

$$= w/(w + 1) \tag{6.3}$$

The determination of m_0 requires care, and should be measured in a laboratory according to the standard procedures for each crop or product. The temperature and time for drying to determine 'oven dry mass' is limited so that other chemical changes do not occur. Some chemically bound water may remain after this process. It is also important to realize that there are limiting temperatures for drying crops for storage, so the product does not crack and allow bacterial attack. Much further detail is available in the references listed at the end of this chapter.

If left for long enough, a moist grain will give up water to the surrounding air until the grain reaches its *equilibrium moisture content* w_e. w_e depends on the crop, and especially on the temperature and humidity of the surrounding air. For rice in air at 30°C and 80% relative humidity (typical of rice growing areas), $w_e \approx 0.16$.

Note that the drying process is not uniform. Much of the moisture present in a crop is 'free water', which is only loosely held in the cell pores, and is therefore quickly lost after harvest. The remaining water (usually 30–40%) is bound to the cell walls by hydrogen bonds, and is therefore harder to remove. It is important that grain be dried quickly, i.e. within a few days of harvest, because fungi thrive in moist or partly moist grain.

6.3.3 *Energy balance and temperature for drying*

If unsaturated air is passed over wet material, the air will take up water from the material as described in the previous section. This water has to be evaporated, and the heat to do this comes from the air and the material. The air is thereby cooled. In particular, if a volume V of air is cooled from T_1 to T_2 in the process of evaporating a mass m_w of water, then

$$m_w \Lambda = \rho c V (T_1 - T_2) \tag{6.4}$$

where Λ is the latent heat of vaporization of water and ρ and c are the density and specific heat of air at constant pressure at the mean temperature, for moderate temperature differences.

The basic problem in designing a crop drier is therefore to determine a suitable T_1 and V to remove a specified amount of water m_w. The temperature T_1 must not be too high, because this would make the grain crack and so allow bacteria and parasites to enter.

Example 6.1
Rice is harvested at a moisture content $w = 0.28$. Ambient conditions are 30°C and 80% relative humidity, at which $w_e = 0.16$ for rice. Calculate how much air at 45°C is required for drying 1000 kg of rice if the conditions are as in Fig. 6.3.

Solution
From (6.2), $m/m_0 = w + 1 = 1.28$, so the dry mass of rice is $m_0 = 780$ kg. The mass of water to be evaporated is therefore

$$m_w = (0.28 - 0.16)(780 \, \text{kg}) = 94 \, \text{kg}$$

For the moist air leaving the drier the exit temperature is found from the humidity data (Table B.2(b)) as follows. 1 m^3 of air at 30°C and 80% relative humidity has absolute humidity

$$(0.8) \times (30.3 \, \text{g m}^{-3}) = 24.2 \, \text{g m}^{-3}$$

(point A in Fig. 6.3).

If the small change in density is neglected, this value will also be the absolute humidity of the same air after heating to 45°C (point B). (The relative humidity will of course be lowered.) After passing through the rice, the exit air will be more moist. If the conditions are according to Fig. 6.3, the exit air is at C, and its temperature will be about 30°C. Then (6.4) gives

$$V = \frac{(94\,\text{kg})(2.4\,\text{MJ}\,\text{kg}^{-1})}{(1.15\,\text{kg}\,\text{m}^{-3})(1.0\,\text{kJ}\,\text{kg}^{-1}\,\text{K}^{-1})(45-30)\,°\text{C}}$$
$$= 13 \times 10^3\,\text{m}^3$$

where the other data come from Appendix B.

More exact calculations would allow for the variations in latent heat, density and humidity of the exit air, but the conclusion would be the same: large scale drying requires the passage of large volumes of warm dry air. Drying with forced convection is an established and complex subject. Drying without forced air flow is even more complex, especially if drying times and temperatures are limited (see Exell, 1980).

6.4 Space heat

A major use of energy in colder climates is to heat buildings in winter. What is perceived as a comfortable air temperature depends on the humidity, the received radiation flux, the wind speed and on how much clothing one is wearing. Therefore we aim to keep the inside (room) temperature T_r in a comfortable range (say 15–20°C) while using the minimum artificial heating P_{boost}, even when the ambient temperature T_a drops to 0°C or less. The heat balance of the inside of a building is given by an equation like (5.1), namely

$$mc\frac{\text{d}T_r}{\text{d}t} = \tau\alpha GA + P_{\text{boost}} - \frac{(T_r - T_a)}{R} \tag{6.5}$$

6.4.1 *Passive solar systems*

Passive solar design consists of arranging the mass m, the sun-facing area A and the loss resistance R to achieve optimum solar benefit by structural design. The first step is to insulate the building properly (high R), including draught prevention and controlled ventilation with heat recovery. For a new building the orientation and window positions should give a maximum product of GA. Remember that we are discussing winter in high latitudes, so the sunshine may be incident more on the vertical walls than on the roof. The sun-facing surfaces should be a dark color with $\alpha > 0.8$ (Fig. 6.4(a)) and the building should be designed to have massive interior walls (high m), thereby limiting the variations in T_r.

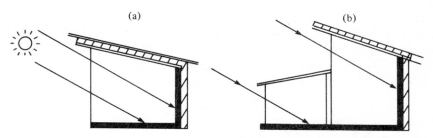

Fig. 6.4 Direct gain passive solar heating: (a) basic system (b) clerestory window (to give direct gain on the back wall of the house). Note the use of massive, dark colored, rear insulated surfaces to absorb and to store the radiation

Example 6.2 Solar heat gain of a house
The Solar Black House shown in Fig. 6.4(a) was designed for Washington DC (latitude 38°N) with a large window on the south side and a massive blackened wall on the north. Assuming that the roof and walls are so well insulated that all heat loss is through the window, calculate the solar irradiance required so that direct solar heating alone maintains room temperature 20°C above ambient.

Solution
If the room temperature is steady, (6.5) reduces to

$$\tau \alpha G = \frac{T_r - T_a}{r}$$

where r is the thermal resistivity from room to outside of a vertical window, single glazed. By the methods of Chapters 3 and 4, we calculate

$$r = 0.07 \, \text{m}^2 \text{K} \text{W}^{-1}$$

Take glass transmittance $\tau = 0.9$ and wall absorptance $\alpha = 0.8$. Then

$$G = \frac{20°C}{(0.07 \, \text{m}^2 \text{K} \text{W}^{-1})(0.9)(0.8)} = 400 \, \text{W} \text{m}^{-2}$$

This irradiance may be expected on a vertical Sun-facing window on a clear day in winter.

This correctly suggests that most of the heating load of a well-designed house can be contributed by solar energy, but the design of practical passive solar systems is more difficult than the above example would suggest. For example, the calculation shows only that the Solar Black House will be adequately heated in the middle of the day. But the heat must also be retained at night and there must be an exchange of air for ventilation.

Example 6.3 Heat loss of a house
The Solar Black House of the previous example measures 2 m high by 5 m wide by 4 m deep. The interior temperature is 20 °C at 4.00 p.m. Calculate the interior temperature at 8.00 a.m. the next day for the following cases:

(1) Absorbing wall 10 cm thick, single window as before
(2) Absorbing wall 50 cm thick, thick curtain covering the inside of the window.

Solution
With $G = P_{boost} = 0$, (6.5) reduces to

$$\frac{dT_r}{dt} = -\frac{(T_r - T_a)}{RC}$$

with $C = mc$.
  The solution is

$$T_r - T_a = (T_r - T_a)_{t=0} \exp[-t/(RC)]$$

assuming T_a is constant (cf Section 16.4). As before, assume all heat loss is through the window, of area $10 \, m^2$. Assume the absorbing wall is made of concrete.

(1) $R = rA^{-1} = 0.007 \, K \, W^{-1}$

$C = mc = [(2.4)(10^3 kg\,m^{-3})(2\,m)(5\,m)(0.1\,m)](0.84 \times 10^3 J\,kg^{-1}K^{-1})$

$\qquad = 2.0 \times 10^6 \, J \, K^{-1}$

$RC = 14 \times 10^3 \, s = 4.0 \, h$

After 16 hours, the temperature excess above ambient is

$(20°C) \exp(-16/4) = 0.4°C.$

(2) A curtain is roughly equivalent to double glazing. Therefore take $r \approx 0.2 \, m^2 K \, W^{-1}$ (from Table 5.1). Hence

$R = 0.02 \, K \, W^{-1}$

$C = 10 \times 10^6 \, J \, K^{-1}$

$RC = 2.0 \times 10^5 \, s = 55 \, h$

$T_r - T_a = (20°C) \exp(-16/55) = 15°C$

This calculation shows the importance of m and R in the heat balance, and also the importance of having parts of the house adjustable to admit heat by day while shutting it in at night (e.g. curtains, shutters).

The analysis for a standard house is more complicated because of the complex absorber geometry, the nonzero losses through the walls, the presence of people in the house and the considerable 'free gains' from lighting etc. Not only can people make adjustments such as drawing the curtains, but also their metabolism can contribute appreciably to the heat balance (say, 100 W per person in the term P_{boost} of (6.5)). A reasonable number of air changes (between one and three per hour) is required for ventilation, and this will usually produce significant heat loss unless heat exchangers are fitted.

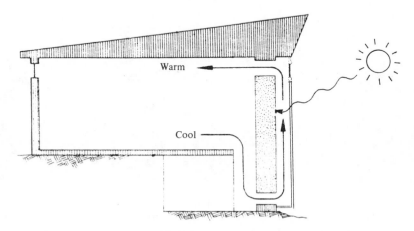

Fig. 6.5 Operation of a Trombe-Michel wall in winter. After Leckie *et al.* (1976)

One drawback of simple direct gain systems is that the building can be too hot during the day – especially during the summer. Although this nuisance can be reduced by suitably large roof overhangs, potentially larger and more controllable heat gain can be achieved by building a so-called 'storage wall' into the sunny side of the building. For example, the Trombe-Michel wall of Fig. 6.5 consists of a concrete slab about 30 cm thick, with a glass cover outside and slots above and below. Air circulates as shown in the winter; the system acts as a built-in air heater with thermosyphon and possible fan assisted circulation. In summer the wall would be shaded, or the flaps can be exchanged so cool air is pulled through the house from the cool side. For esthetic reasons the concrete slab would normally have small windows in it. An example of a solar house is shown in Fig. 6.6.

6.4.2 *Active solar systems*

An alternative space heating method is to use external (separate) collectors, heating either air (Section 6.2) or water (Chapter 5) in an active solar system. Such systems are easier to control than purely passive systems and can be fitted to existing houses. However retrofitting is usually far less satisfactory than

Fig. 6.6 Solar house at Eindhoven, Netherlands, 51°N. Large detached house with 51 m^2 of solar collectors prominently designed into the south wall and yielding 45 GJ y^{-1} of useful heat. The well-insulated house uses water as a working fluid and storage medium, and has a pump to circulate the warm water

correct passive design at the initial constructional stage. In either case a large storage system is needed (e.g. the building fabric, or a rockbed in the basement, or a large tank of water; see Section 16.4). Water based systems require a heat exchanger to heat the room, and air based systems need substantial ducting. A system of pumps or fans is needed to circulate the working fluid.

Like passive systems, active solar systems will work well only if heat losses have been minimized. In practice so-called passive houses are much improved with electric fans controlled to pass air between rooms and heat stores. Thus the term 'passive' tends to be used when the sun's heat is first trapped in rooms or conservatories behind windows, even if controlled ventilation is used in the building. 'Active' tends to be used if the heat is first trapped in a purpose built exterior collector.

6.5 Space cooling

Solar heat can be used not only to heat but also to cool. A mechanical device capable of doing this is the *absorption refrigerator* (Fig. 6.7). All refrigerators depend on the surroundings giving up heat to evaporate a working fluid. In a conventional electrical (or compression) refrigerator, the working fluid is

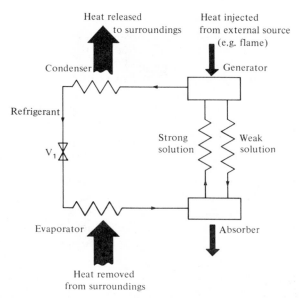

Fig. 6.7 Schematic diagram of an absorption refrigerator. Zigzags represent heat exchangers

recondensed by heat exchange at elevated pressure applied by a motor. In an absorption refrigerator, the required pressure rise is obtained from the difference in vapor pressure of the refrigerant between (1) a part containing refrigerant vapor above a concentrated solution of refrigerant liquid (the generator), and (2) a part containing refrigerant vapor above a dilute solution (the absorber). Instead of an external input of work, as in a compression cycle, the absorption cycle requires an external input of heat. This heat is applied to the generator in order to maintain its temperature at a level such that the vapor pressure there equals the saturation pressure in the condenser.

A suitable combination of chemicals is water as refrigerant and lithium bromide as absorbent. The heat can be applied either by a flame, by waste heat or by solar energy. Although systems are commercially available for use with flat plate collectors, operating at ~80°C, they suffer from high mechanical complexity and low 'efficiency'

$$C_T = \frac{\text{heat removed from cool space}}{\text{heat applied to generator}} \approx 0.7 \qquad (6.7)$$

There is a great variety of solar vapor cycle refrigerators, including some straightforward designs working on a 24 hour cycle.

A better way to cool *buildings* in hot climates is again to use passive designs (cf. Section 6.4.1). These either harness the natural flows of cooling air (in humid areas), store coolness from night time or winter (in dry areas) or in some cases automatically generate a cooling flow by convection. A comprehensive account of the relevant design principles, with examples, is given in the 'Manual of Tropical Building' (Koenigsberger *et al.*, 1974).

For cooling foodstuffs etc., at least in small quantities, there are also available commercial compression refrigerators and freezers powered by solar cells via batteries (see Section 7.5). At present these are economically attractive only in areas remote from conventional electricity supplies.

6.6 Water desalination

To support a community in a desert there is a need to supply fresh water for drinking, crop growth and general purposes. Many desert regions (e.g. central Australia) have a supply of salt water under the ground, and it is usually much cheaper to distil this water than to transport fresh water from afar. Since deserts usually have high insolation, it is reasonable to use solar energy to perform this distillation.

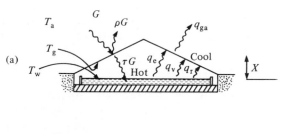

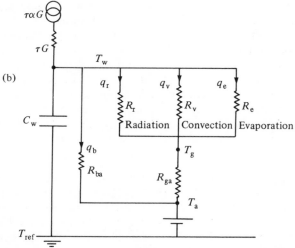

Fig. 6.8 Heat flows in a solar still: (a) schematic (b) heat circuit. Symbols as before, with b base, e evaporation, v convection, r radiation, w water and a ambient

The most straightforward approach is to use a *basin solar still* (Fig. 6.8). It consists of a blackened basin containing impure water at a shallow depth, over which is a transparent, vapor-tight cover that completely encloses the space above the basin. The cover is sloped towards the collection channel. In operation, solar energy passes through the cover and warms the water, some of which then evaporates. The water vapor is carried upwards by thermal convection, where it condenses on the cool cover. The condensed water drops then slide down the cover into the catchment trough.

Example 6.4 Output from an ideal solar still
The insolation in a dry sunny area is typically $20\,MJ\,m^{-2}day^{-1}$. The latent heat of evaporation of water is $2.4\,MJ\,kg^{-1}$. Therefore if all the solar heat goes into evaporation, and all the evaporated water is collected, the output from the still is

$$\frac{20\,MJ\,m^{-2}day^{-1}}{2.4\,MJ\,kg^{-1}} = 8.3\,kg\,day^{-1}m^{-2}$$

These numbers show that substantial areas of glass are required to produce enough fresh water for even a small community.

To calculate the output of a real solar still, we have to determine what proportion of the input solar energy causes evaporation. The heat balance of unit area of water (Fig. 6.8(a)) is

$$mc\frac{dT_w}{dt} = \alpha_w \tau G - q_b - q_r - q_v - q_e \tag{6.8}$$

which differs from similar equations of Chapter 5 by (1) the inclusion of the evaporative heat transfer q_e, and (2) a much smaller value of τ, owing to droplets of water under the cover being highly reflecting. As a simplification assume that the still is well insulated so that the bottom loss q_b is negligible.

The radiative flow is given by (C.18) and (3.44), with the surfaces treated as large and parallel with unit emittance.

$$q_r = 4\sigma_w \left[\frac{T_w + T_g}{2}\right]^3 (T_w - T_g) \tag{6.9}$$

The convective heat flow can be expressed in the form

$$q_v = h_v(T_w - T_g) \tag{6.10}$$

in which the heat transfer coefficient h_v depends on the circulation of air and vapor in the still as described in Section 3.4. The evaporative heat transfer likewise depends on the circulation, from the water surface to the condensing cover. The same flow Q carries heat from hot to cold, and solute from high concentration to low concentration. (Here the solute is water vapor.) The net heat flow per unit area is

$$q_v = 2\rho c(Q/A)\Delta T \tag{6.11}$$

(cf. (3.48)). The factor 2 occurs because the flow carries both hot fluid up and cold fluid down. Similarly the net mass of solute transferred per unit area per unit time in the convective air circulation is

$$W = \dot{m}/A = 2(Q/A)\Delta\chi$$

$$= h_v\rho^{-1}c^{-1}\Delta\chi \tag{6.12}$$

using (6.10) and (6.11). The heat flow per unit area arising from the evaporation and recondensation of the water is

$$q_e = W\Lambda \tag{6.13}$$

For the still shown in Fig. 6.8,

$$q_e = h_v\Lambda\rho^{-1}c^{-1}[\chi(T_w) - \chi(T_g)] \tag{6.14}$$

From Table C.1, with the characteristic dimension X on Fig. 6.8(a),

$$h_v = \mathcal{N}k/X$$

The Nusselt number $\mathcal{N}$ is given by (C.7), so

$$h_v = 0.062(X/k)\mathcal{A}^{1/3} \tag{6.15}$$

where the Rayleigh number

$$\mathcal{A} = g\beta X^3(T_w - T_g)\kappa^{-1}\nu^{-1} \tag{6.16}$$

Strictly speaking, all the fluid properties in this discussion (ρ, k etc.) refer to moist air at the appropriate vapor concentration. However, usually $\chi << \rho$, so the properties of dry air tabulated in Table B.1 are a reasonable approximation.

Fig. 6.9 shows the results of calculations based on the equations given. We see that the fraction of heat going into evaporation is almost independent of $(T_w - T_g)$ but increases strongly with the water temperature T_w. This is to be expected, since the vapor concentration $\chi(T_w)$ increases much faster than T_w (see Fig. 6.3). The results also show that the maximum production achievable with this type of still with basin water at $\sim 50°C$ is 60% of that calculated in Example 6.4, i.e. 5 l day^{-1}m^{-2}.

An alternative approach is the *multiple effect still* in which the heat given off by the condensation of the distilled fresh water is used to evaporate a second mass of saline water. The heat given off by the condensation of the second mass can in turn be used to evaporate a third mass of saline water, and so on. Practical performance is limited by imperfections in heat transfer by the complexity.

The economics of solar distillation depends on the price of alternative sources of fresh water. In an area of high or moderate rainfall (>40 cm y^{-1}) it is almost

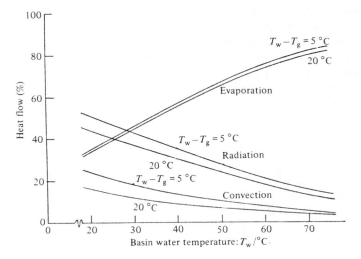

Fig. 6.9 Effect of water temperature on the effectiveness of a basin solar still, as calculated by Cooper (1974)

certainly cheaper to build a water storage system than any solar device. An alternative approach that is becoming cheaper is to purify the water by reverse osmosis, with the water pumped against the osmotic pressure across special membranes which prevent the flow of dissolved material.

6.7 Solar ponds

In applications calling for large amounts of low temperature ($<100\,°C$) heat, the conventional collectors described in Chapter 5 are often too expensive. A solar pond is an ingenious collector which uses water as its top cover. A large 'pond' (say $10^4\,m^2$) can be constructed by simple earthworks for the remarkably low cost of $\sim\$15\,m^{-2}$ (Tabor, 1981). Moreover, it incorporates its own heat storage, which extends the range of its uses.

A solar pond comprises several layers of salty water, with the saltiest layer on the bottom (Fig. 6.10) at 1.5 m depth. Sunshine is absorbed at the bottom so the lowest layer of water heats up. In an ordinary homogeneous pond this warm water would then be lighter than its surroundings and would rise, thus carrying its heat to the air above by free convection (cf. Section 3.4). But in the solar pond the bottom layer was initially made so much saltier than the one above that, even though its density decreases as it warms, it still remains denser than the layer above. Thus convection is suppressed, and the bottom layer remains at the bottom getting hotter and hotter. Indeed there are some solutions that increase density with increase in temperature, so producing very stable solar ponds.

Of course, the bottom layer does not heat up indefinitely but settles to a temperature determined by the heat lost by conduction through the stationary water above. Calculation shows that the resistance to this heat loss is comparable to that in a conventional plate collector (Problem 6.3). Equilibrium temperatures of 90 °C or more have been achieved, with boiling being observed in some exceptionally efficient solar ponds. Note that to set up such a solar pond in practice takes up to several months, because if the upper layers are added too quickly, the resulting turbulence stirs up the lower layers and destroys the desired stratification.

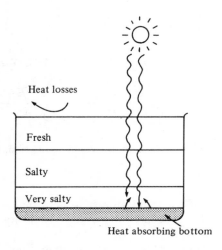

Fig. 6.10 In a solar pond, convection is suppressed and the bottom layer retains the heat from the sun

In a large solar pond, the thermal capacitance and resistance can be made large enough to retain the heat in the bottom layer from summer to winter (Problem 6.3). The pond can therefore be used for heating buildings in the winter. A pond has also many potential applications in industry, as a steady source of heat at a moderately high temperature.

It is also possible to produce electricity from a solar pond by using a special 'low temperature' heat engine coupled to an electric generator. Such systems are conceptually very similar to OTEC systems (Chapter 14). A solar pond at Ein Borek in Israel produces a steady 150 kW (electrical) from 0.74 ha at a busbar cost of about $0.10 kWh^{-1} (Tabor, 1981).

6.8 Solar concentrators

6.8.1 *Basics*

Many potential applications of solar heat require higher temperatures than those achievable by even the best flat plate collectors. In particular a working fluid at 500 °C can drive a conventional heat engine to produce mechanical work and thence (if required) electricity. Higher temperatures still (~2000 °C) are useful in the production and purification of refractory materials.

A concentrating *collector* comprises a *receiver*, where the radiation is absorbed and converted to some other energy form, and a *concentrator*, which is the optical system that directs beam radiation on to the receiver (e.g. Fig. 6.11). Therefore it is usually necessary to continually move the concentrator so that it faces the solar beam. (Section 6.8.4 considers an exceptional case.)

The aperture of the system A_a is the projected area of the concentrator, facing the beam. We define the *concentration ratio* X to be the ratio of the area of aperture to the area of the receiver:

$$X = A_a/A_r \tag{6.17}$$

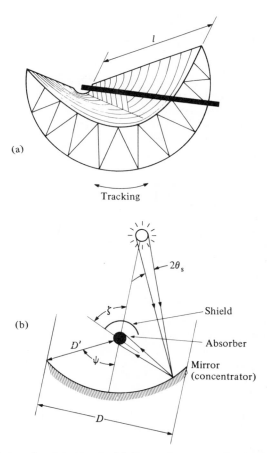

Fig. 6.11 A parabolic trough concentrator. (a) General view, showing the receiver running along the axis. Support struts for the receiver and mirror are also drawn (b) End view of the design discussed in the text (not to scale)

For an ideal collector X would be the ratio of the flux density at the receiver to that at the concentrator, but in practice the flux density varies greatly across the receiver. The temperature of the receiver cannot be increased indefinitely by simply increasing X since by Kirchhoff's laws (Section 3.5.4) receiver temperature T_r cannot exceed the equivalent temperature T_s of the sun. Moreover the sun (radius R_s, distance L) subtends a finite angle at the earth which limits the achievable concentration ratio to

$$X < (L/R_s)^2 = 45\,000 \tag{6.18}$$

(See Problem 6.4.) In the next section we shall see how closely these limits on T_r and X can be approached in practice. In Chapter 7 concentrators are discussed for solar cell arrays (see Fig. 7.24).

6.8.2 *Parabolic trough concentrator*

Fig. 6.11(a) shows a typical trough collector. The concentrator is a parabolic mirror of length l with the receiver running along its axis. This gives concentration only in one dimension, so that the concentration factor is less than for a paraboloid dish. On the other hand, the one-dimensional arrangement is mechanically simpler. Similarly, it is usual to have a trough collector track the sun only in one dimension. The axis is aligned east–west, and the trough rotated (automatically) about its axis to follow the sun in tilt only.

The power absorbed by the absorbing tube is

$$P_{abs} = \rho_c \alpha l D G_b \tag{6.19}$$

where ρ_c is the reflectance of the concentrator, α is the absorptance of the absorber, lD is the area and G_b is the averaged beam irradiance on the trough.

The shield shown in Fig. 6.11(b) is intended to cut down heat losses from the absorber. It also cuts out some unconcentrated direct radiation, but this is insignificant compared with the concentrated radiation coming from the other side. The absorber loses radiation only in directions unprotected by the shield. Therefore its radiated power is

$$P_{rad} = \epsilon(\sigma T_r^4)(2\pi r l)\,(1 - \zeta/\pi) \tag{6.20}$$

where T_r, ϵ and r are respectively the temperature, emittance and radius of the absorber tube. To minimize the losses we want r small, but to gain the full power P_{abs} the tube must be at least as big as the sun's image. Therefore for high temperature we choose

$$r = D'\theta_s \tag{6.21}$$

in the notation of Fig. 6.11(b). Other heat losses can (in principle) be eliminated, but radiative losses cannot. Therefore by setting $P_{rad} = P_{abs}$ we find the stagnation temperature T_r:

$$T_r = \left[\frac{\alpha \rho_c \tau_a G_0 \cos\omega}{\epsilon\sigma}\right]^{1/4} \left[\frac{D}{2\pi r(1 - \zeta/\pi)}\right]^{1/4} \tag{6.22}$$

This will be a maximum when the shield allows outward radiation only to the mirror, i.e. $\zeta \to \pi - \psi$. By trigonometry the geometric term inside the second bracket can be simplified to $1/\theta_s$, so that the maximum obtainable temperature is

$$T_r^{(max)} = \left[\frac{\alpha \rho_c \tau_a G_0 \cos \omega}{\epsilon \sigma \theta_s} \right]^{1/4} = 1160 \, \text{K} \tag{6.23}$$

for the typical conditions $G_0 = 600 \, \text{W m}^{-2}$, $\rho_c = 0.8$, $\alpha/\epsilon = 1$, $\theta_s = R_s/L = 4.6 \times 10^{-3} \, \text{rad}$ and $\sigma = 5.67 \times 10^{-8} \, \text{W m}^{-2} \text{K}^{-4}$.

$T_r = 1160 \, \text{K}$ is a much higher temperature than that obtainable from flat plate collectors (cf. Table 5.1). Practically obtainable temperatures are less than $T_r^{(max)}$ for two main reasons:

(1) Practical troughs are not perfectly parabolic, so that the solar image subtends angle $\theta_s' > \theta_s = R_s/L$.
(2) Use heat is removed by passing a fluid through the absorber, so

$$T_r^4 \, \alpha \, P_{rad} = P_{abs} - P_u < P_{abs}$$

Nevertheless useful heat can be obtained at $\sim 700 \, ^\circ\text{C}$ under good conditions (see Problem 6.5).

Although (6.22) suggests that T_r could be raised even further by using a selective surface with $\alpha/\epsilon > 1$, this approach yields only limited returns because the selectivity of the surface depends on the fact that α and ϵ are averages over different regions of the spectrum (cf. Section 5.6).

Indeed, according to the definitions (3.27),

$$\alpha = \frac{\int_0^\infty \alpha_\lambda \theta_{\lambda, in} \, d\lambda}{\int_0^\infty \phi_{\lambda, in} \, d\lambda} \qquad \epsilon = \frac{\int_0^\infty \epsilon_\lambda \phi_{\lambda, B} \, d\lambda}{\int_0^\infty \phi_{\lambda, B} \, d\lambda} \tag{6.24}$$

As T_r increases the corresponding black body spectrum $\phi_{\lambda, B}(T_r)$ of the emitter becomes more like the equivalent black body spectrum of the sun, $\phi_{\lambda, in} = \phi_{\lambda, B}(T_s)$. Since Kirchhoff's law (3.30) states that $\alpha_\lambda = \epsilon_\lambda$ for each λ, (6.24) implies that as $T_r \to T_s$, $\alpha/\epsilon \to 1$.

6.8.3 *Parabolic bowl concentrator*

Concentration can be achieved in two dimensions by using a bowl shaped concentrator. This requires a more complicated tracking arrangement than the one-dimensional trough, similar to that required for the 'equatorial mounting' of an astronomical telescope. As before the best focusing is obtained with a parabolic shape, in this case a paraboloid of revolution.

Its performance can be found by repeating the calculations of Section 6.8.2 but this time taking Fig. 6.11(b) to represent a section through the paraboloid. The absorber is assumed to be spherical. The maximum absorber temperature is found in the limit $\zeta \to 0$, $\psi \to \pi/2$ and becomes

$$T_r^{(max)} = \left[\frac{\alpha \rho_c \tau_a G_0 \sin^2 \psi}{4 \epsilon \sigma \theta_s^2} \right]^{1/4} \tag{6.25}$$

Comparing this with (6.23) we see that the concentrator now fully tracks the sun, and θ_s has been replaced by $(2\theta_s/\sin\psi)^2$. Thus $T_r^{(max)}$ increases substantially. Indeed for the ideal case, $\sin\psi = \alpha = \rho_c = \tau_a = \epsilon = 1$, we recover the limiting temperature $T_r = T_s$ of Section 6.8.2. Even allowing for imperfections in the tracking and in the shaping of the mirror, and especially for difficulties in designing the receiver, temperatures of up to 3000 K can be achieved in practice.

6.8.4 *Nontracking concentrators*

The previous sections described how high concentration ratios can be achieved with geometric precision and accurate tracking. Nevertheless cheap concentrators of low concentration ratio are useful (e.g. Rabl, 1976; Meinel and Meinel, 1977; Winston, 1974). For example, it may be cheaper and equally satisfactory to use a $5\,m^2$ area concentrator of concentration ratio 5 coupled to a $1\,m^2$ solar cell array than to use $5\,m^2$ of solar cells with no concentration. Such economy can be achieved more readily if the concentrator does not track the sun. However, with solar cells care has to be taken to avoid unequal illumination across the array (see Chapter 7).

6.9 Electric power systems

Collectors with concentrators can achieve temperatures high enough ($\geq 700\,°C$) to operate a heat engine at reasonable efficiency, which can be used to generate electricity. However, there are considerable engineering difficulties in building a single tracking bowl with a diameter exceeding 30 m, and even a single bowl of that size could receive at most a peak thermal power of

$$\pi(15\,m)^2(1\,kW\,m^{-2}) = 700\,kW$$

with subsequent electricity generation of perhaps 200 kW. This would be useful for a small local electricity network, but not for established utility networks.

How then can one build a solar power station large enough to make an appreciable contribution to a local grid, say 10 MW (electric)? Two possible approaches are illustrated in Figs 6.12 and 6.14, namely distributed collectors and a central 'power tower'.

6.9.1 *Distributed collectors*

In Fig. 6.12 many (small) concentrating collectors each individually track the sun. (The collectors need not be parabolic bowls as shown here, though this is the usual choice.) Each collector transfers solar heat to some transfer fluid, and this hot transfer fluid is then gathered from all the collectors at some central power station. The transfer fluid could be steam, to be used directly in a steam turbine, or it could be some *thermochemical storage medium*, such as dissociated ammonia, as illustrated in Fig. 6.13.

The advantage of the latter system, proposed by Carden (1977), is that no heat stored in the chemical is lost between the collectors and the heat engine, so that transmission can be over a long distance or a long time (e.g. overnight, thus

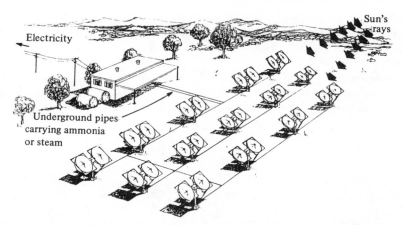

Fig. 6.12 Electricity generation from distributed parabolic collectors. After Electricity Authority of New South Wales (1979)

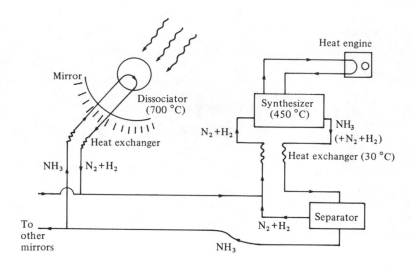

Fig. 6.13 Dissociation and synthesis of ammonia, as a storage medium for solar energy. After Carden (1977)

allowing continuous power generation). In this system the sun's rays are focused on to a receiver in which ammonia gas (at high pressure, ~300 atmospheres) is dissociated into hydrogen and nitrogen. (This reaction is endothermic with $\Delta H = -46 \text{kJ} \text{ (mole NH}_3)^{-1}$; the heat of reaction is provided by the solar energy.) Within the central plant the N_2 and H_2 are (partially) recombined in the synthesis chamber, using a catalyst. The heat from this reaction can be used to drive an external heat engine or other device. The outflow from the synthesis chamber is separated by cooling it, so that the ammonia liquefies. Numerical details are worked out in Problems 6.6, 6.7.

6.9.2 *Power tower*

An alternative approach is to use a large field of sun-tracking plane mirrors, which focus on to a large central receiver. In Fig. 6.14 the mirrors focus beam radiation on to the receiver on top of the tower (hence the name 'power tower').

Fig. 6.14 Solar 'power tower' for generating steam for electricity production. The photograph shows the 10 MW(e) installation of the Southern California Edison Company: 1818 plane mirrors, each 7 m × 7 m, reflect direct radiation to the raised boiler. Reproduced with permission of the company

Problems

6.1 *Theory of chimney*
   A vertical chimney of height h takes away hot air at temperature T_h from a heat source. By evaluating the integral (5.17) inside and outside the chimney, calculate the thermosyphon pressure P_{th} for the following conditions:

   (a) $T_a = 30°C$, $T_h = 45°C$, $h = 4\,m$ (corresponding to a solar crop drier)
   (b) $T_a = 5°C$, $T_h = 300°C$, $h = 100\,m$ (corresponding to an industrial chimney).

6.2 *Flow through a bed of grain*

Flow through a bed of grain is analogous to flow through a network of pipes.

(a) Fig. 6.15(a) shows a cross-section of a solid block pierced by n parallel tubes each of radius a. According to Poiseuille's law, the volume of fluid flowing through *each* tube is

$$Q_1 = \frac{\pi a^4}{8\mu}\left(\frac{dp}{dx}\right)$$

where μ is the dynamic viscosity (see Chapter 2) and dp/dx is the pressure gradient driving the flow. Show that the bulk flow velocity through the solid block of cross-section A_0 is

$$\bar{v} = \frac{Q_{total}}{A_0} = \frac{\epsilon a^2}{8\mu}\frac{dp}{dx}$$

where the porosity ϵ is the fraction of the volume of the block which is occupied by fluid, and Q_{total} is the total volume flow through the block.

(b) The bed of grain in a solar drier has a total volume $V_{bed} = A_0\Delta x$ (Fig. 6.15(c)). The drier is to be designed to hold 1000 kg of grain of bulk volume $V_{bed} = 1.3\,m^3$. The grain is to be dried in four days ($=30$ hours of operation). Show that this requires an air flow of at least

$$Q = 0.12\,m^2 s^{-1}$$

(Refer to Example 6.1.)

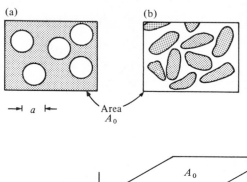

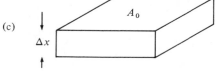

Fig. 6.15 For Problem 6.2. (a) Block pierced by parallel tubes. (b) Pores in a bed of grain. (c) Volume of grain bed

(c) Fig. 6.15(b) shows how a bed of grain can likewise be regarded as a block of area A_0 pierced by tubes whose diameter is comparable to (or smaller than) the radius of the grains. The bulk flow velocity is reduced by a factor $k(<1)$ from that predicted by (a), because of the irregularity and tortuousity of the tubes. If the driving pressure is Δp, show that the thickness Δx through which the flow Q can be maintained is

$$\Delta x = \left(\frac{k\epsilon a^2}{8\mu} \frac{V\Delta p}{Q} \right)^{1/2}$$

For a bed of rice, $\epsilon = 0.2$, $a = 1\,mm$, $k = 0.5$ approximately. Taking Q from (b), and Δp from Problem 6.1(a), calculate Δx and A_0.

6.3 *The solar pond*

An idealized solar pond measures $100\,m \times 100\,m \times 1.2\,m$. The bottom $20\,cm$ (the storage layer) has an effective absorptance $\alpha = 0.7$. The $1.0\,m$ of water above (the insulating layer) has a transmittance $\tau = 0.7$, and its density increases downwards so that convection does not occur. The designer hopes to maintain the storage layer at $80\,°C$. The temperature at the surface of the pond is $27\,°C$ (day and night).

(a) Calculate the thermal resistance of the insulating layer, and compare it with the top resistance of a typical flat plate collector.

(b) Calculate the thermal resistance of a similar layer of fresh water, subject to free convection. Compare this value with that in (a), and comment on any improvement.

(c) The density of NaCl solution increases by $0.75\,gl^{-1}$ for every $1\,g$ of NaCl added to $1\,kg\,H_2O$. A saturated solution of NaCl contains about $370\,g$ NaCl per kg H_2O. The volumetric coefficient of thermal expansion of NaCl solution is about $4 \times 10^{-4}\,K^{-1}$. Calculate the minimum concentration C_{min} of NaCl required in the storage layer to suppress convection. (Assume the water at the top of the pond is fresh.) How easy is is to achieve this concentration in practice?

(d) Calculate the characteristic time scale for heat loss from the storage layer, through the resistance of the insulating layer. If the temperature of the storage layer is $80\,°C$ at sunset (6 p.m.) what is its temperature at sunrise (6 a.m.)?

(e) The molecular diffusivity of NaCl in water is $1.5 \times 10^{-5}\,cm^2s^{-1}$. The pond is set up with the storage layer having twice the critical concentration of NaCl, i.e. double the value C_{min} calculated in (c). Estimate the time for molecular diffusion to lower this concentration to C_{min}.

(f) In the light of your answers to (c)–(e), discuss the practicability of building such a pond, and the possible uses to which it could be put.

6.4 The limiting concentrating system of Section 6.8.3 has as receiver a black body of area A_r, which is in radiative equilibrium at temperature $T_r = T_s$, the equivalent temperature of the sun. By considering the energy balance of the receiver show that these conditions correspond to a limiting concentration ratio

$$X^{(m)} = (L/R_s)^2$$

where R is the radius of the sun and L its distance from the earth.

6.5 Fig. 6.16 shows the key feature of one proposal for the large scale use of solar energy (Meinel and Meinel, 1977). Sunlight is concentrated on to a pipe (running perpendicular to the plane of the diagram), where it is absorbed by the strongly absorbing (selective) surface on the outside of the pipe. The fluid within the pipe is thereby heated to a temperature T_f of about 500 °C.

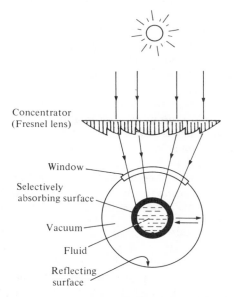

Fig. 6.16 For Problem 6.5. A proposed concentrator system for power generation

    The fluid then passes through a heat exchanger where it produces steam to drive a conventional steam turbine, which in turn drives an electrical generator.

(a) Why is it desirable to make T_f as high as possible?
(b) Suppose the inner pipe is 10 m long and 2 mm thick and has a diameter of 50 mm, and that the fluid is required to supply 12 kW of thermal power to the heat exchanger. If the pipe is made of copper, show that the temperature difference across the pipe is less than 0.1 °C. (Assume that the temperature of the fluid is uniform.)

(c) Suppose the selective surface has $\alpha/\epsilon = 10$ at the operating temperature of 500°C. What is the concentration factor required of the lens (or mirror) to achieve this temperature using the evacuated collector shown? Is this technically feasible from a two-dimensional system?

(d) Suppose the copper pipe was not shielded by the vacuum system but simply sat exposed to the air. Assuming a wind speed of zero, calculate the convective heat loss per second from the pipe.

(e) Suppose the whole system is to generate 50 MW of *electrical* power. Estimate the collector area required.

(f) Briefly discuss the advantages and disadvantages of such a scheme, compared with (i) an oil fired power station of similar capacity, and (ii) small scale uses of solar energy, such as domestic water heaters.

6.6 Suppose the system of Fig. 6.13 is to be used to supply an average of 10 MW of electricity.

(a) Estimate the total collector area this will require. Compare this with a system using photovoltaic cells.

(b) Briefly explain why a chemical (or other) energy store is required, and why the mirrors have to be pointed at the sun. How might this be arranged?

(c) To insure a suitably high rate of dissociation, the dissociator is to be maintained at 700°C. Plumbing considerations (Problem 6.7) require that the dissociator has a diameter of about 15 cm. Assuming (for simplicity) that it is spherical in shape, calculate the power lost from each dissociator by radiation.

(d) Each mirror has an aperture of $10\,\text{m}^2$. In a solar irradiance of $1\,\text{kW}\,\text{m}^{-2}$, what is the irradiance at the receiver? Show that about $2.5\,\text{g}\,\text{s}^{-1}$ of NH_3 can be dissociated under these conditions.

6.7 The system of Fig. 6.13 requires $2.5\,\text{g}\,\text{s}^{-1}$ of NH_3 to pass to each concentrator (see Problem 6.6). Suppose the NH_3 is at a pressure of 300 atmospheres, where it has density $\rho = 600\,\text{kg}\,\text{m}^{-3}$ and viscosity $\mu = 1.5 \times 10^{-4}\,\text{kg}\,\text{m}^{-1}\text{s}^{-1}$.

The ammonia passes through a pipe of length L and diameter d. To keep friction to an acceptable low level, it is required to keep the Reynolds number $\mathcal{R} < 6000$.

(a) Calculate (i) the diameter d (ii) the energy lost to friction in pumping 2.5 g of ammonia over a distance $L = 50\,\text{m}$.

(b) Compare this energy loss with the energy carried. Why is the ammonia kept at a pressure of ~300 atmospheres rather than ~1 atmosphere? (Hint: estimate the dimensions of a system working at ~1 atmosphere.)

Solutions

6.1 (a) $2.7\,\text{N}\,\text{m}^{-2}$ (b) $650\,\text{N}\,\text{m}^{-2}$

6.2 (a) $\epsilon = n\pi a^2/A_0$ (c) 0.13 m, $10\,\text{m}^2$.

6.3 (a) $r_n = 1.7\,\mathrm{m^2 K W^{-1}} > 0.4\,\mathrm{m^2 K W^{-1}}$ (for good flat plate)
   (b) $r_v = 0.0018\,\mathrm{m^2 K W^{-1}} << r_n$ of (a)
   (c) $c \sim 30\,\mathrm{g\ NaCl\ kg^{-1}\ H_2O} <<$ saturated concentration
   (d) $RC \approx 1.3 \times 10^6\,\mathrm{s} = 15$ day (neglected bottom loss) Sunrise temp $=$
       $T_a + (T_{store} - T_a)_0 \exp\left[-t/(RC)\right] = 78.3\,^\circ\mathrm{C}$
   (e) Diffusion time $\approx X^2/D \approx 5\,\mathrm{y}$
   (f) Salt gradient easily established and maintained against diffusion. Heat
       well retained, therefore theoretically viable. Practicalities: see text.

6.5 (a) Higher η
   (b) Use $P = -Ak\,\partial T/\partial r$
   (c) Use $P_u = \alpha GXA - \epsilon\sigma(T_p^4 - T_a^4)A$; $X = 10$, easy (NB $P_u \neq 0$)
   (d) Use (C.4), $\mathscr{A} = 10^5$, $P = 6\,\mathrm{kW}$
   (e) $0.2\,\mathrm{km^2}$ (for $\eta = 0.25$)

6.6 (a) $50 \times 10^3\,\mathrm{m^2}$ (assuming $\eta = \eta_{Carnot}$, and average power $= 1/3$ peak)
       $\approx 40\%$ of area for cells
   (b) Night-time; beam radiation; feedback from sun sensor at focus
   (c) $3\,\mathrm{kW}$
   (d) $560\,\mathrm{kW\,m^{-2}}$

6.7 (a) $4\,\mathrm{mm}$, $2\,\mathrm{kJ}$
   (b) High $T \rightarrow$ high $X \rightarrow$ high (energy) density

Bibliography

General

Duffie, J. A. and Beckman, W. A. (1980) *Solar Engineering of Thermal Processes*, John
   Wiley and Sons.
   Covers most of the topics of this chapter.
Sayigh, A. (ed.) (1977) *Solar Energy Engineering*, Academic Press, London.
   Multiauthor compendium, also covering most of these topics.

Air heaters and crop-drying

Exell, R. (1980) 'A simple solar rice dryer: basic design theory', *Sunworld*, **4**, 186–90.
   Whole issue is about crop drying.
Monteith, J. (1973) *Principles of Environmental Physics*, Edward Arnold.
   Includes full discussion of humidity.
Selcuk, M. K. (1977) 'Solar air heaters and their applications', in Sayigh (1977).

Space heating

Leckie, J., Masters, G., Whitehouse, H. and Young, L. (1976) *Other Homes and
   Garbage: Designs for Self-sufficient Living*, Sierra Club, San Francisco.
Shurcliff, W. (1978) *Solar Heated Buildings of North America*, Brick House, Harrisville,
   NH.
   Descriptions of many solar houses.
Steadman, P. (1975) *Energy, Environment and Building*, Cambridge University Press.
   Advocates energy conservation in buildings by sensible design. Many examples.

Sutton, R. (1982) 'Solar wall passive storage heating', *Solar Progress*, **3**(1), 7–14.
  This journal, published in Australia, has many very readable articles on solar energy generally.
Szokolay, S. V. (1975) *Solar Energy and Building*, 2nd edn, Arnold.
  Good pictorial explanations.

Water desalination
Cooper, P. I. (1974) 'Design philosophy and operating experience for Australian solar stills', *Solar Energy*, **16**, 1–8.
  Summarizes much earlier work.
Howe, E. W. and Tleimat, B. (1977) 'Fundamentals of water desalination', in Sayigh (1977).
  Useful review with basic physics displayed.

Solar cooling
Brinkworth, B. J. (1977) 'Refrigeration and air conditioning', in Sayigh (1977).
Koenigsberger, O. H., Ingersol, T. G., Mayhew, A. and Szokolay, S. V. (1974) *Manual of Tropical Housing and Building. Part I: Climatic Design*, Longmans, London.
  A guide for architects, containing much relevant physics and data. Stresses passive design.

Solar ponds
Savage, S. B. (1977) 'Solar pond', in Sayigh (1977).
  Good review of basic physics.
Tabor, H. (1981) 'Solar ponds', *Solar Energy*, **27**, 181–94.
  Reviews practical details and costs.
Weinberger, H. (1964) 'The physics of the solar pond', *Solar Energy*, **8**, 45–56.
  Pioneering treatment of detailed heat transfer, but has some mistakes (see Savage, 1977).
Wilkins, E. and Pinder, K. (1979) 'Experiments with a model solar pond', *Sunworld*, **3** (4), 111–17.
  Laboratory scale pond.
Winsberg, S. (1981) 'Israel, solar pond innovator', *Sunworld*, **5** (4), 122–5.
  Popular level. Part of a special issue on solar ponds.

Concentrators
Meinel, A. and Meinel, M. (1977) *Applied Solar Energy*, Addison-Wesley, London.
  Good physical descriptions of wide range of designs, but the mathematics needs care.
Rabl, A. (1976) 'Comparison of solar concentrators', *Solar Energy*, **18**, 93.
Sakurai, T. (1977) 'Solar furnaces', in Sayigh (1977).
Winston, R. (1974) 'Solar concentrators of novel design', *Solar Energy*, **16**, 89.

Electric power systems
Carden, P. O. (1977) 'Energy co-radiation using the reversible ammonia reaction', *Solar Energy*, **19**, 365–78.
  First of a long series of articles.
Electricity Authority of New South Wales, Sydney, Australia.
Sunworld, **5** (4), 1981.
  Special issue on solar thermal power. Includes readable articles describing principles and examples of central receiver systems, hemispherical bowl systems and solar pond systems.

7 *Photovoltaic generation*

7.1 Introduction

Photovoltaic generation of power is caused by radiation separating positive and negative charge carriers in absorbing material. If an electric field is present these charges can produce a current for use in an external circuit. Such fields exist permanently at junctions or inhomogeneities in materials as 'built-in' electrostatic fields and provide the EMF for useful power production. The built-in fields of semiconductor/semiconductor and metal/semiconductor photocells produce potential differences of about 0.5 V and current densities of about $200\,\mathrm{A\,m^{-2}}$ in full solar radiation of $1\,\mathrm{kW\,m^{-2}}$. Commercial photocells may have efficiencies of 10 to 20% in ordinary sunshine and can produce electricity of 1 to $2\,\mathrm{kWh\,m^{-2}}$ $\mathrm{day^{-1}}$.

Junction devices are usually named *photovoltaic cells* or *solar cells*, although it is the current that is produced by the radiation photons and not the 'voltage'. The cell itself provides the source of EMF. It is important to appreciate that photo-electric devices are *electrical current sources* driven by a flux of radiation. Solar cells produce current directly linked to the diurnal, seasonal and random variations of the insolation. Efficient power utilization depends not only on efficient generation in the cell, but also on *dynamic load matching* in the external circuit. In this respect photoelectric devices are similar to other renewable sources of power, although the precise methods may vary (e.g. by using DC to DC converters as 'maximum power tracking' interfaces, Section 7.5).

The majority of photovoltaic cells are silicon semiconductor junction devices (Fig. 7.1). These were first produced in 1954 and were rapidly developed to provide power for space satellites, based on semiconductor electronics technology.

The dominant cost of photovoltaic systems used to be the generating cell, but this cost has fallen substantially, and now substrate and encapsulation costs are significant. In the 10 years from 1974 to 1984 the capital cost of cells fell from approximately $(US)100 per peak watt of generating capacity to about $4 per peak watt. The declared aim is about $0.8/$W_p$ (1983 dollar) for international large scale marketing. However at $4/$W_p$ for cells, $2/$W_p$ for ancillary equipment (batteries, control etc.), on a reasonably sunny site of $20\,\mathrm{MJ\,m^{-2}\ day^{-1}}$

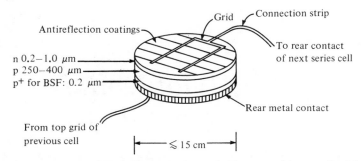

Fig. 7.1 Typical structure of n–p junction solar cell. The cover (glass or plastic) above the cell, and the filler between the cover and the cell, are not shown. BSF: back surface field

insolation, and with a lifetime of 20 years, power can be produced at about 16 cent (US) per kWh (4.4 cent MJ^{-1}). This is competitive with electricity from diesel generators, especially in remote areas where fuel supply and maintenance costs may be large.

The major uses of photovoltaics have been in space satellites, remote radio communication booster stations and marine warning lights. In the next few years it is likely that there will be widespread use of photovoltaics in developing countries for village lighting systems and water pumping.

In the following sections the basic science and technology of photoelectric power systems will be outlined. Preliminary analysis will always refer to the Si p–n junction single crystal solar cell since this is more common and best established. Other developments are analyzed from this basis. Section 7.5 on electrical circuits and loads may be read independently as preparation for practical use of solar cells.

7.2 The silicon p–n junction

The properties of semiconductor materials are described in an ample range of solid state physics and electronics texts. Solar cells are usually considered to be of specialized interest and may be given little attention; however, most books will include the properties of the p–n junction without illumination (dark properties) as summarized in this section. The theory is extended to the illuminated junction for solar applications in sections 7.3 onwards.

7.2.1 *Silicon*

Commercially pure (intrinsic) Si has concentrations of impurity atoms of $<10^{18} m^{-3}$ (<1 in 10^9) and electrical resistivity $\rho_e \approx 2500 \,\Omega m$. Electrical properties are described by the theory of the band gap between conduction and valence bands (Fig. 7.2). The density of charge carrier electrons in the conduction band and holes in the valence band of pure intrinsic material is proportional to $\exp(-E_g/2kT)$ if impurity atoms have no effect. This is equivalent to there being no electron or hole charge carriers with an energy state within the forbidden band gap. Table 7.1 gives basic data for silicon.

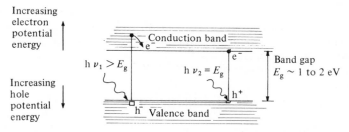

Fig. 7.2 Semiconductor band structure, intrinsic pure material. Photon absorption. $h\nu < E_g$, no photoelectric absorption. $h\nu_1 - E_g$, excess energy dissipated as heat. $h\nu_2 = E_g$, photon energy equals band gap

Table 7.1 Solar cell related properties of silicon

Intrinsic, pure material
Band gap E_g (27°C): 1.14 eV (corresponding $\lambda = 0.97\,\mu m$)
    (~200°C): 1.11 eV (corresponding $\lambda = 1.11\,\mu m$)
Carrier mobility μ, electron $0.14\,m^2\,V^{-1}\,s^{-1}$, hole $0.048\,m^2\,V^{-1}\,s^{-1}$
Carrier diffusion constant $D_e = 35 \times 10^{-4}\,m^2\,s^{-1}$, $D_h = 12 \times 10^{-4}\,m^2\,s^{-1}$
Refractive index at $\lambda = 6\,\mu m$ $n = 3.42$
Extinction coefficient K at $\lambda = 1\,\mu m$ $K = 10^4\,m^{-1}$
        at $\lambda = 0.4\,\mu m$ $K = 10^5\,m^{-1}$
Thermal conductivity $157\,W\,m^{-1}\,K^{-1}$
Specific heat capacity $694\,J\,kg^{-1}\,K^{-1}$
Density; atoms $5.0 \times 10^{28}\,m^{-3}$; $2329\,kg\,m^{-3}$

Typical Si homojunction n–p/p^+ solar cell
n layer, thickness 0.25–0.5 μm; dopant conc. $\lesssim 10^{26}\,m^{-3}$
p layer, thickness 250–350 μm; dopant conc. $\lesssim 10^{24}\,m^{-3}$
p^+ layer, thickness 0.5 μm; dopant conc. $\gtrsim 10^{24}\,m^{-3}$
Surface recombination velocity $10\,m\,s^{-1}$
Minority carrier: diffusion constant $D \sim 10^{-3}\,m^2\,s^{-1}$
        path length $L \sim 100\,\mu m$
        lifetime $\tau \sim 10\,\mu s$

7.2.2 *Doping*

Controlled quantities of specific impurity ions are added to the pure intrinsic material to produce doped (extrinsic) semiconductors. Si is tetravalent in group IV of the periodic table. Impurity ions of less valency (e.g. boron, group III) enter the solid Si lattice and become electron *acceptor* sites which trap free electrons. These traps have an energy level within the band gap, but near to the valence band. The absence of the free electrons produces positively charged states called *holes* that move through the material as free carriers. With such electron acceptor impurity ions, the semiconductor is called *p (positive) type material*, having holes as *majority carriers*.

Conversely, atoms of greater valency (e.g. phosphorus, group V) are electron *donors*, producing *n (negative) type material* with an excess of conduction electrons as the majority carriers.

In each case, however, charge carriers of the complementary polarity also exist in much smaller numbers and are called *minority carriers* (electrons in p type, holes in n type). Holes and electrons may recombine when they meet freely in the lattice or at a defect site.

Both p and n type extrinsic material have higher electrical conductivity than the intrinsic basic material. Indeed the resistivity ρ_e is used to define the material. Common values for silicon photovoltaics are $\rho_e \approx 0.01\,\Omega\,m = 1\,\Omega\,cm$ $(N_d \approx 10^{22}\,m^{-3})$, and $\rho_e \approx 0.1\,\Omega\,m = 10\,\Omega\,cm$ $(N_d \approx 10^{21}\,m^{-3})$, where we use the symbol N_d for dopant ion concentration.

7.2.3 *Fermi level*

n type material has greater conductivity than intrinsic material because electrons easily enter the conduction band by thermal excitation. Likewise p type has

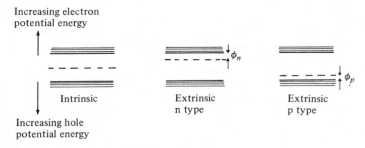

Fig. 7.3 Fermi level in semiconductors (shown by broken line)

holes that easily enter the valence band. The Fermi level is a descriptive and analytical method of explaining this (Fig. 7.3). It is the apparent energy level within the forbidden band gap from which majority carriers (electrons in n type and holes in p type) are excited to become charge carriers. The probability for this varies as $\exp[-e\phi/(kT)]$, where e is the charge of the electron and hole, $e = 1.6 \times 10^{-19}\,C$, and ϕ is the electric potential difference between the Fermi level and the valence or conduction bands as appropriate.

Note that electrons are excited 'up' into the conduction band, and holes are excited 'down' into the valence band. Potential energy increases upwards for electrons and downwards for holes on the conventional diagram.

7.2.4 *Junctions*

p type material can have excess donor impurities added to specified regions so that these become n type in the otherwise continuous material, and vice versa. The region of such a dopant change is a *junction* (which is not formed by physically pushing two separate pieces of material together!). Imagine however that the junction has been formed instantaneously in the otherwise isolated material (Fig. 7.4(a)). Excess donor electrons from the n type material cross to

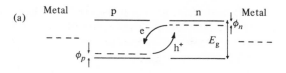

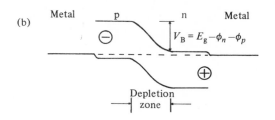

Fig. 7.4 (a) Diagrammatic 'formation' of a p–n homojunction cell. Fermi levels of isolated components shown by broken line. (b) Energy level diagram of a p–n homojunction with metal nonrectifying (ohmic) contacts. Electrons and holes have diffused to reach an equilibrium

the acceptor p type, and vice versa for holes. A steady state is eventually reached. The electric field, caused by the accumulation of charges of opposite sign on each side of the junction, balances the diffusive forces arising from the different concentrations of free electrons and holes. As a result the Fermi level is at constant potential throughout the whole material. However a net movement of charge has occurred, with excess negative charge now on the p side and positive on the n side.

The band gap E_g still exists throughout the material, and so the conduction and valence bands have a step at the junction as drawn in Fig. 7.4(b). The depth of the step is eV_B in energy and V_B in electric potential difference (voltage). $V_{B(I=0)}$ is the band step potential at zero current through the material and is the built-in field potential of the isolated junction. Note that $V_B < E_g$ because

$$V_{B(I=0)} = E_g - (\phi_n + \phi_p) \tag{7.1}$$

$(\phi_n + \phi_p)$ decreases with increase in dopant concentration. For a heavily doped Si p–n junction (dopant ions $\sim 10^{22}\,\mathrm{m}^{-3}$), $E_g = 1.11\,\mathrm{eV}$, and $(\phi_n + \phi_p) \approx 0.3\,\mathrm{V}$. So in the dark, with no current flowing,

$$V_{B(I=0)} \approx 0.8\,\mathrm{V} \tag{7.2}$$

7.2.5 *Depletion zone*

The potential energy balance of carriers each side of the junction (represented by the constancy of the Fermi level across the junction) results in the p type region having a net negative charge ('up' on the energy diagram) and vice versa for the donor region. The net effect is to draw electron and hole carriers out of the junction, leaving it greatly depleted in total carrier density. Let n and p be the electron and hole carrier densities. Then the product $np = C$, a constant, throughout the material. For example,

(1) p region:
$$np = C = (10^{11}\,\text{m}^{-3})(10^{22}\,\text{m}^{-3}) = 10^{32}\,\text{m}^{-6} \tag{7.3}$$

$$n + p = 10^{22}\,\text{m}^{-3}$$

(2) n region:

$$np = C = (10^{22}\,\text{m}^{-3})(10^{11}\,\text{m}^{-3}) = 10^{32}\,\text{m}^{-6} \tag{7.4}$$

$$n + p = 10^{22}\,\text{m}^{-3}$$

(3) Depletion zone: $n = p$ by definition. So

$$n^2 = p^2 = C = 10^{32}\,\text{m}^{-6}$$

$$n = p = 10^{16}\,\text{m}^{-3}$$

$$n + p = 2 \times 10^{16}\,\text{m}^{-3} \tag{7.5}$$

The typical data of this example show that the total charge carrier density at the depletion zone is reduced by 10^5 as compared with the n and p regions each side.

The width w of the junction can be approximated to

$$w \approx \left[\frac{2\epsilon_0\epsilon_\text{f} V_\text{B}}{e\sqrt{(np)}}\right]^{1/2} \tag{7.6}$$

where ϵ_0 is the permittivity of free space, ϵ_r is the relative permittivity of the material, and the other terms have been defined previously.

For Si at $10^{22}\,\text{m}^{-3}$ doping concentration $w \approx 0.5\,\mu\text{m}$ and the electric field intensity V_B/w is $\sim 2 \times 10^6\,\text{V m}^{-1}$. The photovoltaic properties of the junction depend on minority carriers being able to diffuse to the depletion zone and then be pulled across in the large electric field. This demands that $w < L$, where L is the diffusion length for minority carriers, and this is a criterion easily met in solar cell p–n junctions (see (7.11)).

7.2.6 *Biasing*

The p–n junction may be fitted with metal contacts connected to a battery (Fig. 7.5). The contacts are called ohmic contacts – nonrectifying junctions of low resistance compared with the bulk material. In forward bias the positive conventional circuit current passes from p to n material across a reduced band potential difference V_B. In reverse bias the conventional positive current has an increased band potential difference V_B to overcome. The junction therefore acts

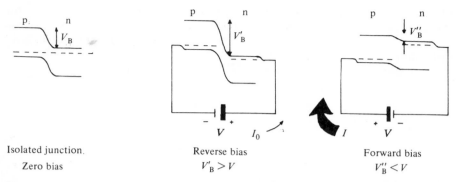

Fig. 7.5 Reverse and forward biasing of a p–n junction. $I_0 << I$, conventional current

as a rectifying diode with an $I–V$ characteristic that will be described later (Fig. 7.8).

7.2.7 Carrier generation

At an atomic scale, matter is in a continuous state of motion. The atoms in a solid oscillate in vibrational modes with quantized energy (phonons). In semi-conductor material electrons and holes are spontaneously generated from bound states for release into the conduction and valence bands as charge carriers. This is a thermal excitation process with the dominant temperature variation given by the Boltzmann probability factor $\exp[-E/(kT)]$, where E is the energy needed to separate the electrons and holes from their particular bound states, k is the Boltzmann constant, and T is the absolute temperature. For pure intrinsic material $2E = E_g$, the band gap. For doped extrinsic material $|E| = |e\phi|$, where ϕ is the potential difference needed to excite electrons in n type material into the conduction band, or holes in p type material into the valence band. Note that ϕ is determined locally at the dopant site and $|e\phi| < E_g$ (Fig. 7.4(a)). In general ϕ decreases with increase in dopant concentration. For heavily doped Si $(\rho_e \approx 0.01\,\Omega\,m, N_d \approx 10^{22}\,m^{-3})$, $|e\phi| \approx 0.2\,eV$.

7.2.8 Relaxation (recombination) time and diffusion length

Thermally or otherwise generated electron and hole carriers recombine after a typical relaxation time τ, having moved a typical diffusion length L through the lattice. In very pure intrinsic material relaxation times can be long $(\tau \sim 1\,s)$, but for commercial doped material relaxation times are much shorter $(\tau \sim 10^{-2}$ to $10^{-8}\,s)$. Lifetime is limited by recombination at sites of impurities, crystal imperfections, surface irregularities and other defects. Thus highly doped material tends to have short relaxation times. Surface recombination is troublesome in solar cells because of the large area and constructional techniques. It is characterized by the surface recombination velocity S_v, typically $\sim 10\,m\,s^{-1}$ for Si, as defined by

$$J = S_v N \tag{7.7}$$

where J is the recombination current number density perpendicular to the surface $(m^{-2}s^{-1})$ and N is the carrier concentration in the material (m^{-3}).

The probability per unit time of a carrier recombining is $1/\tau$. For n electrons the number of recombinations per unit time is n/τ_n, and for p holes is p/τ_p. In the same material at equilibrium these must be equal, so

$$\frac{n}{\tau_n} = \frac{p}{\tau_p}, \quad \tau_n = \frac{n}{p}\tau_p, \quad \tau_p = \frac{p}{n}\tau_n \tag{7.8}$$

In p material, if $p \sim 10^{22} m^{-3}$ and $n \sim 10^{11} m^{-3}$, then $\tau_n \ll \tau_p$ and vice versa. Therefore in solar cell materials minority carrier lifetimes are many orders of magnitude shorter than majority carrier lifetimes (i.e. minority carriers have many majority carriers to recombine with).

Thermally generated carriers diffuse through the lattice down a concentration gradient dN/dx to produce a number current density (in the direction x) of

$$J_x = -D\left(\frac{dN}{dx}\right) \tag{7.9}$$

where D is the diffusion constant. A typical value for Si is $35 \times 10^{-4} m^2 s^{-1}$ for electrons, $12 \times 10^{-4} m^2 s^{-1}$ for holes.

Within the relaxation time τ the diffusion distance L is given by Einstein's relationship

$$L = (D\tau)^{1/2} \tag{7.10}$$

Therefore a typical diffusion length for minority carriers in p type Si $(D \sim 10^{-3} m^2 s^{-1}, \tau \sim 10^{-5} s)$ is

$$L \approx (10^{-3} 10^{-5})^{1/2} m \approx 100 \, \mu m \tag{7.11}$$

Note that $L \gg w$, the junction width of a typical p–n junction (7.6).

7.2.9 *Junction currents*

Electrons and holes may be generated thermally or by light, and so become carriers in the material. Minority carriers, once in the built-in field of the depletion zone, are pulled across electrostatically down their respective potential gradients. Thus minority carriers that cross the zone become majority carriers in the adjacent layer (consider Figs 7.5 and 7.6). The passage of these carriers becomes the *generation current* I_g, which is predominantly controlled by temperature in a given junction without illumination.

In an isolated junction there can be no overall imbalance of current across the depletion zone. A reverse *recombination current* I_r of equal magnitude occurs from the bulk material. This restores the normal internal electric field. Also the band potential V_B is slightly reduced by I_r. Increase in temperature gives increased I_g and so decreased V_B (leading to reduced photovoltaic open circuit voltage V_{oc} with increase in temperature, see later). For a given material, the

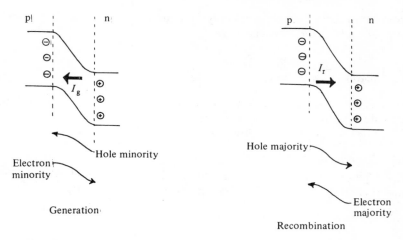

Generation

Recombination

Fig. 7.6 Generation and recombination currents at a p–n junction

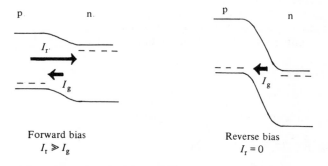

| Forward bias | Reverse bias |
| $I_r \gg I_g$ | $I_r = 0$ |

Fig. 7.7 Recombination and generation junction currents with externally applied bias

generation current I_g is controlled by the temperature. However, the recombination current I_r can be varied by external bias as explained in Section 7.2.6 and in Figs 7.5 and 7.7.

Without illumination, I_g is given by

$$I_g = eN_i^2 \left(\frac{1}{p} \frac{L_p}{\tau_p} + \frac{1}{n} \frac{L_n}{\tau_n} \right) \tag{7.12}$$

where N_i is the intrinsic carrier concentration and the other quantities have been defined before. In practice the control of material growth and dopant concentration is not exact enough to predict how L and τ will vary with material properties and so I_g is not controlled.

Note that recombination is unlikely to occur in the depletion zone, since the transit time across the zone t is

$$t \approx \frac{w}{u} = \frac{w}{\mu(V_B/w)} = \frac{w^2}{\mu V_B} \sim 10^{-12}\,\text{s} \tag{7.13}$$

where u is the carrier drift velocity and μ is the mobility $(\sim 0.1\,\text{m}^2\text{V}^{-1}\text{s}^{-1})$ in the electric field V_B/w $(V_B \sim 0.6\,\text{V}, w \sim 0.5\,\mu\text{m})$.

Thus $t \ll \tau_r$ (τ_r is the recombination time of $\sim 10^{-2}$ to $10^{-8}\,\text{s}$).

7.2.10 *Circuit characteristics*

The p–n junction characteristic (no illumination) is explained by the previous discussion and shown in Fig. 7.8. With no external bias $(V = 0)$,

$$I_r = I_g \tag{7.14}$$

With a positive, forward, external bias of V, the recombination current becomes an increased forward current:

$$I_r = I_g \exp[eV/(kT)] \tag{7.15}$$

as explained in basic solid state physics texts.

The total current (in the dark, no illumination) is

$$I_D = I_r - I_g$$

$$= I_g\{\exp[eV/(kT)] - 1\} \tag{7.16}$$

This is the Shockley equation, usually written

$$I_D = I_0\{\exp[eV/kT] - 1\} \tag{7.17}$$

where $I_0(= I_g)$ is the *saturation current* under full reverse bias before avalanche breakdown. It is also called the leakage or diffusion current. For good solar cells $I_0 \sim 10^{-8}\,\text{A}\,\text{m}^{-2}$.

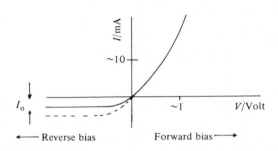

Fig. 7.8 p–n junction dark characteristic. Saturation current I_0 increases with temperature (---)

7.3 Photon absorption

The dominant process causing the absorption of electromagnetic radiation in semiconductors is the generation of electron–hole pairs. This occurs in direct transitions of electrons across the band gap E_g when

$$h\nu \geq E_g \tag{7.18}$$

where h is the Planck constant (6.63×10^{-34} J s) and ν is the radiation frequency. The semiconductor material of solar cells has $E_g \approx 1$ eV. Absorption of photons near this condition occurs with indirect band gap transitions when a lattice vibration phonon (energy $h\Omega \sim 0.02$ eV, where Ω is the phonon frequency) is also involved in the transition. In this case (see (7.36))

$$h\nu \pm h\Omega \geq E_g \tag{7.19}$$

Direct band gap semiconductors (e.g. Ga As) therefore have sharp absorption band transitions with relatively large values of extinction coefficient ($\nu > E_g/h$), whereas *indirect band* gap semiconductors (e.g. Si) have less sharp absorption bands and smaller extinction coefficients (Fig. 7.9).

Band gap absorption for semiconductors occurs at frequencies within the solar spectrum. For Si this occurs when

$$\nu > E_g/h \approx \frac{(1.1\,\text{eV})(1.6 \times 10^{-19}\,\text{J eV}^{-1})}{6.63 \times 10^{-34}\,\text{J s}} = 0.27 \times 10^{15}\,\text{Hz} \tag{7.20}$$

and

$$\lambda \approx \frac{3.0 \times 10^8\,\text{m s}^{-1}}{0.27 \times 10^{15}\,\text{s}^{-1}} = 1.1\,\mu\text{m} \tag{7.21}$$

The number flux of photons in the solar spectrum is large ($\sim 1\,\text{kW m}^{-2}/[(2\,\text{eV})(1.6 \times 10^{-19}\,\text{J eV}^{-1})] \approx 3 \times 10^{21}$ photon m^{-2}s^{-1}). So the absorption of solar radiation in semiconductors can greatly increase electron–hole generation in addition to thermal generation. If this charge carrier creation occurs near a p–n junction, the built-in field across the depletion zone can be the EMF to maintain charge separation and produce currents in an externally connected circuit (Fig. 7.10). Thus the photon generation of carriers in sunlight adds to, and dominates, the thermal generation already present. In dark conditions, of course, only the thermal generation occurs.

The p–n junction with photon absorption is now a DC source of current and power, with positive polarity at the p type material. Power generation from a solar cell corresponds to conditions of diode forward bias. The potential difference across the semiconductor cell (ignoring small contact potentials) is V_B and is due to both the forward biasing and the band displacement. V_B will vary with the external current I between limits of zero (short circuit condition I_{sc}) and V_{oc} (open circuit voltage, $I = 0$). Maximum power is transmitted to an external

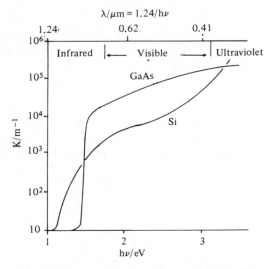

Fig. 7.9 Extinction coefficient K of a direct (GaAs) and indirect (Si) band gap material. Radiation intensity varies as $I(x) = I_0 \exp(-Kx)$ where x is the depth into the surface. Note the logarithmic scale that masks the sharpness of the band gap absorption. After Wilson (1979)

load R_L when R_L equals the internal resistance of the source R_{int}. However, R_{int} is regulated by the absorbed photon flux, so good power matching in a solar cell requires R_L to change in relationship to the insolation.

The solar cell current I is determined by subtracting the photon induced current I_L from the diode dark current I_D (Fig. 7.11; compare with Fig. 7.8). Under reverse bias, I_D is negative and equal to the thermally generated current

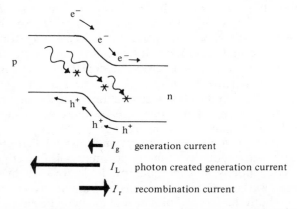

Fig. 7.10 Absorption of active photons ($h\nu \geq E_g$) to create a further current with power generating capability. Currents I are conventional currents

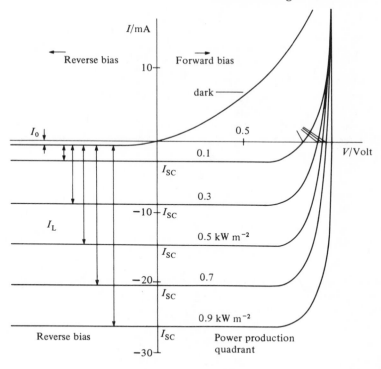

Fig. 7.11 Photovoltaic p–n junction *I–V* characteristic. Drawn in the convention used for rectifying diodes under varying intensity of solar radiation

I_g. Thus I_g and I_L add, since both arise from source generation of electron–hole pairs.

From Fig. 7.11 it is apparent that the external current I is always given by

$$I = I_D - I_L \tag{7.22}$$

and from (7.17),

$$I = I_0 \{\exp[eV/(kT)] - 1\} - I_L \tag{7.23}$$

The details of these characteristics are more fully developed in Section 7.5.

Photocurrent generation depends on photon absorption near the junction region. If the incident solar radiant flux density is G_0, then at depth x the absorbed power per unit area is

$$G = G_0 - G_x = G_0\{1 - \exp[K(\nu)x]\} \tag{7.24}$$

where

$$\frac{dG}{dx} = -K(\nu)G_x \tag{7.25}$$

$K(\nu)$ is the extinction coefficient of Fig. 7.9, and is critically dependent on frequency. Photons of energy less than the band gap are transmitted with zero or very little absorption. At depths of $1/K$ absorption is 63%, at $2/K$ is 86% and at $3/K$ is 95%. For Si at frequencies greater than the band gap, $2/K$ equals $\sim 400\,\mu m$ which gives approximately the minimum thickness for solar cell material.

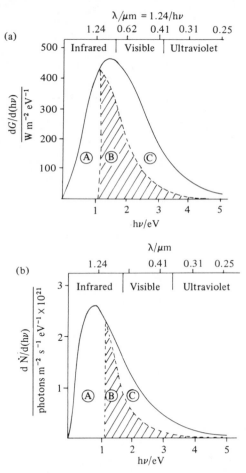

Fig. 7.12 Smoothed spectral distribution of solar radiation. (a) Distribution of solar intensity with photon energy (AM0). (b) Distribution of photon number flux density with photon energy (AM1). A: with photon energy too low for Si photovoltaic generation. B: representing energy going to power production. C: representing excess photon energy not used

7.4 Solar radiation input

Detailed properties of solar radiation were considered fully in Chapter 4. For photovoltaic power generation in a typical solar cell, e.g. Si material, the essential factors are (Fig. 7.12):

(1) The solar spectrum includes frequencies too low for photovoltaic generation ($h\nu < E_g$) (region A). Absorption of these low frequency (long wavelength) photons produces heat, but no electricity.
(2) At frequencies of band gap absorption ($h\nu > E_g$) the excess photon energy ($h\nu - E_g$) is wasted as heat (region C).
(3) Therefore there is an optimum band gap absorption to fit a solar spectrum for maximum electricity production (Fig. 7.13). Note that the spectral distribution of the received solar radiation varies with angular penetration through the atmosphere (see Section 4.6.1 concerning air mass ratio, i.e. AM1 at zenith, AM2 at zenith angle 60°, AM0 in space), and with cloudiness, humidity, air pollution etc. AM1 conditions are usually considered as standard for solar cell design.
(4) Only the energy in region B of Fig. 7.12 is available for photovoltaic power in a basic solar cell. The maximum proportion of total energy $[B/(A + B + C)]$, where A, B, C are the areas of regions A, B, C, is about 47%, but the exact amount varies slightly with spectral distribution. For

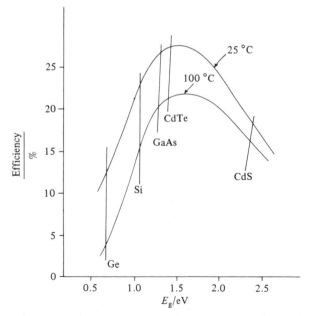

Fig. 7.13 Maximum expected efficiency of homojunction solar cells as a function of band gap. Note the decrease in performance with increase in temperature. Band gaps of common materials are indicated. After Merrigan (1977)

photons with energy greater than the band gap, if the distribution of photon number (N) with photon energy $(E = h\nu)$ is dN/dE, then the maximum theoretical power produced is

$$P = \int_{E_g}^{\infty} \left(\frac{dN}{dE}\right) E_g \, dE \tag{7.26}$$

but

$$dP = h\nu \, dN = E \, dN \tag{7.27}$$

so

$$P = \int_{E_g}^{\infty} \left(\frac{dP}{dE}\right) \frac{E_g}{E} \, dE \tag{7.28}$$

7.5 Photovoltaic circuit properties and loads

With photocells, as with all renewable energy devices, the environmental conditions provide a *current source* of energy. The equivalent circuit portrays the essential macroscopic characteristics for power generation (Fig. 7.14).

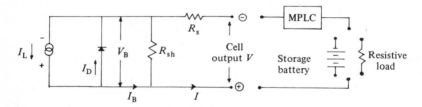

Fig. 7.14 Equivalent circuit of a solar cell. Also drawn are examples of loads with maximum power load control (MPLC) to insure peak power operation

The device current I has positive direction for a power generating cell (NB this would be interpreted as a negative current for a rectifying diode where I_D would be the only component). In a constant radiation flux the circuit asymmetry is dominated by I_D, the dark current diode characteristic, as shown in Fig. 7.15. This relates to the positive V, negative I region of the diode characteristic (Figs 7.8 and 7.11).

Maximum power is obtained by controlling V and I to lie on the *maximum power line*, as the received insolation and load resistance vary. This is nearly at constant voltage, within 25% of V_{oc}. An electrical storage battery remains at nearly constant potential difference whatever the charging current. This is in marked contrast to a purely resistive load where the potential varies directly as

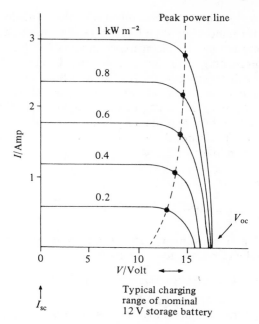

Fig. 7.15 I–V characteristic of a typical 33 cell Si array used for battery charging or other loads. Note that the peak power line, of maximum *IV* product, is a good match with the charging voltage range of the battery even without maximum power load control

the current. Therefore the V/I load line for battery charging can be matched near to the maximum power line.

However, battery EMF does change with the state of charge, within a range of about 10% of maximum EMF. This is shown in Section 16.5 and Fig. 16.4. Thus with all loads, including storage batteries, improved matching is obtained with a maximum power load control. This is essentially a DC to DC converter, or a load switching control. Load management control devices can enable 95% of the maximum output to be used usefully in a load under varying solar conditions.

From the equivalent circuit (Fig. 7.14),

$$I = I_L - I_D - \frac{(V - IR_S)}{R_{sh}} \tag{7.29}$$

Using (7.17),

$$I_D = I_0\{\exp[e(V - IR_s)/(A\,kT)] - 1\} \tag{7.30}$$

For Si material, $I_0 \approx 10^{-7}\,\mathrm{A\,m^{-2}}$.

The *ideality factor A* is introduced at this stage since its inclusion more exactly models actual performance. For all cells $A \gtrsim 1$. Higher values of A produce more curved I–V characteristics, so reducing maximum power. This effect is called the additional curvature factor. It results from increased electron–hole recombination at defects in the junction. Section 7.6 discusses these aspects more fully.

An increase in cell material temperature θ affects performance by decreasing V_{oc} and increasing I_{sc}, with the characteristic changing accordingly. Effectively R_{sh} (often taken as infinitely large) and R_s (made as small as possible) decrease with increase in temperature. Empirical relationships for these effects at $1 \, kW \, m^{-2}$ insolation on Si material are

$$V_{oc}(\theta) = V_{oc}(\theta_1)[1 - a(\theta - \theta_1)] \tag{7.31}$$

$$I_{sc}(\theta) = I_{sc}(\theta_1)[1 + b(\theta - \theta_1)] \tag{7.32}$$

where $\theta_1 = 25\,°C$ is a convenient reference temperature, θ is the material temperature, and the temperature coefficients are

$$a = 3.7 \times 10^{-3}\,°C^{-1}, b = 6.4 \times 10^{-4}\,°C^{-1}$$

The net effect of an increase in temperature is to reduce the power P available. An empirical relationship for Si material is

$$P(\theta) = P(\theta_1)[1 - c(\theta - \theta_1)] \tag{7.33}$$

where $c = 4 \times 10^{-3}\,°C^{-1}$.

The remaining requirements for good power production are obvious from the equivalent circuit, namely:

(1) I_L should be a maximum: for example, through minimum photon losses, absorption near the depletion layer, a low surface reflectance, a small top surface electrical contact area, a high dopant concentration and few recombination centers.
(2) I_D should be a minimum: for example through high dopant concentration.
(3) R_{sh} should be high, such as by careful edge construction.
(4) R_s should be small, for example by ensuring a short path for surface currents to electrical contacts, and by using low resistance contacts and leads.
(5) $R_{load} = R_{internal} = V/I$ for optimum power matching.

Solar cell arrays are often assembled from a combination of individual modules usually connected in parallel. Each module is itself a combination of cells in series. Each cell is a set of surface elements connected in parallel (Fig. 7.16). Maximum open circuit potentials of the modules are usually $\sim 15\,V$, and maximum currents at the module terminals are $\sim 1.5\,A$. It will be obvious that difficulties will arise if one cell or element of a cell becomes faulty, or if the array is unequally illuminated by shading or by unequal concentration of light. Cell or

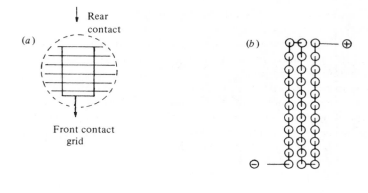

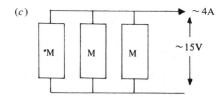

Fig. 7.16 Typical arrangements of commercial Si solar cells: (a) cell (b) module of 33 cells (c) array

cell elements can be driven into forward or reverse bias diode operation, with the danger of overheating. Thus faults can avalanche unless protective bypass diodes are set in parallel with each series linked cell. However, commercial solar arrays can be expected to give trouble-free service so long as elementary abuse is avoided. Lifetimes of at least 20 years are expected.

7.6 Limits to cell efficiency

Photovoltaic cells are limited in efficiency by many losses; some of these are avoidable but others are intrinsic to the system. Some limits are obvious and may be controlled independently, but others are complex and cannot be controlled without producing interrelated effects. For instance, increasing dopant concentration can have both advantageous and harmful effects. Table 7.2 portrays typical losses for Si p–n junction solar cells in AM1 insolation. Unfortunately there is no standard convention for the names of the loss factors. Not included on the table are the economic factors of the cost of manufacture and array durability. When all factors are taken into account, the final strategy for optimum commercial production is exceedingly complex.

Table 7.2 *Limits to efficiency in Si solar cells. Refer to Section 7.6 for explanation of each process*

Text section for process	Data for 1983 cells			Notes	Targets for future cells		
	Energy remaining after process loss (%)	Power loss (%)	Efficiency factor		Efficiency factor	Power loss (%)	Energy remaining after process loss (%)
7.6.3	77	23	0.77	No photovoltaic absorption: $h\nu < E_g$	0.77	23	77
7.6.4	44	33	0.57	Excess photon energy lost as heat: $h\nu - E_g$	0.57	33	44
7.6.2	43	1	0.97	Surface reflection	0.99	1	43
7.6.5	42	0.4	0.99	Quantum efficiency	0.99	0.4	43
7.6.1	39	39	0.92	Top surface contact grid obstruction	0.95	1.6	41
7.6.7	19	20	0.5	Voltage factor $eV_B < E_g$	0.7	12	29
7.6.8	15	4	0.81	Curve factor = (max. power)/$I_{sc} V_{oc}$	0.87	4	25
7.6.9	10	5	0.65	Additional curve factor A, recombination collection losses	0.9	2.4	23
7.6.10	9.7	0.3	0.97	Series resistance	0.97	0.6	22
7.6.11	9.6	0.1	0.99	Shunt resistance	0.99	0.2	21
7.6.12	~10			Delivered power			20

In the following sections the losses are given as a percentage of total incident insolation, AM1 = 100%, and are listed from the top to the base of the cell. The efficiency factors in Table 7.2 refer to the proportion of the remaining insolation that is usefully absorbed at that stage in the photovoltaic generation of electricity.

7.6.1 *Top surface contact obstruction (loss ~3%)*

The electric current leaves the top surface by a web of metal contacts arranged to reduce series resistance losses in the surface (see Section 7.6.10). These contacts have a finite area and so they cover part of the active surface. This loss is not always accounted for in efficiency calculations.

7.6.2 *Reflection at top surface (loss ~1%)*

Without special precautions, the reflectance from semiconductors is high, ~40% of the incident solar radiation. Fortunately this may be dramatically reduced to 3% or less by thin film surface or other treatment.

Consider three materials (air, cover, semiconductor) of refractive index n_0, n_1 and n_2. For dielectric electrically insulating materials, the reflectance between two media is

$$\rho_{ref} = \frac{(n_0 - n_1)^2}{(n_0 + n_1)^2} \tag{7.34}$$

For air ($n_0 = 1$) to plastic (say $n_1 = 1.6$), $\rho = 5.3\%$. Semiconductors have a refractive index represented by a complex number (since they are partly conducting) which is frequency dependent and averages about 3.5 in magnitude over the active spectrum. The reflectance in air varies from $\rho_{ref}(1.1\,eV) = 34\%$ to $\rho_{ref}(5\,eV) = 54\%$.

A thin film (thickness t) of intermediate material between air and semiconductor will largely prevent reflection (Fig. 7.17) if for normal incidence the main reflected components a and b are of equal intensity and differ in phase by π radians ($\lambda/2$ path difference). For this to occur the reflectance at each surface is equal, so $n_1 = \sqrt{(n_0 n_2)}$, and also $t = \lambda/(4n_1)$. There is only one wavelength for which this condition is met exactly; however, broad band reflectance is

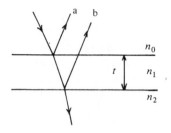

Fig. 7.17 Antireflection thin film

considerably reduced. For Si (if $n_1 = 1.9$, thickness $t = 0.08\,\mu\text{m}$) the broad band reflectance is reduced to $\sim6\%$. Multiple thin layers can reduce broad band reflectance to $<3\%$.

Reflection losses can also be reduced by top surface configurations that reflect the beam for a second opportunity of absorption (Fig. 7.18). The *textured surface* of Fig. 7.18(a) can be produced by chemical etching on Si.

(a)   (b)

Fig. 7.18 Top surfaces for increased absorption: (a) textured (b) structured

7.6.3 *Photon energy less than band gap (loss ~23%)*

Photons of quantum energy $h\nu < E_g$ cannot contribute to photovoltaic current generation. For Si ($E_g \approx 1.1\,\text{eV}$) the inactive wavelengths have $\lambda \gtrsim 1.1\,\mu\text{m}$ and include 23% of AM1 insolation. If this energy is absorbed it causes heating with a temperature rise that lowers power production still further. These photons can theoretically be removed by filters. However a more 'energy efficient' strategy is to make use of the heat in a combined heat and power system.

7.6.4. *Excess photon energy (loss ~33%)*

The excess energy of active photons ($h\nu - E_g$) also appears as heat.

7.6.5 *Quantum efficiency (loss ~0.4%)*

Quantum efficiency – the fraction of incident absorbed active photons producing electron–hole pairs – is usually very high. Thus the design of the cell has to insure that the material is thick enough for at least 95% absorption, as explained in Section 7.3. Reflecting layers at the rear of the cell can return transmitted radiation for a second pass.

7.6.6 ǀ *Collection efficiency*

Collection efficiency is a vague term used differently by different authors. It may be applied to include the losses described in Sections 7.6.3 and 7.6.4 or usually, as here, to electrical collection of charges after carrier generation. Collection efficiency is therefore defined as the proportion of radiation generated electron–hole pairs that produce current in the external circuit. For 10% overall efficiency cells, the collection efficiency factor is usually about 0.7. Increasing this to about 0.9 would produce >20% overall efficiency cells, and so collection efficiency improvement is a major design target.

There are many factors affecting collector efficiency, as outlined in the following. One improvement not otherwise mentioned is back surface field

(BSF). A layer of increased dopant concentration is formed as a further layer beyond the p–n junction (e.g. $1\,\mu m$ of p^+ on p to produce a further junction of $\sim 200\,kV\,m^{-1}$) (Fig. 7.19). Electron minority carriers formed in the p layer near this p^+ region are 'reflected' down a potential gradient back towards the main p–n junction rather than up the gradient to the rear metal contact. Electron–hole recombination at the rear contact is therefore reduced. Similar diode-like layers

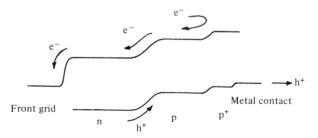

Fig. 7.19 Back surface field (BSF) to lessen diffusion leakage of electron current carriers at the rear of cells, shown here as an n on p cell

can be added to the front surface (e.g. n^+ on n) to produce the same effect for holes, providing optical absorption is not significantly increased.

7.6.7 *Voltage factor F_V (loss ~20%)*

Each absorbed photon produces electron–hole pairs with an electric potential difference of E_g/e (1.1 V in Si). However, only part (V_B) of this potential is available for the EMF of an external circuit. This is made clear in Fig. 7.4, where the displacement of the bands across the junction in open circuit produces the band potential V_B. The voltage factor is $F_V = eV_B/E_g$. In Si F_V ranges from ~ 0.6 ($0.01\,\Omega\,m$ material) to ~ 0.5 ($0.1\,\Omega\,m$ material), so $V_B \approx 0.66\,V$ to $0.55\,V$.

The 'missing' EMF ($\phi_n + \phi_1$) occurs because in open circuit the Fermi level across the junction equates at the dopant n and p levels, and not at the displaced conduction to valence band levels. Increased dopant concentration increases F_V ($0.01\,\Omega\,m$ Si has greater V_B and V_{oc} than $0.1\,\Omega\,m$ Si), but other effects limit the maximum dopant concentrations in Si to $\sim 10^{22}\,m^{-3}$ of $0.01\,\Omega\,m$ materials. In GaAs, F_V is ~ 0.8.

When producing current on load, the movement of carriers under forward bias produces heat as resistive internal impedance heating. This may be included as voltage factor loss, as A factor loss (Section 7.6.9) or, as here, by series resistance heating (Section 7.6.10).

7.6.8 *Curve factor F_c (loss ~4%)*

The solar cell *I–V* characteristic is strongly influenced by the p–n diode characteristic (Fig. 7.8). Thus as the solar cell output voltage is raised towards V_{oc} the diode becomes increasingly forward biased, so increasing the internal recombination current I_r across the junction. This necessary behavior is treated as a fundamental loss in the system. Peak power P_{max} is less than the product $I_{sc} V_{oc}$ owing to the exponential form of the characteristic (7.23).

The curve factor F_c (also called the fill factor) is $P_{max}/(I_{sc} V_{oc})$. The maximum value in Si is 0.88.

7.6.9   *Additional curve factor A (loss ~5%)*

In practice the cell characteristic does not follow (7.23) and is better represented by

$$I = I_0 \exp[eV/(AkT) - 1] - I_L \tag{7.35}$$

See (7.17) and (7.30)

The factor A (>2 for many commercial cells) results from increased recombination in the junction. This effect also tends to change V_{oc} and I_0, so in general optimum output would occur if $A = 1$.

Recombination has already been mentioned for back scatter field (Section 7.6.6). Within the cell, recombination is lessened if:

(1) Diffusion paths are long (in Si > 50 to $100 \mu m$). This requires long minority carrier lifetimes (in Si up to $100 \mu s$).
(2) The junction is near the top surface (within $0.15 \mu m$, as in 'violet' cells, rather than $0.35 \mu m$ as in normal Si cells).
(3) The material has few defects other than the dopant.

Surface recombination effects are influential owing to defects and imperfections introduced at crystal slicing or at material deposition.

7.6.10   *Series resistance (loss ~0.3%)*

The solar cell current has to pass through the bulk material to the contact (ohmic) leads. At the rear the contact can cover the whole cell and the contribution to series resistance is very low. The top surface, however, must be exposed to the maximum amount of insolation. It should therefore be covered by the least area of contacts. This, however, produces relatively long current path lengths with noticeable series resistance. Significant improvements have now been made in forming these contacts and arranging the surface layout to minimize the top surface series resistance to $\sim 0.1 \Omega$ in a cell resistance of $\sim 20 \Omega$ at peak power.

7.6.11   *Shunt resistance (negligible, ~0.1%)*

Shunt resistance is caused by structural defects across and at the edge of the cell. Present technology has reduced these to a negligible effect, so shunt resistance may be considered infinite in single crystal Si cells. This may not be so in polycrystalline cells, however.

7.6.12   *Delivered power (Si cell 10 to 14%)*

Table 7.2 shows the remaining power, after the losses, as the delivered power. This assumes optimum load matching at full insolation, without overheating, to produce peak power on the *I–V* characteristic.

Table 7.2 (right hand side) also shows the planned improvements to produce Si single crystal cells with ~22% efficiency. Most authorities consider 22 to 25% to be the maximum for Si cells, 30% for advanced heterojunction or graded junction cells (see Section 7.7), and 40% for highly specialized systems with concentration or other devices. Note that the spectrum of incident solar radiation sets an absolute limit for the efficiency (Sections 7.6.3, 7.6.4, 7.4). For Si cells this limit is ~47%.

7.7 Solar cell construction

We shall describe briefly the construction of a standard single crystal Si photovoltaic cell. There are many variations, and commercial competition produces continued revision of cell type and fabrication method. A general design is shown in Fig. 7.1.

7.7.1 General design criteria

(1) Initial materials have to be of high chemical purity with consistent properties.
(2) The cells must be mass produced with the minimum cost, but total control of the processes and high levels of precision must be maintained.
(3) The final product has to have a lifetime of at least 20 years in exposed and often hostile environments. Even without concentration of the insolation, the cell temperature may range between -30 and $+200°C$. Electrical contacts must be maintained and all forms of corrosion avoided. In particular water must not be able to enter the fabric.
(4) The design must allow for some faults to occur without failure of the complete system. Thus redundant electrical contacts are useful. The parallel and series connections between the cells must allow for some cells to become faulty without causing an avalanche of further faults.
(5) The completed modules have to be safely transported, often to inaccessible and remote areas.

7.7.2 Crystal growth

High purity electronic grade base material is obtained in polycrystalline ingots. Impurities should be less than 1 atom in 10^9, i.e. less than 10^{18} atoms per m^3. This starter material has to be made into large single crystals.

(1) *Czochralski technique* This well-established crystal growing technique consists of dipping a small seed crystal into molten material (Fig. 7.20(a)). Dopant (e.g. boron acceptors for p type) is added to the melt. Slowly the crystal is mechanically pulled upwards with a large (~15 cm diameter) crystal growing from the seed. As with other single crystal growth techniques, the crystal is then sliced ~300 μm thick with highly accurate diamond saws. About 40 to 50% of crystalline material is lost during this process, which represents a most serious loss.

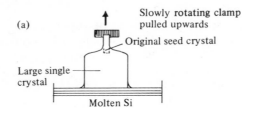

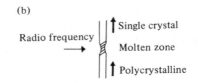

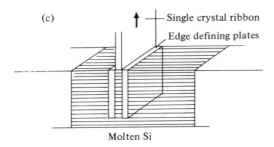

Fig. 7.20 Some crystal growth methods: (a) Czochralski (b) zone recrystallization or laser heating (c) ribbon

(2) *Zone refining*   Polycrystalline material is formed as a rod. A molten zone is passed along the rod by heating with a radio frequency coil or with lasers (Fig. 7.20(b)). This process both purifies the material and forms a single crystal. The single crystal has to be sliced and treated as for other techniques.

(3) *Ribbon growth*   This method avoids slicing and the consequent wastes by growing a continuous thin strip of single crystal up to 10 cm wide and 300 μm thick. The type of process is shown in Fig. 7.20(c). The ribbon may be stored on large diameter rolls, and then later cut for surface treatment to form cells.

(4) *Vacuum deposition*   This technique may be used at different stages of construction, e.g. for the top metal layer of a Schottky diode. Vacuum deposition of Si is difficult and not usually successful.

(5) *Casting*   Polycrystalline material is formed. The cheapness of the process may compensate for the lower efficiency cells produced.

7.7.3 *Slice treatment*

The $300\,\mu$m to $400\,\mu$m thick slices are then chemically etched. A very thin layer of n type material is formed by diffusion of donors (e.g. phosphorus) for the top surface. One method is to heat the slices to $1000\,°$C in a vacuum chamber into which is passed P_2O_5, but more often the slices are heated in nitrogen with the addition of $POCl_3$.

Photolithographic methods may be used to form the grid of electrical contacts. First Ti may be deposited to form a low resistance contact with the Si; then a very thin Pd layer to prevent chemical reaction of Ti with Ag; and then the final Ag deposit for the current carrying grid. Other methods depend on screen printing and electroplating.

The important antireflection layers are afterwards carefully deposited by vacuum techniques; however, the similar properties of textured surfaces are produced merely by chemical etching. The rear surface may be diffused with Al to make a back surface field of p^+ on p (see Section 7.6.6). On to this is laid the rear electrical metal contact as a relatively thick overall layer.

7.7.4 *Modules and arrays*

The individual cells, of size $\sim$10 cm $\times$ $\sim$10 cm, are then connected into modules of about 30 cells. Each module usually has three to five columns of cells in series. Such an arrangement produces an EMF of about 15 V, which is safe and convenient for charging 12 V batteries. (Such batteries are best if especially designed for photovoltaic use.) Close packing of circular cells leaves about 15% of the module area as voids. The cells are sandwiched in an inert filler between a clear front cover (usually ultraviolet resistant plastic) and a backing plate. The cover sealing must be watertight under all conditions including thermal stress. The rear plate must be strong and yet present a low thermal resistance.

7.8 Types and adaptations of photovoltaics

Although the flat plate Si solar cell is the most common commercial product, there is a great variety of alternative types and constructions. These seek to improve efficiency, and to decrease the cost of the power produced by reducing capital cost. This section is a brief summary of a complex and continually changing scene (see Table 7.3).

7.8.1 *Solid state structure variations*

(1) *Homojunctions* If the base semiconductor material remains the same across the p–n junction, and only changes in type or concentration of dopant, it is a homojunction. The Si cells so far discussed are homojunctions. The band gap is constant across the junction (Fig. 7.21(a)).

(2) *Heterojunctions* The band gap changes across the junction because of a significant change in base material (Fig. 7.21(b)). The advantage is that band gap photon absorption is now possible at two frequencies. This increases the total proportion of photons that may be absorbed, and so decreases the

Table 7.3 Solar cell base material parameters, AM1 conditions*

Material base	Band gap E_g† eV	Direct D or indirect I	Example of cell	V_{oc} V	$\dfrac{I_{sc}}{A\,m^{-2}}$	Practical efficiency 1983 (%)	Target efficiency (%)
Ge	0.6		Not used				
Si	1.1	I	n/p 0.1 Ωm	0.55	340	14	20
			n/p 0.01 Ωm	0.6		15	22
GaAs	1.4	D	p/n	0.9		12	25
CdTe	1.4	D					
$Ga_{1-x}Al_xA_s$ $0 < x < 0.34$	1.4–1.9	D	Heterojunction with GaAs base	0.95		16	25
$Ga_{1-x}Al_xA_s$ $0.34 < x < 1$	1.9–2.2	I	Heterojunction with GaAs base	0.95		16	25
CdS	2.4		Thin film complex structure	0.5		10	15

*The optimum band gap in AM1 radiation is between 1.4 and 1.5 eV (Fig. 7.13)
†Band gap decreases with temperature increase. Data here for ambient temperature [e.g. Si 1.14 eV (30°C), 1.09 (130°C)]

excess photon energy loss ($h\nu - E_g$). Normally the wider band gap material is on the top surface, so the less energetic photons cross to the narrower band gap material.

The lattice structure on each side of the heterojunction has to be compatible for continuous growth of the complete cell. Examples are $Ga_{1-x}Al_xAs$ on GaAs (operational efficiency 12%), and SnO_2 on n type Si (operational efficiency 10%).

A continuously decreasing band gap (the *graded* band gap cell) has been proposed. Manufacture is difficult yet possible (e.g. with $Ga_{1-x}Al_x$ material), but V_B would be low.

(3) *Direct and indirect band gap* Consider the Brillouin energy (E), momentum (k) diagrams of Fig. 7.22. We wish photons to be able to cause electron–hole transitions across the minimum band gap. With direct band gap materials the transition can be caused by photons alone, with $h\nu = E_g$. However, with indirect band gap materials the minimum gap can only be crossed with low energy photons if the absorption coincides with phonon absorption. Thus if the phonon energy is $h\Omega$, the condition for absorption at the minimum between the bands is

$$h\nu \pm h\Omega = E_g \tag{7.36}$$

As explained in section 7.3, indirect band gap material (e.g. Si) has a lower extinctive coefficient than direct band gap material (e.g. GaAs), so requiring thicker cells.

(4) *Schottky barrier, MS and MIS* A p–n junction may be formed at a metal–semiconductor (MS) interface as shown in Fig. 7.21(c). The advantage is simplicity of construction, since the metal can be deposited as a thin film on the one base material. The disadvantages are the increased reflectance of the metal (and hence loss of input), and the increased recombination losses at the junctions.

In fabrication it is difficult to avoid a thin insulating layer of oxide forming between the metal and semiconductor (MIS cell). Control of this insulating layer can, however, lead to improved cells (Fig. 7.21(d)) through suppression of surface recombination.

(5) *Polycrystalline* The cells described so far assume single crystal base material. This is not necessary, and considerable savings in production costs can be avoided by using polycrystalline material. This is not necessarily structurally weak, but the presence of boundaries between the crystal grains increases recombination of electron–hole pairs. Consequently poly-crystalline solar cells have lower efficiencies than single crystal material.

Polycrystalline cells may be made by low cost thin film techniques (see later), which may however alter the solid state properties significantly.

(6) *Amorphous* Amorphous materials are solids that have only a short range of order (glass has an amorphous structure). Materials that are considered semiconductors (e.g. Si) may under particular conditions remain with semi-conductor properties in the amorphous state. Electrical resistivity may be similar, and in particular n and p type dopants may be added to produce

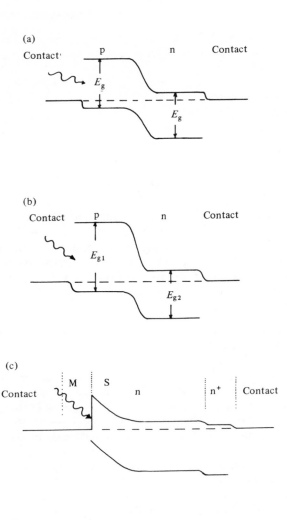

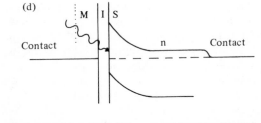

Fig. 7.21 Solar cell junction types. (a) Homojunction: base material and band gap constant across junction. (b) Heterojunction: base material and band gap change across junction. (c) Schottky metal semiconductor (MS) cell, e.g. Au/Si, shown here with n/n^+ back surface field. The important antireflection layer is not shown. (d) Schottky metal insulator semiconductor (MIS) cell

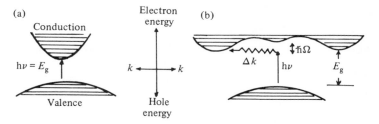

Fig. 7.22 Brillouin zone, energy band/lattice vibration momentum diagrams. Absorption near the band gap requires the creation of a lattice phonon (energy $\pm \hbar\Omega$) with indirect band gap semiconductors. (a) Direct band gap e.g. GaAs. (b) Indirect band gap e.g. Si. See Smith (1978) for further explanation

similar effects to that in crystalline material. Normally however the amorphous structure produces a very high proportion of unsatisfied 'dangling' chemical bonds that readily act as electron or hole traps in an uncontrollable manner. However, if a high proportion of hydrogen is present as the material is formed, the number of such bonds is dramatically reduced.

Development of amorphous cells has raised efficiency to about 10% and promises low cost construction.

(7) *Liquid interface*   The top surface of a solar cell may be a liquid electrolyte. The advantages are potentially good electrical contact, and the possibility of chemical change in the liquid for energy storage. The disadvantages are the operational difficulties, generally low efficiency and easy contamination.

(8) *Organic material*   Carbon based organic materials may be semiconductors (carbon is in the same periodic group as silicon but not itself a semiconductor). The extensive knowledge of organic chemistry and the possible cheapness of organic materials make developments in this area of great interest. Efficiencies of 1% have been achieved, which should be improved to about 10% with present research.

(9) *Intermediate transitions (phosphors)*   In principle, the front surface of a photovoltaic cell could be coated with a fluorescent or phosphorescent layer to absorb photons of energy significantly greater than the band gap ($h\nu_1 \gg E_g$). The emitted photons would have to be actively absorbed ($h\nu_2 \geq E_g$). Thus the excess energy of the original photons ($h\nu_1 - h\nu_2$) would be dissipated in the surface, hopefully with less temperature increase of the cell. Other similar ideas have been put forward for either releasing two active photons from each original photon, or absorbing two inactive photons ($h\nu < E_g$) to produce one active photon in a manner reminiscent of photosynthesis. Commercial exploitation of these ideas has not been tried.

7.8.2 *Important substrate materials*

Table 7.3 lists important parameters for a small range of solar cell materials, and gives some typical properties of p–n junction cells made with the material as bottom layer. Si is commercially and historically the most important. Comment will be made on two other materials of commercial importance.

(1) *Gallium arsenide (GaAs)*   GaAs is a direct band gap semiconductor with $E_g = 1.43\,\text{eV}$. This is near the optimum band gap of $1.5\,\text{eV}$ for a solar cell in AM1 insolation. Heterojunctions with $Ga_{1-x}Al_xAs$ can be made commercially. Theoretical target efficiencies for cells are high at about 25%, and GaAs devices have reached practical efficiencies of 16%. The high extinction coefficient necessitates accurate control of layer depths, and surface recombination can be high.

(2) *Cadmium sulphide (CdS)*   CdS has been considered important because cells can be made by thin film vacuum deposition or by chemical spraying. The lower CdS layer is made p type, and the n type layer of Cu_xS is formed by chemical dipping. The p–n zone is a heterojunction, which unfortunately easily degrades by the diffusion of Cu ions into the CdS and by various chemical changes in manufacture.

7.8.3 *Variations in cell construction*

(1) *Stacked cells*   These are a series of physically separated p–n junctions of decreasing band gap, with light incident on the largest band gap material. Photons of energy less than the first gap ($h\nu < E_{g1}$) are transmitted to the next junction, and so on.

(2) *Vertical multijunction cells (VMJ)*

    (a) *Series linked*   A series of perhaps 100 similar p–n junctions are made in a pile (Fig. 7.23(a)). Light is incident through the edges of the junctions, so the output potential ($\sim 50\,\text{V}$) is the series sum of many junctions. The current is related only to the absorbed radiation flux per junction edge area, and so is not large.

    (b) *Parallel linked* arrangements are also possible. This is a form of *grating cell*, usually made with the aim of absorbing photons more efficiently in the region of the junction (Fig. 7.23(b)).

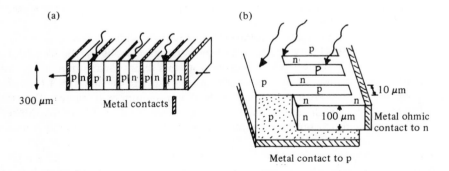

Fig. 7.23 Vertical multijunction cells (VMJ): (a) series linked. (b) parallel linked

(3) *Thin film cells*   The great disadvantage of conventional Si cells has been the cost of single crystal growth, slicing, and surface layer preparation. It is therefore attractive to consider construction by the deposition of layers from vapors or sprays. Silicon has a high boiling point (2620°C) and as a vapor is chemically very active and difficult to handle. It is not an attractive element for thin film deposition from a hot vapor, requiring extreme vacuums ($<10^{-8}$torr) to exclude oxygen and contaminants. Nevertheless the deposition of amorphous silicon compounds by electric glow discharge in gaseous silane (SiH_4) has been possible (see Section 7.8.1(6)) and other alternative techniques to vacuum deposition are proposed.

The most common thin film solar cell has been the CdS cell (see Section 7.8.2(2)). In principle all types of cells could be manufactured by thin film techniques, but the difficulties of forming efficient cells seem very great. However, technical improvements should occur in the future to make such cells commercially viable.

(4) *Reflecting or textured surfaces*   The top surface of the solar cells can be designed to pass reflected radiation back into the surface (Fig. 7.18). Some systems have to be made mechanically; others may be textured by etching with chemicals that give asymmetric activity according to crystal axis orientation.

7.8.4 *Variation in system arrangement*

(1) *Concentrators* (see Section 6.8)   Since the active solar cell material is usually the most expensive component of an array, it is sensible strategy to concentrate the insolation into the cell (Fig. 7.24). Performance is not impaired in the increased radiation flux if the cell temperature remains near ambient, by active or passive cooling. The heat removed in active cooling may be used to increase the total energy efficiency of the system.

The *concentration ratio X* of a perfectly focused system is the ratio of the concentrator input aperture to the surface area of the cell. In practice energy concentration reaches 80 to 90% of this geometric factor. Low X systems ($X \lesssim 5$) do not have to be oriented through the day to follow the sun and so make use of some diffuse as well as direct insolation. Increased X systems have to track the sun, and are only sensible in places with a large proportion ($> 70\%$) of direct radiation.

There is a wide variety of concentrator systems based on lenses (usually Fresnel flat plane lenses), mirrors and various novelties such as prism internal reflection (Fig. 7.24). (See also Section 6.8.)

(2) *Spectral splitting*   Separate solar cells with increasing band gap could be laid along a solar spectrum (say from a prism, and ranging from infrared to ultraviolet) to obtain excellent frequency matching. The dominant losses (totalling ~50%) from the mismatch of photon energy and band gap could therefore be greatly decreased. Final efficiencies of about 60% might be possible. Economically this would only be possible if combined with a high concentration ratio before dispersion. Systems of this principle using three dichroic mirrors have been made.

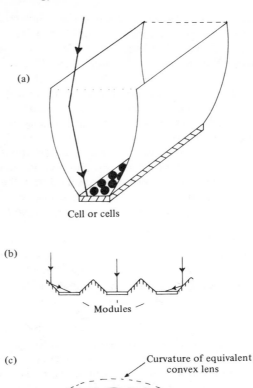

Fig. 7.24 Some concentrator systems. Beware: grossly unequal illumination of cells or modules can cause cell damage. (a) Linear parabolic reflector: may be constructed as a solid block of transparent plastic. (b) side reflectors. (c) Fresnel lens

(3) *Thermophotovoltaic* Insolation may be highly concentrated on to an absorbing surface which then reradiates according to its temperature. Effectively the peak frequency of the radiation is shifted into the infrared to obtain a better match with a low band gap photovoltaic cell. Efficiencies of 40% have been claimed in laboratory systems.

7.9 Other types of photoelectric and thermoelectric generation

Photoelectric generation is defined in this text as the generation of electrical current by the action of absorbed photons directly creating electron–hole pairs. Current can only be obtained in an external circuit if a (voltage) source of EMF exists. In *photovoltaic devices* the EMF is obtained from the built-in field of p–n or metal–semiconductor junctions. Photoelectric generation could still, however, occur if the source of EMF is obtained externally, as with the emission of an electron

(7.37)

work function of the metal or sion surfaces in series along an produce power generation. The w, however, and commercially onsidered unlikely.

photon absorption, but *thermionic* ectricity from the solar heat flux. A an electron current I that may be surface. I is given by Richardson's

(7.38)

dson's constant for the material, T is the function. A good thermionic emitter is $^{-2}K^{-2}$, $\phi \sim 2\,eV$). An external EMF is not necessary, other metal of lower work function, and at lower temperature, the heated surface. In effect the system is similar to a p–n junction with generation current being used for power production. Such devices using solar heating have a short lifetime and a low efficiency, however.

Electrical power can be obtained from *thermoelectric devices*, based on the Seebeck (thermocouple) effect. Electric carrier excitation occurs by the thermal energy of high temperature, and not by photon absorption directly. Thermoelectric generation is most efficient with heated semiconductor junctions. Solar systems have been constructed with low efficiency. Essentially the hot junction of the thermocouple is heated at the focus of a concentrating system with direct solar radiation. However, energy efficiency is low and the heating and cooling systems complex. The theory of these devices is well covered in thermodynamics texts and, since the systems do not rely particularly on renewable energy supplies, they will be considered no further here.

Problems

7.1 The band gap of GaAs is 1.4 eV. Calculate the optimum wavelength of light for photovoltaic generation in a GaAs solar cell.
7.2 (a) Give the equation for the I–V characteristic of a p–n junction diode in the dark.
   (b) If the saturation current is $10^{-8}\,A\,m^{-2}$, calculate and draw the I–V characteristic as a graph to 0.2 V.

7.3 (a) What is the *approximate* photon flux density (photon $s^{-1}m^{-2}$) for AM1
          solar radiation at $0.8\,kW\,m^{-2}$?
     (b) AM1 insolation of $0.8\,kW\,m^{-2}$ is incident on a single Si solar cell of area
          $100\,cm^2$. Assume 10% of photons cause electron–hole separation across
          the junction leading to an external current. What is the short circuit
          current I_{sc} of the cell? Sketch the I–V characteristic for the cell.

7.4 A small household lighting system is powered from a $8\,V$ storage battery
     having a $30\,Ah$ supply when charged. The lighting is used for $4\,h$ each night
     at $3\,A$.
         Design a suitable photovoltaic power system that will charge the battery
     from an arrangement of Si solar cells.
     (a) How will you arrange the cells?
     (b) How will the circuit be connected?
     (c) How will you test the circuit and performance?

Solutions

7.1 $h\nu = E_g$; $\lambda = 0.88\,\mu m$

7.2 Graph of $I = (10^{-8}\,A\,m^2)\,[\exp(eV/kT) - 1]$.

7.3 (a) From Fig. 7.12(b), the photons with average photon number flux
          density have energy ~1 eV. Thus

$$N \approx \frac{(0.8\,kW\,m^{-2})(10^3\,W\,kW^{-1})(1\,\text{photon}/1\,eV)}{(1.6 \times 10^{-19}\,J\,eV^{-1})}$$

$$\approx 0.5 \times 10^{22}\,\text{photon}\,m^{-2}s^{-1}$$

     (b) Assume 10% of the photons each produce one electron and one hole as
          charge carriers. Then

$$I = (0.5 \times 10^{21}\,\text{photon}\,m^{-2}s^{-1})(2\,\text{carrier/photon})(10^{-2}\,m^{-2})$$

$$\times (1.6 \times 10^{-19}\,C/\text{carrier})$$

$$= 1.6\,A$$

     (NB These are approximate 'ballpark' answers. Mathematical rigor could
        produce accurate answers.)

7.4 Cells should produce about $10\,V$ to charge an $8\,V$ storage battery. Each cell
     on peak load has EMF ~$0.5\,V$, so we need parallel arrangements of 20 cells
     in series. Each night $12\,Ah$ is discharged, so each day $12\,Ah/0.8 = 15\,Ah$ is
     needed from the cells. Assume $3\,h$ of direct sunlight each day, so $5\,A$
     charging current is needed. In series connection this could be obtained from

$5\,\mathrm{A}/(200 \times 10^{-4}\,\mathrm{A\,cm^{-2}}) = 250\,\mathrm{cm^2}$. With a series arrangement of 20 cells, each cell needs an area of $12.5\,\mathrm{cm^2}$ (radius $\sim 2.5\,\mathrm{cm}$). Thus a series arrangement of 20 cells each of radiu $2.5\,\mathrm{cm}$ is reasonable for charging the batteries with the *assumed* solar radiation flux.

Test for short circuit current, open circuit voltage, in direct sunshine normal to the cells.

Bibliography

Chamberlain, G. A. (1983) 'Organic solar cells – a review', *Solar Cells*, **8**, 47–83.

Green, M. A. (1982) *Solar Cells: Operating Principles, Technology and System Application*, Prentice-Hall.
Excellent text by an outstanding researcher.

Merrigan, J. A. (1977) *Sunlight to Electricity*, MIT Press, Boston.
Summary of photovoltaic and other systems. Not theoretical.

Open University (1973 and later) *Solar Cell*, unit TS251 (10 and 11), Open University Press, Milton Keynes.
Clear introduction of physical and materials requirements.

Smith, R. A. (1978) *Semiconductors*, 2nd edn, Cambridge University Press.
Basic text on semiconductor properties, including the p–n junction. Very little on solar cells explicitly.

Sogesta Publications (International School on Solar Energy and Other Renewable Energy Sources), Sogesta SpA, Casella Postale 65, 61029, Urbino, Italy.
Excellent series of graduate level books associated with summer schools, e.g. Galluzzi, F. 'Photovoltaic conversion of solar energy', in Barra, O. A. (ed.) *Small Scale Power Conversion*.

Spirito, P. and Vitale, G. (1981) *Photovoltaic Solar Devices*, Workshop on System Design, Instituto Electrotecnico, University of Naples.
Very practical guide for real installations, including battery charging, lighting, telecommunications, water pumping etc.

Wilson, J. I. B. (1979) *Solar Energy*, Wykeham, London.
Well written, thoughtful text on solar thermal and electric systems. Particularly strong on photovoltaics. Covers fundamental theory and practical construction. Strongly recommended.

8 *Hydro-power*

8.1 Introduction

The term hydro-power is usually restricted to the generation of shaft power from falling water. The power is then used for direct mechanical purposes or, more frequently, for generating electricity. Other sources of water power are waves and tides (Chapters 12 and 13).

Hydro-power is by far the most established renewable resource for electricity generation and commercial investment. The early generation of electricity from about 1880 often derived from hydro-turbines, and the capacity of total worldwide installations has grown at about 5% per year ever since, i.e. doubling about every 15 years. In 1980 the worldwide generating capacity was about 500 000 MW (0.5 TW), mostly in installations above about 10 MW. The estimated total potential for such installations is about 1.5 TW (Sørensen, 1979), with much of this potential in Africa, China and South America. However, global estimates can be completely misleading for local hydro-power planning, since (1) small scale (1 MW to 10 kW) applications are often neglected, despite the sites for such installations being the most numerous, and since (2) the economics of generation are sensitive to particular environmental and consumer use factors not recognized in large surveys. Thus the potential for hydro generation from rivers has been greatly neglected, and the value of hydro-power for the lucky few with local streams is not recognized in large scale studies. Environmental factors are also important, and these too cannot be judged by global surveys but only by evaluating local conditions.

Hydro installations and plant are long lasting, e.g. turbines for about 50 years. This is due to continuous steady operation without high temperature or other stress. Consequently established plant often produces electricity at low cost (~4 US cent kWh^{-1}) with consequent economic benefit. For instance Norway with 90% of electricity from hydro-power receives a significant economic benefit (see Fig. 1.3). Tables 8.1 and 8.2 review the importance of hydroelectric generation for various countries and regions. In general the best sites are developed first on a national scale, so the rate of exploitation of total generating capacity tends to diminish with time.

Hydro turbines have a rapid response for power generation and so the power

Table 8.1 Total hydroelectric resources of the world from installations ≳5 MW. 'Total' includes resources not harnessed. After United Nations, 1979, and other sources known to the authors.

Region	Total maximum capacity GW	Total maximum production at 0.5 capacity factor GWh	Fraction of world total capacity %	Fraction of each region's resource *not* harnessed %	Fraction of each region's resource harnessed 1983 %
Asia (excluding USSR)	630	2720	28	90	10
South America	440	1940	20	83	17
Africa	350	1550	16	95	5
North America (with Mexico)	350	1550	16	54	46
USSR	240	1070	11	79	21
Europe (excluding USSR)	150	680	7	35	65
Oceania	40	190	2	80	20
World	2200	9700	100	79	21

Table 8.2 Hydroelectric resource (total potential and 1985 production) for countries of UN category ECE. Based on the references of Table 8.1. For installations $\gtrsim 5$ MW. Such data are always subject to debate

Country	Total potential recoverable $TWh\,y^{-1}$	Production ~ 1985 $TWh\,y^{-1}$	Approximate fraction harnessed %
Europe			
Austria	46	28	60
Belgium	0.5	0.5	100
Denmark	0.2	0.0	0
Finland	13	12	100
France	65	64	100
Germany (Fed.)	21	19	90
Greece	16	4.1	25
Iceland	18	2.4	10
Ireland	0.7	0.7	100
Italy	50	46	90
Luxembourg	0.6	0.5	80
Netherlands	0.0	0.0	—
Norway	160	100	60
Portugal	14	10	70
Spain	48	42	90
Sweden	95	65	70
Switzerland	33	33	100
Turkey	72	25	35
UK	7.5	5	60
Yugoslavia	45	35	80
Total	~ 700	~ 500	70
Eastern Europe			
Bulgaria	16	3.5	20
Czechoslovakia	7.2	3.9	50
Germany (Dem.)	1.3	1.3	100
Hungary	4.9	1.2	25
Poland	6.0	2.7	50
Romania	27	20	70
USSR (all)	1100	220	20
Total	~ 1200	~ 250	20
North America			
Canada	540	300	55
USA	700	360	50
Total	~ 1240	~ 660	50

may be used to supply both base load and peak demand requirements on a grid supply. Power generation efficiencies may be as high as 90%. Turbines are of two types:

(1) Reaction turbines, where the turbine is totally embedded in the fluid and powered from the pressure drop across the device
(2) Impulse turbines, where the flow hits the turbine as a jet in an open environment, with the power deriving from the kinetic energy of the flow.

Reaction turbine generators may be reversed so water can be pumped to high levels for energy storage at an overall efficiency of about 80%.

The main disadvantages of hydro-power are associated with effects other than the generating equipment, particularly for large systems. These are the difficult problems of environmental impact, silting of dams, corrosion of turbines in certain water conditions, and the relatively high capital costs compared with those of fossil power stations. For instance the benefits and disadvantages of the Aswan Dam for Egypt and the Sudan (and indeed for the Nile delta) are continually debated.

This chapter considers certain fundamental aspects of hydro-power and does not attempt to be comprehensive in such a developed subject. In particular we have considered small scale application, and we refer readers to the bibliography for comprehensive works at established engineering level.

8.2 Principles

A volume per second Q of water falls down a slope. The density of the fluid is ρ. Thus the mass falling is ρQ, and the potential energy lost by the falling fluid in each second is

$$P_0 = \rho Q g H \tag{8.1}$$

where g is the acceleration due to gravity, P_0 is the energy change per second (a power measured in watts), H is the vertical component of the water path.

The purpose of a hydro-power system is to convert this power to shaft power. Unlike some other power sources, there is no fundamental thermodynamic or dynamic reason why the output power of a hydro-system should be less than the input power P_0, apart from the loss of energy required to remove the water from the turbine. The advantages of hydro-power can be inferred from (8.1). For a given site, H is fixed and Q can usually be held fairly constant by insuring that the supply pipe is kept full. Hence the actual output is close to the design output, and it is not necessary to install a machine of capacity greater than normally required.

The main disadvantage of hydro-power is also clear from (8.1): the site must have sufficiently high Q and H. In general this requires a rainfall $\gtrsim 40 \, \text{cm} \, \text{y}^{-1}$ dispersed through the year, a suitable catchment and, if possible, a water storage site. Where these are available, hydro-power is almost certainly the most suitable electricity generating source. However, considerable civil engineering (in the form of dams, pipework, etc.) is always required to direct the flow through the turbines. These civil works often cost more than the mechanical and electrical components.

8.3 Assessing the resource for small installations

Suppose we have a stream available which *may* be useful for hydro-power. At first only approximate data, with an accuracy of about ±50%, are needed to estimate the power potential of the site. If this survey proves promising, then a detailed investigation will be necessary involving data taken over at least several years. It is clear from (8.1) that to estimate the input power P_0 we have to measure the flow rate Q and the available vertical fall H (usually called the head, cf. Section 2.2). For example with $Q = 40 \text{ ls}^{-1}$ and $H = 20 \text{ m}$, the maximum power available at source is 8 kW. This might be very suitable for a household supply.

8.3.1 *Measurement of head H*

For nearly vertical falls, trigonometric methods (perhaps even using the lengths of shadows) are suitable; whereas for more gently sloping sites, the use of level and pole is straightforward.

Note that the power input to the turbine depends not on the geometric (or total) head H_t as measured this way, but on the available head H_a:

$$H_a = H_t - H_f \tag{8.2}$$

where H_f allows for friction losses in the pipe and channels leading from the source to the turbine (see Section 2.6).

By a suitable choice of pipework it is possible to keep $H_f \lesssim H_t/3$, but according to (2.12) H_f increases in proportion to the total length of pipe, so that the best sites for hydro-power have steep slopes.

8.3.2 *Measurement of flow rate Q*

The flow through the turbine produces the power, and this flow will usually be less than the flow in the stream. However the flow in the stream varies with time, for example between drought and flood periods. For power generation we usually want to know the *minimum* (dry season) flow, since a turbine matched to this will produce power all the year without overcapacity of machinery. We also need to know the *maximum* flow and flood levels to avoid damage to installations.

The measurement of Q is more difficult than the measurement of H. The method chosen will depend on the size and speed of the stream concerned. As in Section 2.2,

$$\text{flow rate } Q = (\text{volume passing in time } \Delta t)/\Delta t \tag{8.3a}$$

$$= (\text{mean speed } \bar{u})(\text{cross-sectional area } A) \tag{8.3b}$$

$$= \int \mathbf{u} \cdot \hat{\mathbf{n}} \, dA \tag{8.3c}$$

where $\hat{\mathbf{n}}$ is the unit vector normal to the elemental area dA. The methods that follow from each of these equations we call basic, refined and sophisticated. In addition, if water falls freely over a ledge or weir, then the height of the flow at the ledge relates to the flow rate. This provides a further method.

(1) *Basic method* (Fig. 8.1(a))   The whole stream is either stopped by a dam or diverted into a containing volume. In either case it is possible to measure the

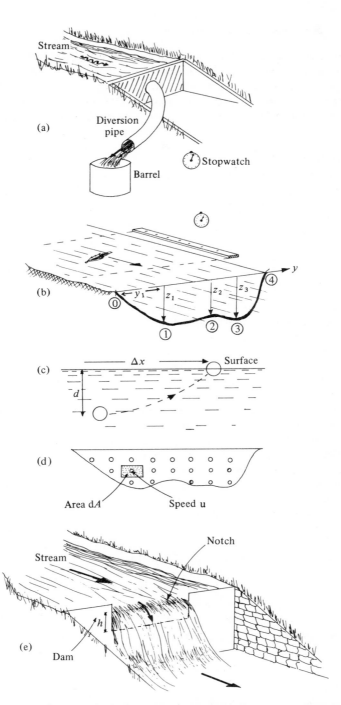

Fig. 8.1 Measuring water flow: (a) basic method (b) refined method I (c) refined method II (d) sophisticated method (e) weir method

flow rate from the volume trapped (8.3a). This method makes no assumptions about the flow, is accurate and is ideal for small flows, such as those at a very small waterfall.

(2) *Refined method I* (Fig. 8.1(b))   Equation (8.3b) defines the mean speed $\bar{u}$ of the flow. Since the flow speed is zero on the bottom of the stream (because of viscous friction), the mean speed will be slightly less than the speed u_s on the top surface. For a rectangular cross-section, for example, it has been found that

$$\bar{u} \approx 0.8 u_s$$

u_s can be measured by simply placing a float (e.g. a leaf) on the surface and measuring the time it takes to go a certain distance along the stream. For best results the measurement should be made where the stream is reasonably straight and of uniform cross-section.

The cross-sectional area A can be estimated by measuring the depth at several points across the stream and integrating across the stream in the usual way (Fig. 8.1(b)):

$$A \approx \tfrac{1}{2}y_1 z_1 + \tfrac{1}{2}(y_2 - y_1)(z_1 + z_2) + \tfrac{1}{2}(y_3 - y_2)(z_2 + z_3) + \tfrac{1}{2}(y_4 - y_3)z_3$$

(3) *Refined method II* (Fig. 8.1(c))   A refinement which avoids the need for accurate timing can be useful on fast flowing streams. Here a float (e.g. a table tennis ball) is released from a standard depth below the surface. The time for it to rise to the surface is independent of its horizontal motion and can easily be calibrated in the laboratory. Measuring the horizontal distance required for the float to rise gives the speed in the usual way. Moreover, what is measured *is* the mean speed (although averaged over depth rather than over cross-section: the difference is small).

(4) *Sophisticated method* (Fig. 8.1(d))   This is the most accurate method for large streams and is used by professional hydrologists. Essentially the forward speed u is measured with a small flow meter at the points of a two-dimensional grid extending across the stream. The integral (8.3c) is then evaluated by summation.

(5) *Using a weir* (Fig. 8.1(e))   If Q is to be measured throughout the year for the same stream, measurement can be made by building a dam with a specially shaped calibration notch. Such a dam is called a weir. The height of flow through the notch gives a measure of the flow. The system is calibrated against a laboratory model having the same form of notch. The actual calibrations are tabulated in standard handbooks. Problem 8.2 shows how they are derived.

8.4 An impulse turbine

Impulse turbines are easier to understand than reaction turbines, so we shall consider a particular impulse turbine – the Pelton wheel.

8.4.1 *Forces*

To operate an impulse turbine, we first turn the potential energy of the water in the reservoir into kinetic energy of one or more jets. Each jet then hits a series of buckets or 'cups' placed on the perimeter of a wheel, as shown in Fig. 8.2. The resulting deflection of the fluid constitutes a change in momentum of the fluid. The cup has exerted a force on the fluid, and therefore the fluid has likewise exerted a force on the cup. This tangential force applied to the wheel causes it to rotate. After the water leaves the nozzle, the flow process takes place at atmospheric pressure.

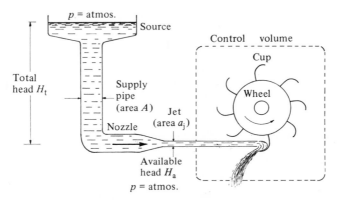

Fig. 8.2 Schematic diagram of a Pelton wheel impulse turbine

Fig. 8.3(a) shows the jet hitting the cup as seen in the 'laboratory' (i.e. earthbound) frame. The cup moves to the right with steady speed u_t. As seen by the cup, the jet enters tangentially and is deflected smoothly through almost 180°. In the ideal case of 180° deflection the *speed* (i.e. magnitude of velocity) of the fluid relative to the cup can be decreased only by friction, which can be made small by suitably polishing the cup. Thus in vector notation (Fig. 8.3(b)), with conservation of momentum in a fluid stream of constant cross-section,

$$\mathbf{u}_{r1} - \mathbf{u}_{r2} = 2(u_j - u_t)\hat{\mathbf{x}}$$

where $\hat{\mathbf{x}}$ is a unit vector in the jet direction.

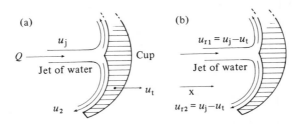

Fig. 8.3 Speed of cup and fluid, as seen in (a) the laboratory frame (b) the frame of the cup

This will also be the *change* in velocity seen in the laboratory frame. Thus, taking a control volume, indicated in Fig. 8.3, we find the change in momentum of the fluid going through the volume to be

$$\mathbf{F} = 2\rho Q(u_j - u_t)\hat{\mathbf{x}} \tag{8.4}$$

where Q is the total flow through the control volume, i.e. the total flow through all the jets. Ideally, at each instant, this force is pushing some blade forward at velocity $u_t\hat{\mathbf{x}}$, thus transferring power

$$P(u_t) = Fu_t$$

$$= 2\rho Q(u_j - u_t)u_t \tag{8.5}$$

This is a maximum for fixed u_j when

$$u_t/u_j = 0.5 \tag{8.6}$$

in which case the output power of the turbine is

$$P_j = \tfrac{1}{2}\rho Q u_j^2 \tag{8.7}$$

We note that P_j equals the total kinetic energy leaving the jet per second, so that the ideal turbine efficiency is 100%. For this ideal case, the velocity of the water

Fig. 8.4 Impulse turbine runner (Pelton type) with buckets cast integrally with the hub

leaving the cup would have zero component in the direction of the jet, i.e. $u_2 = 0$ in the laboratory frame. Therefore the water from a horizontal jet would fall vertically from the lowest cup.

The design of a practical Pelton wheel (Fig. 8.4) aims to match the ideal performance described. For instance, nozzles are adjusted so that the water jets hit the moving cups perpendicularly at the optimum relative speed for maximum momentum transfer. The ideal cannot be achieved in practice, because an incoming jet would be disturbed both by the reflected jet and by the next cup revolving into place. Pelton made several improvements in the turbines of his time (1860) to overcome these difficulties. Notches in the tops of the cups gave the jets better access to the turbine cups. The shape of the cups incorporated a central splitter section so that the water jets were reflected away from the incoming water.

The mechanical efficiency η_{jm} of an actual Pelton wheel arrangement is estimated in Problems 8.3 and 8.4. Experimental values range from 50% for small units to 90% for accurately machined large commercial systems.

8.4.2 *Jet velocity and nozzle size*

As indicated in Fig. 8.2, the pressure is atmospheric both at the top of the pipe and at the jet. Therefore Bernoulli's theorem (2.3) implies that, in the absence of friction in the pipe, $u_j^2 = 2gH_t$. Pipe friction can be allowed for by replacing the total head H_t by the available head H_a to give

$$u_j^2 = 2gH_a \tag{8.8}$$

In practice the size of the pipes is chosen so that u_j is independent of the nozzle area. If there are n_j nozzles, each of area a_j, then the total flow is

$$Q = n_j a_j u_j \tag{8.9}$$

The mechanical power output from the turbine is, from (8.6) and (8.7),

$$P_m = \eta_{jm} P_j = \eta_{jm} \tfrac{1}{2} \rho Q u_j^2$$
$$= \tfrac{1}{2} \eta_{jm} n_j a_j \rho (2gH_a)^{3/2} \tag{8.10}$$

This shows the importance of obtaining the maximum available head H_a between turbine and reservoir. The output power is proportional to the total jet cross-sectional area $A_j = n_j a_j$. However a_j is limited by the size of cup, so if a_j is to be increased a larger turbine is needed. It is usually easier to increase the number of nozzles n_j than to increase the overall size of the turbine, but the arrangement becomes unworkably complicated for $n_j \approx 4$. For small wheels, $n_j = 2$ would be most common.

Of course, the total flow Q through the turbine must be less than the flow in the stream (Q_{stream}) measured in Section 8.3.2:

$$n_j a_j \leq Q_{stream}/(2gH_a)^{1/2} \tag{8.11}$$

8.4.3 *Angular velocity and turbine size*

Suppose we have chosen the nozzle size and number in accordance with (8.9) and (8.10) to give the maximum power available. The nozzle size has fixed the size of the cups, but not the overall size of the wheel. The latter is determined by geometric constraints, and also by the required rotational speed. For electrical generation the output characteristics (voltage, frequency, efficiency) depend on the angular speed of the generator. Most electric generators have highest efficiency at high rotational frequency ($\sim$1500 rpm). To avoid complicated and lossy gearing, it is important that the turbine should also operate at high frequency, and the Pelton wheel is particularly suitable in this respect.

If the wheel has radius R and turns at angular velocity ω then, by (8.4) and (8.5),

$$P = FR\omega \tag{8.12}$$

Thus for a given output power, the larger the wheel the smaller its angular velocity. Since $u_t = R\omega$, by (8.6),

$$R = 0.5(2gH_a)^{1/2}/\omega \tag{8.13}$$

The nozzles usually give circular jets of radius r_j. Then $a_j = \pi r_j^2$, and with (8.9),

$$r_j^2 = \frac{P_m}{\eta_{jm}\rho n_j\pi(gH_a)^{3/2}\sqrt{2}} \tag{8.14}$$

Combining (8.13) and (8.14), we find

$$\frac{r_j}{R} = 0.68\,(n_j\eta_{jm})^{-1/2}\,\mathscr{S} \tag{8.15}$$

where

$$\mathscr{S} = \frac{P_m^{1/2}\omega}{\rho^{1/2}(gH_a)^{5/4}} \tag{8.16}$$

is a nondimensional measure of the operating conditions, called the *shape number* of the turbine. Equation (8.15) relates the shape of a Pelton wheel (measured by the nondimensional parameters r_j/R and n_j) to the nondimensional parameter $\mathscr{S}$ characterizing the operating conditions, and hence to the efficiency η_{jm} under these conditions.

Implicit in (8.15) is the relation (8.6) between the speed of the moving parts u_t and the speed of the jet u_j. If the ratio u_t/u_j is the same for two wheels of different size but the same shape, then the whole flow pattern is also the same for both. It follows that all nondimensional measures of hydraulic performance, such as η_{jm} and $\mathscr{S}$, are the same for impulse turbines with the same ratio of u_t/u_j. Moreover, for a particular shape of Pelton wheel (specified here by r_j/R and n_j), there is a particular combination of operating conditions (specified by $\mathcal{S}$) for maximum efficiency.

Example 8.1
Determine the dimensions of a single jet Pelton wheel to develop 160 kW under a head of (1) 81 m (2) 5 m. What is the angular speed at which these wheels will perform best?

Solution

Assume that water is the working fluid. It is difficult to operate a wheel with $r_j > R/10$, since the cups would then be so large that they would interfere with each other's flow. Therefore assume $r_j = R/12$. Figure 8.6 (see Section 8.5) suggests that at the optimum operating conditions $\eta_{jm} \approx 0.9$. From (8.15), the characteristic shape number for such a wheel is

$$\mathcal{S} = 0.11 \tag{8.17}$$

(1) From (8.16), the angular speed for best performance is

$$\omega_1 = \mathcal{S} \rho^{1/2}(gH_a)^{5/4}P^{-1/2}$$

$$= \frac{0.11(10^3 \, \text{kg m}^{-3})^{1/2}[(9.8 \, \text{m s}^{-2})(81 \, \text{m})]^{5/4}}{(16 \times 10^4 \, \text{W})^{1/2}}$$

$$= 36 \, \text{rad s}^{-1}$$

From (8.8),

$$u_j = (2gH_a)^{1/2} = 40 \, \text{m s}^{-1}$$

Therefore

$$R = \tfrac{1}{2}u_j/\omega = 0.55 \, \text{m}$$

(2) Similarly, with $H_a = 5$ m,

$$\omega_2 = \omega_1(5/81)^{5/4} = 1.1 \, \text{rad s}^{-1}$$

$$u_j = 10 \, \text{m s}^{-1}$$

$$R = 4.5 \, \text{m}$$

It can be seen, by comparing cases (1) and (2), that Pelton wheels to produce power from low heads should rotate slowly. Such wheels would be unwieldy and costly, especially because the size of framework and housing increases with size of turbine. In practice therefore Pelton wheels are confined to low flow/high head installations.

8.5 Reaction turbines

It is clear from the fundamental formula (8.1) that, to have the same power from a lower head, we have to maintain a greater flow Q through the turbine. This can also be expressed in terms of the shape number $\mathscr{S}$ of (8.16). In order to maintain the same ω and P with a lower H, we require a turbine with higher $\mathscr{S}$. One way of doing this is to increase the number of nozzles on a Pelton wheel (see (8.16) and Fig. 8.5(a)). However the plumbing becomes unduly complicated for $n_j \gtrsim 4$, and too many jets interfere with each other, so decreasing efficiency.

To allow a larger flow through the turbine, it is necessary to make a large change in the design. The entire periphery of the wheel is made into one large 'slot' jet which flows into the rotating wheel, as in Fig. 8.5(b). Such turbines are called *reaction* machines because the fluid pushes (or 'reacts') continuously against the blades. This contrasts with impulse machines, where the blades (cups) receive a series of impulses. The wheel, called the runner, must however be adapted so that the fluid enters radially and leaves parallel to the turbine axis. One design that accomplishes this is the *Francis* turbine, shown in Fig. 8.5(b). It can be seen that while the fluid is in the turbine it has a radial component of velocity in addition to the tangential velocity. This complicates analysis, which is, however, given in standard textbooks (Francis, 1974; Barna, 1970; Moody and Zowski, 1969).

A larger water flow can be obtained by making the 'jet' almost as large as the wheel itself. This concept leads to a turbine in the form of a *propeller*, with the flow mainly along the axis of rotation (Fig. 8.5(c)). Note however that the flow is not exactly axial. Guide vanes on entry are used to give the fluid a whirl (rotary) component of velocity, like that in the Francis turbine. It is the tangential momentum from this whirl component which is transferred to the propeller, thus making it rotate.

Since an axial flow machine is the most compact way of converting fluid power into mechanical power, it might be asked why designs of lower $\mathscr{S}$ (such as the Pelton or Francis turbines) should ever be used. The main reason is the great pressure changes which take place in the fluid as it moves, sealed off from the outside air, through a reaction turbine. Bernoulli's equation (2.2) can be used to show that the lowest water pressure in the system will be well below atmospheric (Problem 8.7). Indeed the lowest pressure may even be less than the vapor pressure of water. If this happens, bubbles of water vapor will form within the fluid – a process called *cavitation*. Downstream from this, the water pressure might suddenly increase, so causing the bubble to collapse. The resulting rush of liquid water can cause considerable mechanical damage to nearby mechanical parts. These effects increase with flow speed and head, and so axial machines are restricted in practice to low H. Moreover, the performance of reaction turbines in general, and the propeller turbine in particular, is very sensitive to changes in

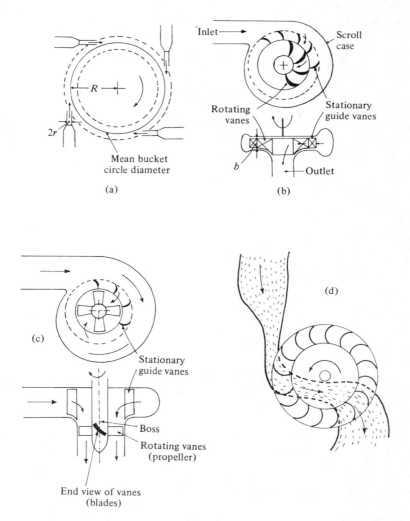

Fig. 8.5 Methods of increasing the power from a given size of machine, working at the same water pressure. (a) A four-jet Pelton wheel, the power of which is four times greater than that from a one-jet wheel of the same size and speed. (b) The jets supplying water to the rotor now exist all round the circumference as a slot, the water leaving the rotor axially. A reaction or radial flow turbine: e.g. Francis turbine. (c) The greatest $\mathscr{S}$ is obtained if the jet is made the same size as the rotor, and there is no radial flow over the rotor. A propeller turbine; e.g. Kaplan turbine. (d) the Banki cross-flow turbine is an intermediate type, in which the flow goes across an open, case-like, structure. Adapted from Francis (1974)

the flow rate. The efficiency drops off rapidly if the flow diminishes, because the slower flow no longer strikes the blade at the correct angle. It is possible to allow for this by automatically adjusting the blade angle, but this is complicated and expensive. Propeller turbines with automatically adjustable blade pitch are usually considered only worth while on large scale installations, e.g. the *Kaplan* turbine. However, smaller propeller turbines with adjustable blades are now available commercially for small scale operation.

The operation of a Pelton wheel (and also the intermediate *Banki* cross-flow turbine; Fig. 8.5(d)) is not so sensitive to flow conditions as a propeller turbine. The Banki turbine is particularly easy to construct with limited facilities (Saunders, 1979; Hamm, 1967). A technical description is given by Haimerl (1960).

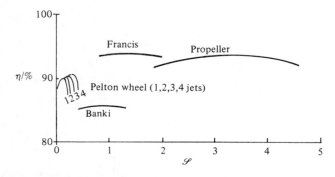

Fig. 8.6 Peak efficiencies of various turbines in relation to shape number. After Moody and Zowski (1969)

As a guide to choosing the appropriate turbine for given Q and H, Fig. 8.6 shows the range of shape number $\mathscr{S}$ over which it is possible to build an efficient turbine. In addition, for each type of turbine there will be a relationship between the shape number $\mathscr{S}$ (characterizing the operating conditions under which the turbine performs best) and another nondimensional parameter characterizing the form of the turbine. One such parameter is the ratio r_j/R of (8.15). These relationships may be established theoretically or experimentally, and are used to optimize design. Details are given in the recommended texts.

8.6 Hydroelectric systems

Most modern hydro-power systems are used to drive electric generators, although some special purpose devices are still useful (notably the hydraulic ram pump of Section 8.7). A complete hydroelectric system, such as that shown in Fig. 8.7, must include the water source, the pipe (penstock), flow control, the turbine, the electric generator, fine control of the generator, and wiring for electricity distribution (reticulation). The dam insures a steady supply of water to the system without fluctuations, and enables energy storage in the reservoir. It may also be used for purposes other than generating electricity, e.g. for roads or water supply. Small run-of-the-river systems from a moderately large and steady stream may require only a retaining wall of low height, but this does not produce any storage.

The supply pipe (penstock) is a major construction cost. It will be cheaper if it is thin walled, short and of small diameter D. Unfortunately it is seldom possible to meet these conditions. In particular the diameter cannot be decreased because of the head loss $H_f(H_f \propto D^{-5}$; see Problem 2.6). Therefore if the pipe diameter is too small, almost all the power will be lost in the pipe. The greater

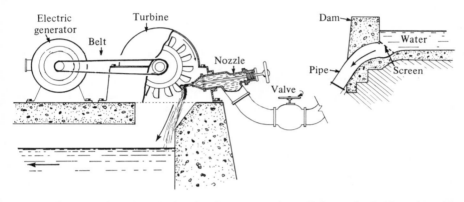

Fig. 8.7 Layout of a micro hydroelectric system using a Pelton wheel. Note that this diagram does not indicate the water head H required

cost of a bigger pipe has to be compared with the continued loss of power by using a small pipe. A common compromise is to make $H_f \lesssim 0.1 H_t$.

The material of the pipe is required to be both smooth (to lessen friction losses) and strong (to withstand the static pressures, and the considerably higher dynamic pressures that arise when the flow is turned suddenly on or off). For small installations, PVC plastic is suitable for the main length of the pipe, with a short steel section at the bottom to withstand the higher pressures there. A screen is needed at the top of the supply pipe to intercept rubbish (e.g. leaves) before it blocks the pipe. This screen has to be regularly cleaned. A settling chamber is essential to deposit suspended material before the water goes down the pipe.

The turbine speed is chosen (as described in the previous sections) to be suitable for the electric generator. However, turbine speeds are never very high by generator standards, and therefore the initial generation is usually at no more than 400 V AC, since voltage is proportional to frequency. Large (megawatt) systems usually have a specially built generator running from the same shaft as the turbine; this minimizes power losses between the turbine and the generator. Small systems (~10 kW) probably will use off-the-shelf generators. These operate at high rotational frequency and therefore require gearing from the turbine. A V-belt is the most common coupling in this case, which may give power losses of 10–20%.

8.6.1 *Power regulation and control*

Much of the material for this section is also considered in the discussions of wind power (Section 9.8) and of renewable energy control (Section 1.4.4). With a large hydro installation feeding electricity into a national grid, it is important that the voltage and frequency of the output closely match that of the rest of the grid. Although the primary generation is always at a relatively low voltage, the voltage can easily be transformed to a higher level, both to match the rest of the grid and to minimize I^2R losses in transmission. It is important that the voltage and frequency be maintained within about ±2% to maintain common standards and consumer requirements. This is done traditionally by mechanical feedback systems which control the flow through the turbine, so that it maintains constant frequency. For example, with a Pelton wheel a spear valve is made to move in and out of the nozzle (as indicated in Fig. 8.7), thus regulating Q. For reaction turbines it is necessary to adjust the blade angles as well. All these systems are mechanically complicated and expensive, especially for smaller scale application.

Small hydroelectric systems intended to supply electricity to a village or farm also require some regulation, but the devices they are required to run, such as lights and small electric motors (e.g. in refrigeration) can tolerate a wider range of voltage and frequency (e.g. ±10%). Moreover the currents involved are low enough that they can easily be switched by electronic devices, such as thyristors. This gives the possibility of a much cheaper control than the conventional mechanical systems.

With an *electronic load control* system, major variations in output are accomplished by manually switching nozzles completely in or out, or by manually controlling the total flow through the turbine. Finer control is achieved by an electronic feedforward control which shares the output of the generator between the main loads (e.g. house lights) and a ballast (or 'off peak') heating circuit which can tolerate a varying or intermittent supply (see Fig. 1.5c). The generator thus always sees a constant total load (= main + ballast); therefore it can run at constant power output, and so too can the turbine from which the power comes. The flow through the turbine does not therefore have to be continually automatically adjusted, which greatly simplifies its construction. In one common type of system the electronic control box is based on a thyristor which responds to the difference between the actual and nominal voltage in the main load.

8.6.2 *System efficiency*

Even though the efficiency of each individual step is high, there is still a substantial energy loss in passing from the original power P_0 of the stream to the electrical output P_e from the generator. Considering the successive energy transformations, we obtain

$$\frac{P_e}{P_0} = \frac{P_j}{P_0}\frac{P_m}{P_j}\frac{P_e}{P_m}$$

$$= \frac{H_a}{H_t}\,\eta_{jm}\eta_{me} \tag{8.18}$$

$$\approx (0.8)(0.8)(0.8)$$

$$\approx 0.5$$

Further losses occur in the distribution and use of the electricity.

8.7 The hydraulic ram pump

This mechanical hydro-power device is well established for water pumping at remote sites, where there is a steady flow of water at a low level. The momentum of the stream flow is used to pump some of the water to a considerably higher level. For example, a stream falling 2 m can be made to pump 10% of its flow to a height 12 m above. This is clearly a useful way of filling a header tank for piped water, especially in rural areas.

Fig. 8.8 shows the general layout of a pumping system using a hydraulic ram. The water supply flows down a strong, inclined pipe called the drive pipe. The potential energy MgH of the supply water is first converted into kinetic energy and subsequently into potential energy mgh. The kinetic energy is obtained by a mass of water M falling through a head H, and out through the impulse valve V_1. Operation is as follows:

(1) The speed of flow increases under the influence of the supply head so a significant dynamic pressure term arises for the flow through the valve (see Bernoulli's equation (2.2)).
(2) The static pressure acting on the underside of the impulse valve overcomes the weight of the valve, so it closes rapidly.
(3) Consequently the water at the bottom becomes compressed by the force of the water still coming down the pipe.
(4) The pressure in the supply pipe rises rapidly, forcing open the delivery valve V_d, and discharging a mass of water m into the delivery section.
(5) The air in the air chamber is compressed by the incoming water, but its compressibility cushions the pressure rise in the delivery pipe.
(6) The combined pressure of air and water forces the mass m up the delivery pipe.
(7) As soon as the momentum of the supply column is exhausted the delivery valve closes, and the water contained in the drive pipe recoils towards the supply.
(8) This recoil removes the pressure acting on the underside of the impulse valve, which thereupon falls and again allows the escape of water.

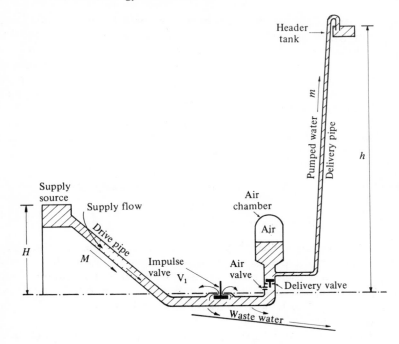

Fig. 8.8 General layout of a hydraulic ram pumping system. After Watt (1975)

(9) Simultaneously, the recoil causes the small air charging valve to open, admitting a small amount of air into the impact chamber of the ram. This air is carried along with the water into the air chamber to compensate for that absorbed by the water.

The whole cycle repeats indefinitely at a rate which is usually set to be about 1 Hz.

A theory for calculating all these quantities is given by Krol (1951) which uses only one main empirical parameter, the drag coefficient of the impulse valve. The efficiency of the device over a period equals mh/MH. Very solid and reliable rams are available commercially. Their efficiency is about 60%. It is also possible to build a ram (of slightly lower performance) from commercial high pressure pipe fittings (Inverson, 1978; Watt, 1975).

Problems

8.1 Use an atlas to estimate the *hydro-potential of your country* or state, as follows:

(a) Call the place in question X. What is the lowest altitude in X? What area of X lies more than 300 metres above the lowest level? How much rain falls per year on this high part of X? What would be the potential energy per year given up by this mass of water if it all ran down to the lowest level? Express this in megawatts.

(b) Refine this power estimate by allowing for the following: (i) not all the rain that falls appears as surface runoff; (ii) not all the runoff appears in streams that are worth damming; (iii) if the descent is at too shallow a slope, piping difficulties limit the available head.

(c) If a hydroelectric station has in fact been installed at X, compare your answer with the installed capacity of X, and comment on any large differences.

8.2 The flow over a U-weir can be idealized into the form shown in Fig. 8.9. In region 1 before the weir the stream velocity u_1 is uniform with depth. In region 2 after the weir, the stream velocity increases with depth h in the water.

(a) Use Bernoulli's theorem to show that for the streamline passing over the weir at a depth h below the surface,

$$u_h = (2g)^{1/2}(h + u_1^2/2g)^{1/2}$$

Hints: assume that p_h in the water = atmospheric pressure, since this is the pressure above and below the water. Assume also that u_1 is small enough that p_1 is hydrostatic.

(b) Hence show that the discharge over the idealized weir is

$$Q_{th} = (8g/9)^{1/2}LH^{3/2}$$

(c) By experiment, the actual discharge is found to be

$$Q_{exp} = C_w Q_{th}$$

where $C_w \approx 0.6$. (The precise value of C_w varies with H/L' and L/b.) Explain why $C_w < 1$.

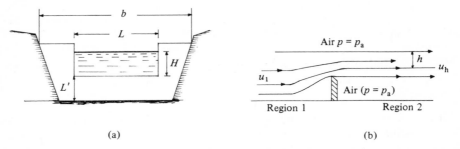

(a) (b)

Fig. 8.9 A U-weir. (a) Front elevation. (b) Side elevation of idealized flow (u_h is the speed of water over the weir where the pressure is p_h)

(d) Calculate Q_{exp} for the case $L' = 0.3\,\text{m}$, $L = 1\,\text{m}$, $b = 4\,\text{m}$, $H = 0.2\,\text{m}$. Calculate also u_1 and justify the assumptions about u_1 used in (a) and (b).

8.3 A Pelton wheel cup is so shaped that the exit flow makes an angle θ with the incident jet, as seen in the cup frame. As in Fig. 8.3, u_t is the tangential velocity of the cup, measured in the laboratory frame. The energy lost by friction between the water and the cup can be measured by a loss coefficient k such that

$$u_{r1}^2 = u_{r2}^2 (1 + k)$$

Show that the power transferred is

$$P = Q\rho u_t (u_j - u_t) \left[1 + \frac{\cos \theta}{\sqrt{(1+k)}} \right]$$

How does this compare with the ideal power when $\theta = 7°$, $k = 0.1$? What is the angle of deflection seen in the laboratory frame?

8.4 A Pelton wheel is to be installed in a site with $H = 20\,\text{m}$, $Q_{min} = 0.05\,\text{m}^3\text{s}^{-1}$.

(a) Neglecting friction for the moment, find (i) the jet velocity (ii) the maximum power available (iii) the radius of the nozzles (assuming there are two nozzles).
(b) Assuming further that the wheel has shape number

$$\mathscr{S} = \frac{\omega P_1^{1/2}}{\rho^{1/2}(gH)^{5/4}} = 0.1$$

where P_1 is the power per nozzle, find (iv) the number of cups (v) the diameter of the wheel (vi) the angular speed of the wheel in operation.
(c) If the main pipe – the penstock – had a length of $100\,\text{m}$, how would your answers to (a) and (b) be modified by fluid friction using: (vii) PVC pipe with a diameter of $15\,\text{cm}$? (viii) Common plastic hosepipe with a diameter of $5\,\text{cm}$? In each case determine the Reynolds number in the pipe.

8.5 Verify that $\mathscr{S}$ defined by (8.16) is dimensionless. What are the advantages of presenting performance data for turbines in dimensionless form?

8.6 A propeller turbine has shape number $\mathscr{S} = 4$ and produces $100\,\text{kW}$ (mechanical) at a working head of $6\,\text{m}$. Its efficiency is about 70%. Calculate

(a) The flow rate
(b) The angular speed of the shaft

(c) The gear ratio required if the shaft is to drive a four-pole alternator to produce a steady 50 Hz.

Solutions

8.1 Viti Levu (Fiji). The catchment area taken has in fact been developed as Fiji's first major hydroelectric station at Monosavu.

(a) $P_1 = (1000\,\mathrm{kg\,m^{-3}})(4\,\mathrm{m\,y^{-1}})(2000\,\mathrm{km^2})(9.8\,\mathrm{m\,s^{-2}})(300\,\mathrm{m}) = 800\,\mathrm{MW}$.
(b) Guesses by A. D. Weir: (i) ~50% (ii) ~50% (iii) ~50%. Hence $P_2 = 100\,\mathrm{MW}$.
(c) Monasavu stage 1 is 40 MW.

8.2 (a) Note hints in question.

(b) $Q = \int_{h=0}^{H} u_h L\,\mathrm{d}h$

Assume $u_1^2/2g \ll h$.
(c) Turbulence on entry and exit implies loss of KE (cf. Bernoulli).
(d) $Q = 0.16\,\mathrm{m^3\,s^{-1}}$, $u_1 = 0.03\,\mathrm{m\,s^{-1}}$.
Hence $[u_1^2/(2g) \ll h]$ for $h > 1\,\mathrm{mm}$

8.3 Follow the derivations of Section 2.3. Power reduced by 3%. Laboratory angle $\approx \frac{1}{2}$ cup angle.

8.4 (a) (i) $20\,\mathrm{m\,s^{-1}}$ (ii) $9.8\,\mathrm{kW}$ (iii) 2.0 cm.
(b) Consider for example $\eta = 0.9$: (iv) 25 cups (v) 52 cm (vi) $33\,\mathrm{rad\,s^{-1}}$.
(c) (vii) $\mathcal{R} = 4 \times 10^5$, $H_a = 16\,\mathrm{m}$, $P = 7.2\,\mathrm{kW}$. (viii) First approx $H_f \approx 800\,\mathrm{m}$ (see Problem 2.5). Therefore $P = 0$ (all potential energy goes into friction).

8.5 Simple algebra. See text.

8.6 (a) $2.4\,\mathrm{m^3\,s^{-1}}$ (b) $65\,\mathrm{rad\,s^{-1}}$ (c) 2.4, smaller pulley on generator.

Bibliography

General articles and books on hydro-power

Basset, (1949) 'There's power in that stream', reprinted in *Handbook of Home-made Power, Mother Earth News*, 1977.
Still the best layman's guide to the measurement of Q and H.
Brown, J. Guthrie (1958) (ed.) *Hydroelectric engineering practice*, Blackie, Glasgow.
Covers civil, mechanical, electrical and economic aspects.

Cotillon, J. (1979) 'Micro-power: an old idea for a new problem', *Water Power and Dam Construction*, January. Part of a special issue on mini-hydro.

Davis, C. V. and Sorensen, K. E. (ed.) (1969) *Handbook of Applied Hydraulics*, 3rd edn, McGraw-Hill.
A technical manual of US engineering practice.

Doland, J. J. (1954) *Hydro power engineering*, Ronald Press, New York.
A well-written practical text at a professional level.

Hamm, H. W. (1967) *Low Cost Development of Small Water Power Sites*, Volunteers in Technical Assistance, Maryland, USA.
Includes construction plans.

Hammond, R. (1958) *Water Power Engineering*, Heywood, London.
A more concise version of Guthrie Brown (1958), aimed at students.

Leckie, J., Masters, G., Whitehouse, H. and Young, L. (1976) *Other Homes and Garbage*, Sierra Club, San Francisco.
A useful chapter on hydro-power, including measurement techniques.

National Academy of Sciences (1976) *Energy for Rural Development*, Washington, DC.
The chapter on hydro-power is a copiously illustrated nontechnical survey of small systems suitable for rural use.

Saunders, R. (1979) in Merril and Gage (eds), *Energy Primer*, 2nd edn, Dell, New York.
Useful chapter on hydro-power for homesteaders. Extensive bibliography.

Sørensen, B. (1979) *Renewable Energy*, Academic Press, London.

United Nations (1979) *Energy Reserves and Supplies in the ECE Region*, UN ref. E/ECE/984.

Mechanics of turbines

Barna, P. S. (1970) *Fluid Mechanics for Engineers*, 3rd edn, Butterworths, London.
Longer account of turbomachinery than Francis, but still at student level.

Francis, J. R. (1974) *A Textbook of Fluid Mechanics*, 4th edn, Edward Arnold, London.
Has a clear chapter on hydraulic machinery, with thorough physics but not too much technical detail.

Haimerl, L. A. (1960) 'The cross flow turbine', in *Water Power*, **12**, 5–13.
Technical account of the Banki turbine (which is not dealt with in the previous three references).

Moody, L. and Zowski, T. (1969) 'Hydraulic machinery', in Davis and Sorensen (1969).
Engineering detail, but still clear.

Turton, R. K. (1984) *Principles of Turbomachinery* E. and F. N. Spon, London.
Technical details.

Hydraulic ram

Inverson, A. R. (1978) *Hydraulic Ram Pump*, Volunteers in Technical Assistance, Maryland, USA, Technical Bulletin no. 32.
Construction plans of the ram itself.

Krol, J. (1951) 'The automatic hydraulic ram', *Proc. Inst. Mech. Eng.*, **165**, 53–65.
Mathematical theory and some supporting experiments. Clumsy writing makes the paper look harder than it is.

Watt, S. B. (1975) *A Manual on the Hydraulic Ram for Pumping Water*, Intermediate Technology Publications, London.
Plans for an alternative design of home-made ram, plus details of installation and operation.

9 *Power from the wind*

9.1 Introduction

The extraction of power from the wind with modern turbines and energy conversion systems is an established industry. Machines are manufactured with a capacity from a few kilowatt to several megawatt in Europe, the United States and, increasingly, in other parts of the world. Most machines are built for electricity production, either linked to a grid or in an autonomous mode.

Later sections will show that in a wind of speed u_0 and density ρ, a turbine intercepting a cross-section A of wind front will produce power given by

$$P_T = C_P A \frac{\rho u_0^3}{2} \tag{9.1}$$

Here C_P is an efficiency factor called the power coefficient. Note that the power P_T is proportional to A and the cube of wind speed u_0. Thus whereas doubling A may produce twice the power, a doubling of wind speed produces eight times the power potential. The power coefficient C_P also varies with wind speed for individual machines. Since wind speeds above average are less likely than those below, and since power is usually needed for extended periods, optimum design for particular uses is complex. Often the average annual power per unit area from a wind turbine approximates to the product of C_P, air density and the mean wind speed cubed: $P_T \sim C_P \rho(\bar{u})^3$, see (9.73).

The maximum rated power capacity of a wind energy conversion system (WECS) is given for a specified wind speed. Commonly this rated wind speed is about $12 \, \mathrm{m \, s^{-1}}$, in which case power production of about $300 \, \mathrm{W \, m^{-2}}$ of cross-section would be expected with power coefficients C_P between 35 and 45%. Tables 9.1 and 9.2 give outline details of wind speeds and machine size. At good wind sites, annual average power production is expected to be between 25 and 33% of the rated capacity. Machines would be expected to last for at least 15 to 20 years and cost about $(US) 1000 to $(US) 1500 per kW rated capacity.

Wind power for mechanical purposes, milling and water pumping, has been established for many hundreds of years. Modern designs for electricity production date from about 1930 to about 1955. At this time development subsided owing to the availability of cheap oil, but interest reawakened and increased rapidly from about 1973. A few of the older machines are still able to operate (e.g. the Gedser 100 k W, 24 m diameter machine in Denmark, built in 1957), but most machines have been built since about 1980 and incorporate modern engineering design and materials. Of particular importance has been the use of microelectronic monitoring and control. A major design criterion is the need to protect the machine against damage in the infrequent very high gale force gusts. Wind forces tend to increase as the square of the wind speed, and since the 1 in 50 year gale speed will be 5 to 10 times the average wind speed, considerable overdesign has to be incorporated for structural strength. In addition wind speed is highly fluctuating, so considerable fatigue damage can occur, especially in blades, from the frequent stress cycles of gravity loading (about 10^7 cycles over 20 years of operation).

Wind results from expansion and convection of air as solar radiation is absorbed on earth. On a global scale these thermal effects combine with dynamic effects from the earth's rotation to produce prevailing wind patterns (Fig. 9.1). In addition to this general or synoptic behavior of the atmosphere there is important local variation caused by geographical and environmental factors. Wind velocities increase with height, and the horizontal components are significantly greater than the vertical components. The latter are however important in causing gusts and short term variations. The kinetic energy stored in the winds is about 0.7×10^{21} J, and this is dissipated by friction, mainly in the air but also by contact with the ground and the sea. About 1% of absorbed solar radiation, 1200 TW (1200×10^{12} W), is dissipated in this way.

The ultimate world use of wind power cannot be estimated in any meaningful way, since it is so dependent on the success and acceptance of machines and suitable energy end-use systems. However, without suggesting any major changes in electrical infrastructure, official estimates of wind power potential for the electrical supply of the United Kingdom and Denmark are at least 20% of the total supply. With changes in the systems, e.g. by connection with hydro storage, significantly greater penetration would be possible. Autonomous wind power systems have great potential as substitutes for oil used in heating or for the generation of electricity from diesel engines. These systems are particularly applicable for remote and island communities.

9.2 Turbine types and terms

The names of different types of wind turbine depend on their geometry, and the way wind passes over the airfoils or blades. Fig. 9.2 shows wind arriving at a moving airfoil, and indicates the essential forces and relative motions that occur.

The airstream of velocity u has velocity v_r relative to the blade section of velocity v. As the air is perturbed by the blade, a force acts which is resolved into two components. The main factors are:

(1) The *drag force* F_D is the component in line with the relative velocity v_r.
(2) The *lift force* F_L is the component perpendicular to F_D. The use of the word

Table 9.1 Wind speed relationships based on the Beaufort scale

Beaufort number	Wind speed range at 10 m height				Description	Wind turbine effects	Power generation possibility for *average* speed in range	Observable effects of wind	
	($m\,s^{-1}$)	($km\,h^{-1}$)	($mi\,h^{-1}$)	(knot)				Land	Sea
0	0.0 → 0.4	0.0 → 1.6	0.0 → 1	0.0 → 0.9	Calm	None	–	Smoke rises vertically	Mirror smooth
1	0.4 → 1.8	1.6 → 6	1 → 4	0.9 → 3.5	Light	None	–	Smoke drifts but vanes uneffected	Small ripples
2	1.8 → 3.6	6 → 13	4 → 8	3.5 → 7.0	Light	None	Useless	Wind just noticeable across the skin; leaves move slightly; vanes unaffected	Definite waves
3	3.6 → 5.8	13 → 21	8 → 13	7.0 → 11	Light	Start-up by few-turbines for light winds e.g. pumping	Water pumping and minor electrical power	Leaves in movement; flags begin to extend	Occasional wave crests break, glassy appearance of whole sea
4	5.8 → 8.5	21 → 31	13 → 19	11 → 17	Moderate	Start-up be few-bladed turbines for power generation	Useful electrical power production	Small branches move; dust raised; pages of books loosened	Larger waves, white crests common
5	8.5 → 11	31 → 40	19 → 25	17 → 22	Fresh	Useful power generation ~1/3 capacity	Extremely good prospects for power	Small trees in leaf sway, wind noticeable for comment	White crests everywhere
6	11 → 14	40 → 51	25 → 32	22 → 28	Strong	Rated range at full capacity	Only for the strongest small machines	Large branches sway, telephone lines whistle	Larger waves appear, foaming crests extensive

Beaufort									
7	14 → 17	51 → 63	32 → 39	28 → 34	Strong	Full capacity reached	Life not worth living here	Whole trees in motion	Foam begins to break from crests in streaks
8	17 → 21	63 → 76	39 → 47	34 → 41	Gale	Shutdown or self-stalling initiated		Twigs break off. Walking difficult	Dense streaks of blown foam
9	21 → 25	76 → 88	47 → 55	41 → 48	Gale	All machines shut down or stalled		Slight structural damage, e.g. chimneys	Blown foam extensive
10	25 → 29	88 → 103	55 → 64	48 → 56	Strong gale	Design criteria against damage		Trees uprooted. Much structural damage	Large waves with long breaking crests
11	29 → 34	103 → 121	64 → 75	56 → 65	Strong gale	Only strengthened machines would survive		Widespread damage	
12	>34	>121	>75	>65	Hurricane	Serious damage certain unless precollapse		Only occurs in tropical cyclones. Countryside devastated. Disaster conditions	Ships hidden in wave troughs. Air filled with spray

$1\,\mathrm{m\,s^{-1}}$ $= 3.6\,\mathrm{km\,h^{-1}}$ $= 2.237\,\mathrm{mi\,h^{-1}}$ $= 1.943\,\mathrm{knot}$
$0.278\,\mathrm{m\,s^{-1}}$ $= 1\,\mathrm{km\,h^{-1}}$ $= 0.658\,\mathrm{mi\,h^{-1}}$ $= 0.540\,\mathrm{knot}$
$0.447\,\mathrm{m\,s^{-1}}$ $= 1.609\,\mathrm{km\,h^{-1}}$ $= 1\,\mathrm{mi\,h^{-1}}$ $= 0.869\,\mathrm{knot}$
$0.515\,\mathrm{m\,s^{-1}}$ $= 1.853\,\mathrm{km\,h^{-1}}$ $= 1.151\,\mathrm{mi\,h^{-1}}$ $= 1\,\mathrm{knot}$

Table 9.2 Typical wind turbine generating characteristics at rated power in $12\,\mathrm{m\,s^{-1}}$ wind speed. Data calculated assuming power coefficient $C_P = 30\%$, air density $\rho = 1.2\,\mathrm{kg\,m^{-3}}$, tip speed ratio $\lambda = 6$. Rated power $P_T = \frac{1}{2}\rho[\pi D^2/4]u_0^3\,C_P$. Hence $D = (2.02\,\mathrm{m})\sqrt{(P/1\,\mathrm{kW})}$, $T = (0.0436\,\mathrm{s\,m^{-1})}\,D$.

Class	Small		Intermediate			Large			Very large		
Rated power P_T kW	10	25	50	100	150	250	500	1000	2000	3000	4000
Diameter D m	6.4	10	14	20	25	32	49	64	90	110	130
Period (T) s	0.3	0.4	0.6	0.9	1.1	1.4	2.1	3.1	3.9	4.8	5.7

'lift' does not mean F_L is necessarily upwards, and derives from the equivalent force on an airplane wing.

(3) *Rotational movement* of the air occurs as the airstream flows around the blade. This may be apparent as distinct vortexes and eddies (whirlpools of air) created near the surface. Vortex shedding occurs as these rotating masses of air break free from the surface and move away, still rotating, with this airstream.

(4) The air is disturbed by the blade movement, and the flow becomes erratic and perturbed. This *turbulence* (see Section 2.5) occurs before and after the rotating blades, so each individual blade may often be moving in the turbulence created by other blades.

(5) The wind turbine presents a certain *solidity* to the airstream. This is the ratio of the total area of the blades at any one moment in the direction of the airstream to the swept area across the airstream. Thus, with identical blades, a four-bladed turbine presents twice the solidity of a two-bladed turbine.

The characteristics of a particular wind turbine are described by the answers to a number of questions (see also Fig. 9.3). The theoretical justification for these criteria will be given in later sections.

(1) *Is the axis of rotation parallel or perpendicular to the airstream?*   The former is a horizontal axis machine, the latter usually a vertical axis machine in a cross-wind configuration.

(2) *Is the predominant force lift or drag?*   Drag machines can have no part moving faster than the wind, but lift machines can have blade sections moving considerably faster than the wind speed. This is similar to a keeled sail boat which can sail faster than the wind.

(3) *What is the solidity?*   For many turbines this is described by giving the number of blades. High solidity machines start easily with high initial torque, but soon reach maximum power at low rotational frequency. Low solidity devices may require starting, but reach maximum power at high rotational frequency. Thus high solidity machines are used for water pumping even in light winds. Low solidity turbines are used for electricity generation, since high shaft rotational frequency is needed.

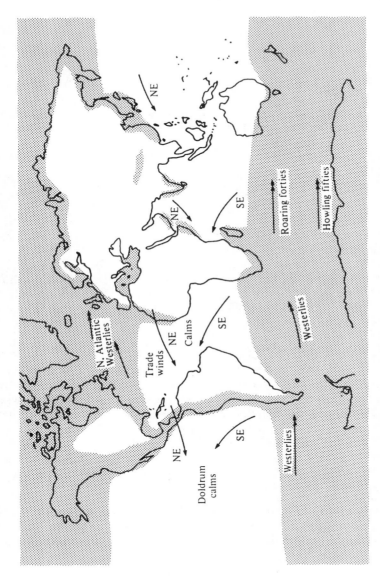

Fig. 9.1 Prevailing strong winds. The shaded areas indicate regions of wind attractive for wind power development, with average wind speed $\geq 5\,\mathrm{m\,s^{-1}}$ and average generation $\geq 33\%$ of rated power. Note the importance of marine situations, and beware of non site related generalizations

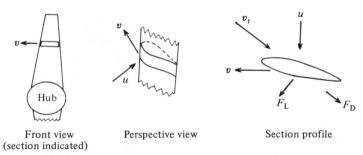

Front view   Perspective view   Section profile
(section indicated)

Fig. 9.2 Velocities and forces at a moving airfoil section. Velocities: u wind, v blade section, v_r relative of section in wind. Forces: F_D drag in line with v_r, F_L perpendicular to F_D

(4) *What is the purpose of the turbine?*   A device for direct mechanical work is often called a 'windmill' or just a 'wind turbine'. If electricity is produced, the combination of turbine and generator may be called a 'wind generator' or an 'aerogenerator'. Because of the confusion of these terms, the acronym WECS is increasingly used (wind energy conversion system). Small machines may be called SWECS (small WECS).

(5) *Is the frequency of rotation controlled to be constant, or does it vary with wind velocity?*   A WECS linked directly to a strong AC electrical grid will be controlled by the grid to rotate at nearly constant frequency, and will not require other expensive control equipment. However a turbine of variable frequency can be matched more efficiently to the varying wind speed than a constant frequency machine.

(6) *Is the turbine of the WECS directly coupled to its generator, or is there an intermediate energy store that acts as a smoothing device?*   A decoupling of this kind filters out high frequency turbine fluctuations, and allows better matching of turbine to wind, and generator to load, than direct coupling. Partial decoupling of the turbine from the generator is a *soft coupling*. Since wind velocities fluctuate rapidly (see Section 9.6) the inertia of the wind turbine and the 'softness' of the turbine–generator coupling are used to prevent these fluctuations appearing in the electricity output. Similar effects occur if the blades are independently hinged against a spring, or hinged together (teetered).

A classification of WECS can now be given in association with Fig. 9.3. This includes the main types, but numerous other designs and adaptations occur. In particular we can expect future designs to include many features to concentrate wind into the turbine.

9.2.1 *Horizontal axis machines*

We consider here the propeller type.

The dominant driving force is lift. Blades on the rotor may be in front (upwind) or behind (downwind) of the tower. Upwind turbines need a tail or

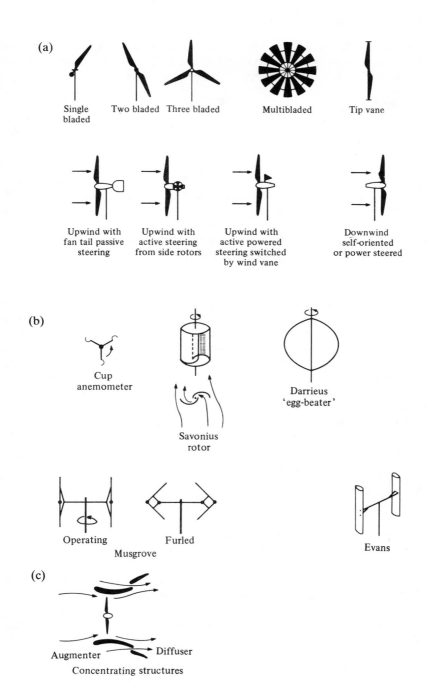

Fig. 9.3 Classification of wind turbines: (a) horizontal axis (b) vertical axis (c) concentrators

some other mechanism to maintain orientation, such as side-facing fan tail rotors. Downwind turbines may be quite seriously affected by the tower, which produces wind shadow and turbulence in the blade path. Perturbations of this kind cause cyclic stresses on the structure, noise and output fluctuations. Wind may be expected to veer frequently in a horizontal plane, and the rotor must turn (yaw) to follow the wind without oscillation. Upwind and downwind machines of capacity greater than about 50 kW are turned by electric motors in a controlled mode.

Two and three-bladed rotors are common for electricity generation. The three-bladed rotors operate smoothly and may be cross-linked for greater rigidity. Gearing and generators are usually at the top of the tower in a nacelle. It is possible to run a shaft down the tower for power generation at ground level, but the complications are usually thought to outweigh the advantages. Multi-blade rotors, having high starting torque in light winds, are used for water pumping and other low frequency mechanical power.

9.2.2 *Vertical axis machines*

By turning with a vertical axis, a machine can accept wind from any direction without adjustment. The other main benefit is that gearing and generators can be directly coupled to the axis at ground level. The principal disadvantages are: (1) many vertical axis machines have suffered from fatigue failures arising from the many natural resonances in the structure, and (2) the rotational torque from the wind varies periodically within each cycle, and thus unwanted power periodici-ties appear at the output. As a result the great majority of working machines are horizontal axis, not vertical. Nevertheless research and development on several types of vertical axis machine continue, for both small and large commercial systems.

(1) *Cup anemometer*   This device rotates by drag force. The shape of this cup produces a nearly linear relationship between rotational frequency and wind speed.

(2) *Savonius rotor (turbo machine)*   There is a complicated motion of the wind through and around the two curved sheet airfoils. The driving force is principally drag. The construction is simple and inexpensive. The high solidity produces high starting torque, so Savonius rotors are used for water pumping.

(3) *Darrieus rotor*   This rotor has two or three thin curved blades with an airfoil section. The driving forces are lift, with maximum torque occurring when a blade is moving across the wind at a speed much greater than the wind speed. Uses are for electricity generation. The rotor is not usually self-starting. Therefore movement may be initiated with the electrical generator used as a motor.

(4) *Musgrove rotor*   The blades of this form of rotor are vertical for normal power generation, but tip or turn about a horizontal point for control or shutdown. There are several variations (see Fig. 9.3), which are all designed to have the advantage of failsafe shut down in strong winds.

(5) *Evans rotor*   The vertical blades twist about a vertical axis for control and failsafe shut down.

9.2.3 *Concentrators*

Turbines draw power from the intercepted wind, and it may be advantageous to funnel or concentrate wind into the turbine from outside the rotor section. Various systems have been developed or suggested for horizontal axis propeller turbines (Fig. 9.3).

(1) *Blade tips*   Various blade designs and adaptations are able to draw air into the rotor section, and hence harness power from a cross-section greater than the rotor area.
(2) *Concentrating structures*   Funnel shapes and deflectors fixed statically around the turbine draw the wind into the rotor. Concentrators are not yet generally used for commercial machines.

9.3 Linear momentum and basic theory

In this section we shall discuss important concepts for wind machines. Basic coefficients concerning power, thrust and torque will be defined. The analysis proceeds by considering the linear momentum loss of the wind. More rigorous treatment will be outlined in later sections.

9.3.1 *Energy extraction*

In the unperturbed state (Fig. 9.4) a column of wind upstream of the turbine, with cross-sectional area A_1 of the turbine disk, has kinetic energy passing per unit time of

$$P_0 = \tfrac{1}{2}(\rho A_1 u_0) u_0^2 = \tfrac{1}{2}\rho A_1 u_0^3 \tag{9.2}$$

Here ρ is the air density and u_0 the unperturbed wind speed. This is the *power in the wind* at speed u_0.

Mass of column $\rho A u$, kinetic energy $\tfrac{1}{2}(\rho A u)u^2$

Fig. 9.4 Power in wind

Air density ρ is a function of height and meteorological condition. Wind speed increases with height, is affected by local topography, and varies greatly with time. These effects are considered fully in Section 9.6, and for the present we consider u_0 and ρ constant with time and over the area of the air column. Such incompressible flow has been treated in Chapter 2 on fluid mechanics. A typical value for ρ is $1.2\,\mathrm{kg\,m^{-3}}$ at sea level, and useful power can be harnessed in moderate winds when $u_0 \sim 10\,\mathrm{m\,s^{-1}}$. In this condition $P_0 = 600\,\mathrm{W\,m^{-2}}$. In gale force conditions $u_0 \sim 25\,\mathrm{m\,s^{-1}}$, so $P_0 \sim 10\,000\,\mathrm{W\,m^{-2}}$. Tables 9.1 and 9.2 give further details of meteorological wind conditions related to wind turbine size, and Section 9.7 considers wind properties related to power extraction.

The theory proceeds by considering constant velocity airstream lines passing through and by the turbine (Fig. 9.5). The rotor is treated as an 'actuator disk', across which there is a change of pressure as energy is extracted and a consequent decrease in the linear momentum of the wind. Perturbations to the smooth laminar flow are not considered, although they undoubtedly occur. Yet despite these severe limitations, the model is extremely useful.

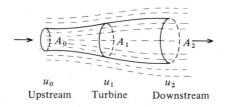

u_0 u_1 u_2
Upstream   Turbine   Downstream

Fig. 9.5 Betz model of expanding airstream

Area A_1 is the rotor swept area, and areas A_0 and A_2 enclose the stream of constant air mass passing through A_1. A_0 is positioned in the oncoming wind front unaffected by the turbine, and A_2 at the position of minimum wind speed before the wind front reforms downwind. A_0 and A_2 can be located experimentally for wind speed determination. Such measurement at A_1 is not possible because of the rotating blades.

The force or thrust on the turbine is the reduction in momentum per unit time from the air mass flow rate $\dot{m}$:

$$F = \dot{m}u_0 - \dot{m}u_2 \tag{9.3}$$

This force is applied by an assumed uniform air flow of speed u_1. The power extracted by the turbine is

$$P_T = Fu_1 = \dot{m}(u_0 - u_2)u_1 \tag{9.4}$$

The loss in energy per unit time by that airstream is the power extracted from the wind:

$$P_W = \tfrac{1}{2}\dot{m}(u_0^2 - u_2^2) \tag{9.5}$$

Equating (9.4) and (9.5)

$$(u_0 - u_2)u_1 = \tfrac{1}{2}(u_0^2 - u_2^2) = \tfrac{1}{2}(u_0 - u_2)(u_0 + u_2) \tag{9.6}$$

Hence

$$u_1 = \frac{u_0 + u_2}{2} \tag{9.7}$$

Thus according to this linear momentum theory, the air speed through the activator disk cannot be less than half the unperturbed wind speed.

The mass of air flowing through the disk per unit time is given by

$$\dot{m} = \rho A_1 u_1 \tag{9.8}$$

So in (9.4),

$$P_T = \rho A_1 u_1^2 (u_0 - u_2) \tag{9.9}$$

Now substitute for u_2 from (9.7)

$$P_T = \rho A_1 u_1^2 [u_0 - (2u_1 - u_0)] = 2\rho A_1 u_1^2 (u_0 - u_1) \tag{9.10}$$

The *interference factor a* is the fractional wind speed decrease at the turbine. Thus

$$a = (u_0 - u_1)/u_0 \tag{9.11}$$

and

$$u_1 = (1 - a)u_0 \tag{9.12}$$

Using (9.7),

$$a = (u_0 - u_2)/(2u_0) \tag{9.13}$$

Other names for a are the induction or the perturbation factor.

In (9.10),

$$P_T = 2\rho A_1 (1 - a)^2 u_0^2 [u_0 - (1 - a)u_0]$$
$$= (\tfrac{1}{2}\rho A_1 u_0^3)[4a(1 - a)^2] \tag{9.14}$$

Comparing this with (9.2),

$$P_T = C_P P_0 \tag{9.15}$$

where P_0 is the power in the unperturbed wind, and C_P is the fraction of power extracted, the *power coefficient*:

$$C_P = 4a(1 - a)^2 \tag{9.16}$$

Analysis could have proceeded in terms of the ratio $b = u_2/u_0$, sometimes also called an interference factor (see Problem 9.2).

The maximum value of C_P occurs when $a = 1/3$ (see Problem 9.1 and Fig. 9.6):

$$C_{Pmax} = 16/27 = 0.59 \tag{9.17}$$

Only about half the power in the wind can be extracted because the air has to have kinetic energy to leave the turbine region. The criterion for maximum power extraction ($C_{Pmax} = 16/27$) is called the *Betz criterion*, and may be applied

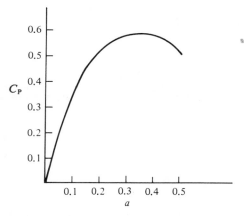

Fig. 9.6 Power coefficient C_P as a function of interference factor a. $C_P = 4a(1-a)^2$; $a = (u_0 - u_1)/u_0$; $C_{Pmax} = 16/27 = 0.59$; $u_1 = 0$ when $a = 0.5$

to all turbines set in an extended fluid stream. Thus it applies to power extraction from tidal and river currents (see Chapter 13). With conventional hydro-power (Chapter 8) the water reaches the turbine from a pipe and is not in extended flow, so other criteria apply.

In practical operation, a good commercial WECS may have a maximum power coefficient of about 0.4, as discussed in Section 9.4. This may be described as having an efficiency relative to the Betz criterion of $0.4/0.59 = 68\%$.

The power coefficient C_P is the efficiency of extracting power from the mass of air in the supposed stream tube passing through the activator disk, area A_1. This incident air passes through area A_0 upstream of the turbine. The power extracted per unit area of a cross-section equal to A_0 upstream is greater than per unit area of A_1, since $A_0 < A_1$. It can be shown (see Problem 9.3) that the maximum power extraction per unit of A_0 is 8/9 of the power in the wind, and so the turbine has a maximum efficiency of 89% considered in this way. Effects of this sort are important for arrays of wind turbines in a wind 'farm'.

9.3.2 *Thrust on turbines*

The motion of incompressible fluids in streamlined frictionless flow is treated by Bernoulli's equation (see (2.2) and (2.3)). We shall use this to calculate the thrust on a wind turbine when treated as an activator disk in streamlined flow (Fig. 9.7). The effect of the turbine is to produce a pressure difference Δp between the near upwind and near downwind parts of the flow.

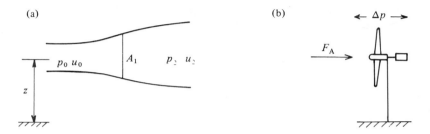

Fig. 9.7 Thrust on turbines: (a) air flow speed u, pressure p, height z (b) axial thrust F_A, pressure difference Δp

From (2.2),

$$p_0/\rho_0 + gz_1 + u_0^2/2 = p_2/\rho_2 + gz_2 + u_2^2/2 \tag{9.18}$$

The changes in z and ρ are negligible compared with the other terms, so if ρ is the average air density then

$$\Delta p = p_0 - p_2 = (u_0^2 - u_2^2)\rho/2 \tag{9.19}$$

Δp is called the static pressure difference, and the terms in $u^2\rho/2$ are the dynamic pressures. The maximum value of static pressure difference occurs as u_2 approaches zero (compare (9.13) and Fig. 9.6), so

$$\Delta p_{max} = \rho u_0^2/2 \tag{9.20}$$

and the maximum thrust on the turbine is

$$F_{A max} = \rho A_1 u_0^2/2 \tag{9.21}$$

On a horizontal axis machine this thrust is centered on the turbine axis and is called the axial thrust F_A.

The thrust equals the rate of loss of momentum of the airstream:

$$F_A = \dot{m}(u_0 - u_2) \tag{9.22}$$

Using (9.8), (9.11) and (9.13),

$$F_A = (\rho A_1 u_1)(2u_0 a)$$
$$= \rho A_1 (1-a) u_0 (2u_0 a)$$
$$= \frac{\rho A_1 u_0^2}{2} 4a(1-a) \tag{9.23}$$

The term $\rho A_1 u_0^2/2$ would be the force given by this model for wind hitting a solid disk. The fraction of this force experienced by the actual turbine is the *axial force coefficient* C_F:

$$F_A = C_F \rho A_1 u_0^2/2$$

where

$$C_F = 4a(1-a) \tag{9.24}$$

and from (9.13)

$$a = (u_0 - u_1)/u_0 = (u_0 - u_2)/2u_0 \tag{9.25}$$

The maximum value of C_F would be 1 when $a = 1/2$, equivalent to $u_2 = 0$. Maximum power extraction by the Betz criterion occurs when $a = 1/3$ (Fig. 9.6 and (9.17)), corresponding to $C_F = 8/9$.

In practice the maximum value of C_F on a solid disk approaches 1.2 owing to edge effects. Nevertheless the linear momentum theory shows that the turbine appears to the wind as a near solid disk when extracting power. It is quite misleading to estimate the forces on a rotating wind turbine by picturing the wind passing unperturbed through the gaps between the blades. If the turbine is extracting power efficiently, these gaps are not apparent to the wind and extremely large thrust forces occur.

The term $A_1 u_0^2/2$ of (9.23) increases rapidly with increase in wind speed, and in practice normal wind turbines become unable to accept the thrust forces for wind speeds above about 15 to $20\,\mathrm{m\,s^{-1}}$. The solutions to overcome this are (1) to turn the turbine out of the wind, (2) to lessen power extraction and hence thrust by rotating the airfoils or extending spoil flaps, (3) to design fixed blades so they become extremely inefficient and self-stalling in high wind speed, (4) to stop the rotation by braking. Method (3) is perhaps the best, giving failsafe operation at low cost without severe stresses on the machinery. However, self-stalling blades may have a low power coefficient and not give optimum power extraction in normal conditions.

9.3.3 *Torque*

The previous calculation of thrust on a wind turbine provides a convenient opportunity to introduce definitions for the torque causing rotational shaft power. At this stage no attempt is made to analyze angular momentum exchange between the air and the turbine. However, it is obvious that if the turbine turns one way the air must turn the other, and full analysis must eventually consider the vortices of air circulating downwind of the turbine (see Section 9.5).

The maximum conceivable torque Γ on a turbine rotor would occur if the maximum thrust could somehow be applied at the blade tip furthest from the axis. For a propeller turbine of radius R,

$$\Gamma_{max} = F_{max} R \tag{9.26}$$

From (9.21),

$$F_{max} = \rho A_1 u_0^2/2 \tag{9.27}$$

So

$$\Gamma_{max} = \rho A_1 u_0^2 R/2 \tag{9.28}$$

For a working machine producing a shaft torque Γ, the torque coefficient C_Γ is defined by

$$\Gamma = C_\Gamma \Gamma_{max} \tag{9.29}$$

As will be discussed in Section 9.4, the tip speed ratio λ is defined as the ratio of the outer blade tip speed v_t to the unperturbed wind speed u_0:

$$\lambda = v_t/u_0 = R\omega/u_0 \tag{9.30}$$

where R is the outer blade radius and ω is the rotational frequency.

From (9.28), substituting for R

$$\Gamma_{max} = \rho A_1 u_0^2(u_0\lambda)/2\omega$$
$$= P_0\lambda/\omega \tag{9.31}$$

where P_0 is the power in the wind from (9.2).

The shaft power is the power derived from the turbine P_T, so

$$P_T = \Gamma\omega \tag{9.32}$$

Now from (9.15) $P_T = C_P P_0$. Thus, using (9.29) and (9.31), (9.32) becomes

$$C_P P_0 = C_\Gamma \Gamma_{max} \omega$$
$$C_P P_0 = C_\Gamma P_0 \lambda$$
$$C_P = \lambda C_\Gamma \tag{9.33}$$

Note that in practice C_P and C_Γ will both be functions of λ and are not constants.

By the Betz criterion (9.17) the maximum value of C_P is 0.59, so in the 'ideal' case

$$(C_\Gamma)_{max} = 0.59/\lambda \qquad (9.34)$$

Fig. 9.8 shows the characteristics of practical turbines. High solidity turbines operate at low values of tip speed ratio and have high starting torque. Conversely low solidity machines (e.g. with narrow two-bladed rotors) have low starting torque and may indeed not be self-starting. At high values of λ the torque coefficient, and hence the torque, drops to zero and the turbines 'freewheel'. Thus with all turbines there is a maximum rotational frequency in high winds despite there being large and perhaps damaging axial thrust. Note too that maximum torque and maximum power extraction are not expected to occur at

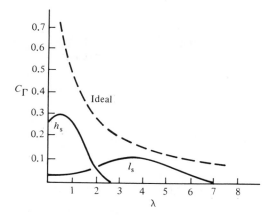

Fig. 9.8 Torque coefficient C_Γ versus tip speed ratio λ, sketched for high solidity h_s, low solidity l_s, and the 'ideal' criterion

the same values of λ. The relationship of power coefficient C_p to tip speed ratio λ is discussed in Section 9.4.

9.3.4 *Drag machines*

The ideal drag machine consists of a device with wind driven surfaces or flaps moving parallel to the undisturbed wind of speed u_0 (Fig. 9.9). The pressure difference across a stationary flap held perpendicular to the wind velocity is given by (9.20), if edge effects are neglected. For a flap of cross-section A moving with a speed v, the maximum driving drag force is

$$F_{max} = \rho A(u_0 - v)^2/2 \qquad (9.35)$$

A dimensionless drag coefficient C_D is used to describe devices departing from the ideal, so the drag force becomes

$$F_D = C_D \rho A (u_0 - v)^2 / 2 \qquad (9.36)$$

The power transmitted to the flap is

$$P_D = F_D v = C_P \rho A (u_0 - v)^2 v / 2 \qquad (9.37)$$

This is a maximum with respect to v when $v = u_0/3$, so

$$P_{D\,max} = \frac{4}{27} C_D \frac{\rho A u_0^3}{2} \qquad (9.38)$$

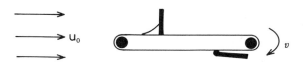

Fig. 9.9 Simplified drag machine with hinged flaps on a rotating belt

The power coefficient C_P is defined from (9.15) as

$$P_{D\,max} = C_P \rho A u_0^3 / 2$$

so

$$C_{P\,max} = \frac{4}{27} C_D \qquad (9.39)$$

Values of C_D range from nearly zero for a pointed object, to a maximum of about 1.5 for a concave shape as used in standard anemometers. Thus maximum power coefficient for a drag machine is

$$C_{P\,max} \approx \left(\frac{4}{27} \right) (1.5) = \frac{6}{27} = 22\% \qquad (9.40)$$

This may be compared with the Betz criterion for an 'ideal' turbine of $C_p = 16/27 = 59\%$ (9.17). In Section 9.4 we show that lift force turbines may have power coefficients of 40% and higher, and it is possible for this Betz criterion to be approached. Therefore drag-only devices have power extraction efficiencies of only about 33% that of lift force turbines for the same area of cross-section. Moreover returning drag flaps move against the wind and power is reduced even more. Power extraction from drag machines may be increased slightly by incorporating more flaps or by arranging concentrated air flows. However in practice a flap may easily meet the wakes of other flaps and power is reduced. The only way to improve drag machines is to incorporate lift forces, as happens in some forms of the Savonius rotor.

9.4 Dynamic matching

9.4.1 *Tip speed ratio*

Wind power devices are placed in wide fluxes of air movement. The air that passes through a wind turbine cannot therefore be deflected into regions where there is no air already, unlike a water turbine, and so there are distinctive limits to wind machine efficiency. Essentially the air must remain with sufficient energy to move away downwind of the turbine. The Betz criterion provides the accepted standard of 59% for the maximum extractable power, but the derivation of Section 9.3 tells us nothing about the dynamic rotational state of a turbine necessary to reach this criterion of maximum efficiency. This section explores this dynamic requirement with a qualitative analysis.

Power extraction efficiency will decrease from an optimum (see Fig. 9.10) if:

(1) The blades are so close together, or rotating so rapidly, that a following blade moves into the turbulent air created by a preceding blade; or
(2) The blades are so far apart or rotating so slowly that much of the air passes through the cross-section of the device without interfering with a blade.

It therefore becomes important to match the rotational frequency of the turbine to particular wind speeds so that the optimum efficiency is obtained.

Power extraction is a function of the time t_b for one blade to move into the position previously occupied by the preceding blade, as compared with the time t_w between the disturbed wind moving past that position and the normal airstream becoming re-established. t_w varies with the size and shape of the blades and inversely as the wind speed.

For an n-bladed turbine rotating at angular velocity Ω,

$$t_b \approx \frac{2\pi}{n\Omega} \tag{9.41}$$

A disturbance at the turbine disk created by a blade into which the following blade moves will last for a time t_w, where

$$t_w \approx d/u_0 \tag{9.42}$$

Here u_0 is the speed of the oncoming wind and the equation defines a distance d. This distance is a measure of the length of the strongly perturbed air stream upwind and downwind incident on the blade.

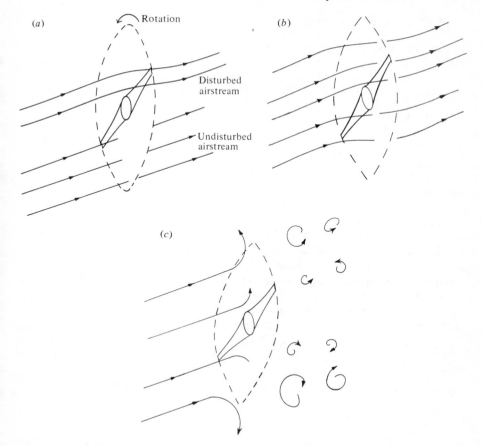

Fig. 9.10 Turbine frequency and power output. (a) Rotational frequency too low: some wind passes unperturbed through the actuator disk. (b) rotational frequency optimum: whole airstream affected. (c) Rotational frequency too high: energy is dissipated in turbulent motion and vortex shedding

Maximum power extraction occurs when $t_w \approx t_b$ at the blade tips, where maximum incremental area is swept by the blades. From (9.41) and (9.42), therefore,

$$\frac{n\Omega}{u_0} \approx \frac{2\pi}{d} \qquad (9.43)$$

If each side of this equation is multiplied by R, the blade tip radius of rotation, and if the tip speed ratio λ is defined as in (9.30) by

$$\lambda = \frac{\text{speed of tip}}{\text{speed of oncoming wind}} = \frac{R\Omega}{u_0} \qquad (9.44)$$

then at optimum power extraction

$$\lambda \approx \frac{2\pi}{n}\left(\frac{R}{d}\right) \tag{9.45}$$

We might expect $d \approx kR$, where $k \sim 1$, so in (9.45) the tip speed ratio for maximum power extraction is

$$\lambda_0 \approx \frac{2\pi}{kn} \tag{9.46}$$

Practical results show that $k \sim 1/2$, so for an n-bladed turbine

$$\lambda_0 \approx 4\pi/n \tag{9.47}$$

For example, for a two-bladed turbine C_{Pmax} occurs for $\lambda_0 \approx 4\pi/2 \approx 6$, and for a four-bladed turbine C_{Pmax} occurs for $\lambda_0 \approx 4\pi/4 \approx 3$.

The preceding calculation is not rigorous, but it does describe the most important phenomena. With carefully designed airfoils, optimum tip speed ratio λ_0 may be $\sim 30\%$ above these values.

We shall see in Section 9.5 that λ is related to the angle that air is incident on the moving blade. Thus in Fig. 9.13(a) $\lambda = \tan\phi$. The criterion for constant optimum tip speed ratio λ_0 can be interpreted as the need to maintain ϕ at the optimum for the particular blade configuration whatever the wind speed.

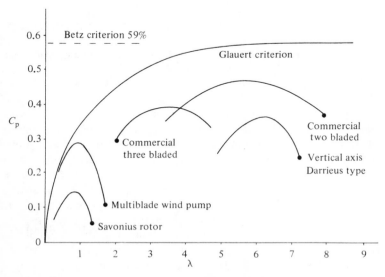

Fig. 9.11 Power coefficient C_p as a function of tip speed ratio λ

The Betz efficiency criterion of Section 9.3 takes no account of any dynamic effects. There are several approaches to an attempt at dynamic calculation (see the review by Shepherd, 1978). Glauert's criterion describing the variation of power coefficient C_P with tip speed ratio λ becomes of value. Both criteria and the relationship of C_P and λ for a variety of wind turbine types are shown in Fig. 9.11. A further constraint on the design of high angular velocity turbines is that the tip speed should not reach the speed of sound ($330\,\mathrm{m\,s^{-1}}$), so creating shock waves. This is possible for well-matched two-bladed turbines in strong winds of speed $\sim 50\,\mathrm{m\,s^{-1}}$.

Tip speed ratio is probably the most important parameter of an aero-generator. It is a function of the three most important variables: blade swept radius, wind speed and rotor frequency. Being dimensionless, it becomes an essential scaling factor in design and analysis.

9.4.2 *Extensions of linear momentum theory*

Fig. 9.6 is a graph of power coefficient C_P against interference factor a in the range $0 < a < 0.5$, as given by simple linear momentum theory. Thus from (9.16),

$$C_P = 4a(1-a)^2 \tag{9.48}$$

where, from (9.11),

$$a = 1 - u_1/u_0 \tag{9.49}$$

Extensions to the simple theory extend analysis into other regions of the interference factor, and link turbine driven performance with aircraft propeller characteristics. In Fig. 9.12, the airstreams are sketched on the graph for specific regions that may be associated with actual air flow conditions:

(1) $a < 0$, C_P *negative*   This describes airplane propeller action where power is added to the flow to obtain forward thrust. In this way the propellers pull themselves into the incoming airstream and hence propel the airplane forward.

(2) $0 < a < 0.5$, C_P *positive and peaking*   At $a = 0$, $u_1 = u_0$ and $C_P = 0$; the turbine rotates freely in the wind and is not coupled to a load to perform work. As a load is applied, power is abstracted so C_P increases as u_1 decreases. Maximum power is removed from the airstream when $a = 1/3$ and $u_1 = 2u_0/3$ ((9.17) and (9.12)). At $a = 1/2$, basic linear momentum theory predicts maximum thrust on the turbine (9.24) with torque coefficient $C_\Gamma = 1$.

(3) $0.5 < a < 1$, C_P *decreasing to zero*   From (9.25), $a = (u_0 - u_2)/2u_0$. When $u_2 = 0$, $a = 0.5$ and the simple model breaks down as, erroneously, no wind is predicted to be leaving the airstream. In practice it is possible to consider this region as equivalent to the onset of turbulent downwind air motion. It is equivalent to a turbine operating in high wind speed when the power extraction efficiency decreases, owing to a mismatch of rotational frequency and wind speed.

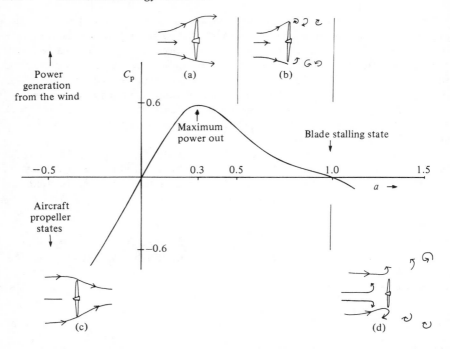

Fig. 9.12 Power coefficient C_p versus interference factor $a = 1 - u_1/u_0$, showing air motion and turbine/propeller states. (a) As Fig. 9.6, normal energy abstraction by a WECS. (b) Turbulent wake reduces efficiency, as met at high wind speeds. (c) Normal airflow of an aircraft propeller. Energy is added to the airstream. (d) Equivalent to aircraft propeller reverse thrust for breaking on landing

At $a = 1$, $C_P = 0$, the turbine is turning and causing extensive turbulence in the airstream, but no power is extracted. Real turbines may reach this state in a stall condition.

(4) $a > 1$ This implies negative u_1, and is met when an airplane reverses thrust by changing blade pitch on landing. Intense vortex shedding occurs in the airstream as the air passes the propellers. In the airplane, additional energy is being added to the airstream and is apparent in the vortices, yet the total effect is a reverse thrust to increase braking.

9.5 Streamtube theory

The complete analysis of horizontal axis turbine performance must consider the forces from the deflected airstream on each element of the rotating blades. Fig. 9.13 shows the important initial parameters. At radius r from the axis, the blade element has a speed $r\omega$ perpendicular to the unperturbed oncoming wind of speed u_1. The relative speed of this element to the moving air is however v_r,

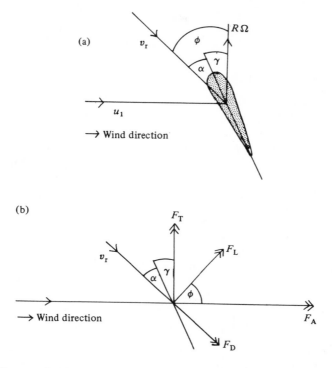

Fig. 9.13 Streamtube theory parameters. (a) Unperturbed wind speed u_0. Wind speed at the turbine disk u_1. Relative speed of air to blade v_r. Angle of attack α. Blade setting angle γ. $\tan \phi = u_1/(R\Omega) = \lambda$, tip speed ratio. (b) As for (a) with the drag force F_D parallel to v_r: the lift force F_L along the rotational axis is F_A, and along the tangent to the blade rotation is F_T

where usually $v_r \gg u_0$. At this position on the blade, the blade setting angle is γ, the angle of attack of the relative wind speed to the blade is α, and $\phi = \gamma + \alpha$. Note that $\tan \phi = u_1/(R\Omega) = \lambda$, the tip speed ratio.

By definition, F_D is the drag force parallel to v_r, and F_L is the lift force perpendicular to v_r.

If F_A is the increment of axial force, and F_T that of tangential force, then

$$F_A = F_L \cos \phi + F_D \sin \phi \tag{9.50}$$

$$F_T = F_L \sin \phi - F_D \cos \phi \tag{9.51}$$

Streamtube theory considers a cylinder or tube of the oncoming wind incident towards elements of the blades at radius r from the axis, cord (width) $c = c(r)$, and incremental length dr. One such cylinder can be treated independently of others both upstream and downstream of the turbine disk. The theory is developed from these basic principles and gives a more physically accurate interpretation of wind turbine operation. Advanced texts should be consulted for the further development, e.g. Shepherd (1978).

9.6 Characteristics of the wind

9.6.1 *Basic meteorological data*

All countries have national meteorological services that record and publish weather related data, including wind speeds and directions. The methods are well established and coordinated within the World Meteorological Organization in Geneva, with a main aim of providing continuous runs of data for many years. Consequently only the most basic data tend to be recorded at a few permanently staffed stations using robust and trusted equipment. Unfortunately for wind power prediction, measurements of wind speed tend to be measured only at the one standard height of 10m, and at stations near to airports or towns where shielding from the wind might be a natural feature of the site. Thus standard meteorological wind data from the nearest station are useful to provide first order estimates of wind power at a particular site, but are hardly ever sufficient for detailed planning or for a particular machine. Usually careful measurements around the nominated site are needed at several locations and heights for several months to a year. These detailed measurements can then be related to the standard meteorological data, and this provides a long term base for comparison.

Classification of wind speeds by meteorological offices is linked to the historical Beaufort scale, which itself relates to visual observations. Table 9.1 gives details together with the relationship between various units of wind speed.

A standard meteorological measurement of wind speed measures the 'length' or 'run' of the wind passing a 10m high cup anemometer in 10 minutes. Such measurements may be taken hourly, but usually less frequently. Such data give little information about fluctuations in the speed and direction of the wind necessary for accurately predicting WECS performance. Continuously reading anemometers are better, but these too will have a finite response time. A typical continuous reading (Fig. 9.14(a)) shows the rapid and random fluctuations that occur. Transformation of such data into the frequency domain gives the range and importance of these variations (Fig. 9.14(b)).

The direction of the wind refers to the compass bearing from which the wind comes. Meteorological data are usually presented as a wind rose (Fig. 9.15(a)) showing the average speed of the wind within certain ranges of direction. It is also possible to show the distribution of speeds from these directions on a wind rose (Fig. 9.15(b)). Such information is of great importance when siting a wind machine in hilly country, near buildings, or in arrays of several machines where shielding could occur.

9.6.2 *Variation with height*

Wind speed varies considerably with height above ground. A machine with a hub height of, say 30m will experience far stronger winds than a person at ground level.

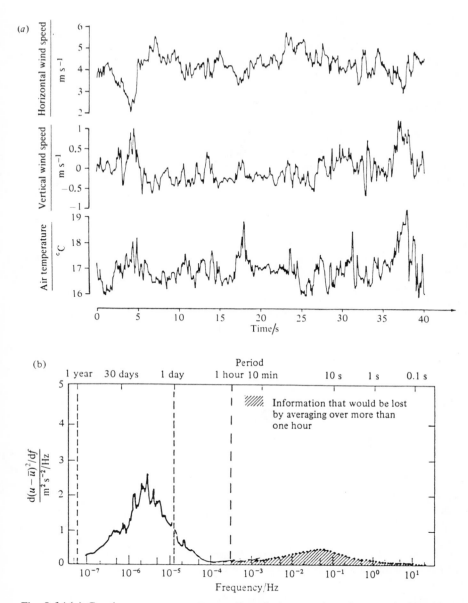

Fig. 9.14 (a) Continuous anemometer reading. A short section of a record of horizontal wind speed, vertical wind speed and temperature at a height of 2 m at the meteorological field, Reading University, UK. Note the positive correlations between vertical wind speed and temperature, and the negative correlations between horizontal and vertical wind speeds. (b) Frequency domain variance spectrum. From Petersen (1975). The graph is a transformation of many time series measurements in Denmark, which have been used to find the square of the standard deviation (variance) of the wind speed u from the mean speed $\bar{u}$. Thus the graph relates to the energy in wind speed fluctuations as a function of their frequency

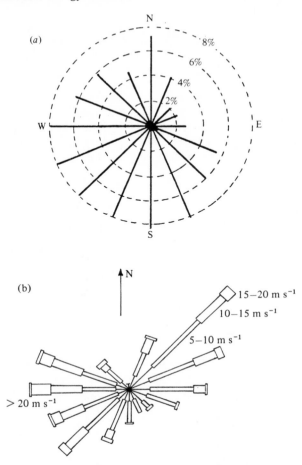

Fig. 9.15 Wind rose from accumulated data. (a) Direction. Station on the island of Tiree in the Outer Hebrides. The radial lines give percentages of the year during which the wind blows from each of 16 directions. The values are 10 year means and refer to an effective height above ground of 13 m. (b) Direction and distribution of speed. Malabar Hill on Lord Howe Island, New South Wales. The thicker sections represent the proportion of time the wind speed is between the specified values, within 16 directional sectors. From Bowden *et al.* (1983b)

Fig. 9.16 shows the form of wind speed variation with height z in the near-to-ground boundary layer up to about 100 m. At $z = 0$ the air speed is always zero. Within the height of local obstructions wind speed increases erratically, and violent directional fluctuations can occur in high winds. Above this erratic region, the height/wind speed profile is given by expressions of the form

$$z - d = z_0 \exp(u_z/V) \tag{9.52}$$

Hence

$$u_z = V \ln\left(\frac{z - d}{z_0}\right) \tag{9.53}$$

Here d is the zero plane displacement with magnitude a little less than the height of local obstructions, z_0 is the roughness length and V is a characteristic speed. On Fig. 9.16 the function is extrapolated to negative values of u to show the form of the expression. Readers should consult texts on meteorology and micro-meteorology for correct detail and understanding of wind speed boundary layer

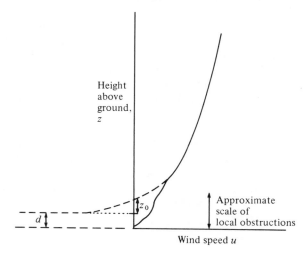

Fig. 9.16 Wind speed variation with height (see (9.52))

profiles. However, the most important practical aspect is the need to place a turbine well above the height of local obstructions to insure that the turbine disk receives a strong uniform wind flux across its area without erratic fluctuations.

The best sites for wind power are at the top of smooth dome shaped hills that are positioned clear of other hills. In general the wind should be incident across water surfaces or smooth land for several hundred meters, i.e. there should be a good fetch.

Most wind turbines operate at hub heights between 5 m and 50 m. It is common for standard meteorological wind speed measurements u_s to be taken at a height of 10 m, and so an approximate expression often used to determine the wind speed u_z at height z is

$$u_z = u_s\left(\frac{z}{10\,\text{m}}\right)^{b'} \tag{9.54}$$

It is often stated that $b' = 1/7 = 0.14$ for open sites. Good sites should have low values of b' to avoid changes in stress across the turbine disk, and high values of mean wind speed u_m to increase power extraction. There is increasing evidence that b' varies with season and time through the day. Great care should be taken with this formula, especially for $z > 50$ m (see Sisterson *et al.*, 1982).

9.6.3 *Variation with time*

For most applications of wind power, it is more important to know about the continuity of supply than the total amount of energy available in a year. In practice when the wind blows strongly, e.g. more than $12 \, \text{m s}^{-1}$, there is no shortage of power and often generated power has to be dumped. Difficulties appear, however, if there are extended periods of light or zero wind. A rule of thumb for electricity generation is that sites with average wind speed less than $5 \, \text{m s}^{-1}$ will have unacceptably long periods without generation, and that sites of average $8 \, \text{m s}^{-1}$ or above can be considered very good. In all cases it will be necessary to carefully match the machine characteristics to the local wind regime to give the type of supply required. The following example outlines the method, and explains the types of probability function needed for later theoretical analysis.

Example 9.1   Wind speed analysis for the island of North Ronaldsay, Orkney
A ten-minute 'run of the wind' anemometer was installed at 10 m height on an open site near a proposed WECS. Five readings were recorded each day at 9 a.m., 12 noon, 3 p.m., 6 p.m. and 9 p.m., throughout the year. Table 9.3 gives a selection of the total data and analysis.

(1) Readings were classed within intervals of $\Delta u = 1 \, \text{m s}^{-1}$, i.e. 0.0 to 0.9; 1.0 to 1.9, etc. A total of $N = 1763$ readings were recorded, with 62 missing owing to random faults.
(2) The number of occurrences of readings in each class were counted to give $\Delta N(u)/\Delta u$, with units of number per speed range (dN/du in Table 9.3).
   Note: $\Delta N(u)/\Delta u$ is a number per speed range, and so is called a frequency distribution of wind speed. Take care, however, to clarify the interval of the speed range Δu (in this case $1 \, \text{m s}^{-1}$, but often a larger interval).
(3) $\Delta N(u)/N = \Phi_u$ is a normalized probability function, often called the probability distribution of wind speed. Φ_u is plotted against u in Fig. 9.17. The unit of Φ_u is the inverse of speed interval, in this case $(1 \, \text{m s}^{-1})^{-1}$. $\Phi_u \Delta u$ may be interpreted as the proportion of time in a year that the wind speed is in the class defined by u (i.e. between u and $u + \Delta u$).
(4) The average or mean wind speed u_m is calculated from $u_m \sum \Phi_u = \sum \Phi_u u$, and a check is made that $\sum \Phi_u \Delta u = 1$. This mean speed $u_m = 8.2 \, \text{m s}^{-1}$ is indicated on Fig. 9.17. Notice that u_m is greater than the most probable wind speed of $6.2 \, \text{m s}^{-1}$ on this distribution.
(5) The cumulative total of the values of $\Phi_u \Delta u$ is tabulated to give the probability, $\Phi_{u > u'}$, of speeds greater than a particular speed u'. The units are number per speed range multiplied by speed range, i.e. dimensionless. This function is

Table 9.3 Wind speed analysis for the example of North Ronaldsay from Barbour (1984). This is a selection of the full data to show the method of calculation

u'	dN/du	Φ_u	$\Phi_{u \geq u'}$	$\Phi_u u$	u^3	$\Phi_u u^3$	P_u	$P_u \Phi_u$
$\overline{\text{ms}^{-1}}$	$\overline{(\text{ms}^{-1})^{-1}}$	$\overline{(\text{ms}^{-1})^{-1}}$			$\overline{(\text{ms}^{-1})^3}$	$\overline{(\text{ms}^{-1})^2}$	$\overline{\text{kW m}^{-2}}$	$\overline{(\text{W m}^{-2})(\text{ms}^{-1})^{-1}}$
>26	1	0.000	0.000	0.000	17576	0.0	11.4	0.0
25	1	0.001	0.001	0.025	15625	15.6	10.2	10.2
24	1	0.001	0.002	0.024	13824	20.7	9.0	9.0
23	2	0.002	0.004	0.046	12167	18.3	7.9	15.8
22	4	0.002	0.006	0.044	10648	21.3	6.9	13.8
.	.	.	.	.	.	.	.	.
.	.	.	.	.	.	.	.	.
.	.	.	.	.	.	.	.	.
8	160	0.091	0.506	0.728	512	46.6	0.3	27.3
7	175	0.099	0.605	0.693	343	340	0.2	19.8
6	179	0.102	0.707	0.612	216	22.0	0.1	10.2
5	172	0.098	0.805	0.805	125	12.3	0.1	9.8
4	136	0.077	0.882	0.882	64	4.9	0.0	0.0
.	.	.		.	.	.	.	.
.	.	.		.	.	.	.	.
0	12	0.007		0	0		0	0
Totals	1763	1.000		8.171		1044.8		
Comment		Peaks at $6.2\,\text{m s}^{-1}$		$u_m = 8.2\,\text{m s}^{-1}$		$(\overline{u^3})^{1/3} = 10.1$		

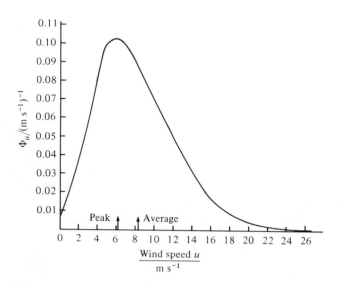

Fig. 9.17 Probability distribution of wind speed against wind speed. Data for North Ronaldsay from Barbour (1984)

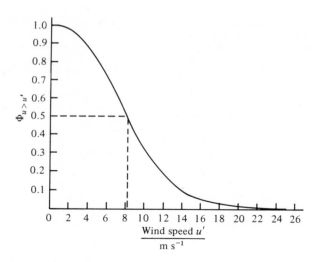

Fig. 9.18 Probability of wind speeds greater than a particular speed u', for example of North Ronaldsay

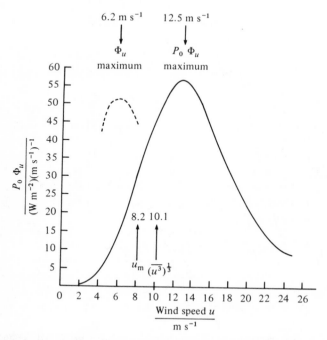

Fig. 9.19 Distribution of power in the wind, for example of North Ronaldsay

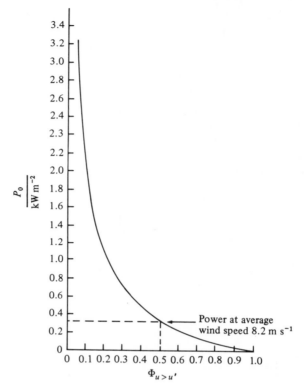

Fig. 9.20 Power per unit area in the wind against probability of wind speeds greater than a particular speed u'

plotted in Fig. 9.18, and may be interpreted as the proportion of time in the year that u exceeds u'.

(6) The power unit area of wind cross-section is $P_0 = \frac{1}{2}\rho u^3$. So if, say, $\rho = 1.3\,\text{kg}\,\text{m}^{-3}$ then $P_0 = Ku^3$ where $K = 0.65\,\text{W}\,\text{m}^{-2}\,(\text{m}\,\text{s}^{-1})^{-3}$. $P_0\Phi_u$ is the distribution of power in the wind (Fig. 9.19). Notice that the maximum of $P_0\Phi_u$ occurs on North Ronaldsay at $u = 12.5\,\text{m}\,\text{s}^{-1}$, about twice the most probable wind speed of $6.2\,\text{m}\,\text{s}^{-1}$.

(7) Finally, Fig. 9.20 plots the power per unit area in the wind against $\Phi_{u>u'}$, to indicate the likelihood of obtaining particular power levels.

The analysis of Example 9.1 is entirely in terms of the probability of wind characteristics as obtained from records of wind speed. Such information over short time intervals can be used to investigate the frequency of wind changes e.g. Fig. 9.14(b).

Note the importance of fluctuations ~10 s. These not only contain significant energy, but lead to damaging stresses on wind machines. A measure of these time variations is the turbulence intensity, equal to the standard deviation of the instantaneous wind speed divided by the mean value of the wind speed. Turbulence intensity is a useful measure over time intervals of several minutes.

9.6.4 *Weibull, Rayleigh and chi-squared distributions*

The analysis of Example 9.1 depended solely on field data and repetitive numerical calculation. It would be extremely useful if the important function Φ_u, the probability distribution of wind speed, could be given an algebraic form that accurately fitted the data. Two advantages follow: (1) fewer data need be measured, and (2) analytic calculation of wind machine performance could be attempted.

Using the symbols of the previous section,

$$\Phi_{u>u'} = \int_{u=u'}^{\infty} \Phi_u(u)\,\mathrm{d}u = 1 - \int_0^{u'} \Phi_u\,\mathrm{d}u \tag{9.55}$$

Therefore, by the principles of calculus,

$$\frac{\mathrm{d}\Phi_{u>u'}}{\mathrm{d}u'} = -\Phi_u \tag{9.56}$$

It is commonly found that a two-parameter exponential function can be closely fitted to measured data. One such function, often used in wind speed analysis, is the Weibull function Φ_u, obtained from

$$\Phi_{u>u'} = \exp\left[-\left(\frac{u'}{c}\right)^k\right] \tag{9.57}$$

so (Weibull):

$$\Phi_u = \frac{k}{c}\left(\frac{u}{c}\right)^{k-1} \exp\left[-\left(\frac{u}{c}\right)^k\right] \tag{9.58}$$

Quite excellent fits to experimental data are found with values of the number k ranging from about 1.8 to 2.3, and the speed c being near to the mean wind speed for the site (see Bowden *et al.*, 1983a).

For many sites it is adequate to reduce (9.58) to the one-parameter Rayleigh distribution (also called the chi-squared distribution), by setting $k = 2$. So (Rayleigh)

$$\Phi_u = \frac{2u}{c^2}\exp\left[-\left(\frac{u}{c}\right)^2\right] \tag{9.59}$$

The mean wind speed is then

$$\bar{u} = \frac{\int_0^\infty \Phi_u u\,du}{\int_0^\infty \Phi_u\,du} \tag{9.60}$$

For the Weibull distribution of (9.58), this becomes

$$\bar{u} = \frac{\int_0^\infty uu^{k-1}\exp[-(u/c)^k]\,du}{\int_0^\infty u^{k-1}\exp[-(u/c)^k]\,du} \tag{9.61}$$

Let $(u/c)^k = v$, so $dv = (k/c^k)\,u^{k-1}\,du$. Equation (9.61) becomes

$$\bar{u} = \frac{c\int_0^\infty v^{1/k}\exp[-v]\,dv}{\int_0^\infty \exp[-v]\,dv} \tag{9.62}$$

The denominator is unity, and the numerator is in the form of a standard integral called the factorial or gamma function, i.e.

$$\Gamma(z+1) = z! = \int_{v=0}^\infty v^z e^{-v}\,dv \tag{9.63}$$

Note that the gamma function is unfortunately usually written, as here, as a function of $z+1$ and not z (refer to Jeffreys and Jeffreys, 1966).

Thus

$$\bar{u} = c\Gamma(1+1/k) = c[(1/k)!] \tag{9.64}$$

Using the standard mathematics of the gamma function it is easy to calculate the mean value of u^n, where n is an integer or fractional number, since in general for the Weibull function

$$\overline{u^n} = c^n\,\Gamma(1+n/k) \tag{9.65}$$

For instance the mean value of u^3 becomes

$$\overline{u^3} = c^3\Gamma(1+3/k) \tag{9.66}$$

from which the mean power in the wind is obtained.

Values for c and k for any particular wind distribution are found by fitting the distribution to meteorological measurements. For instance if $\bar{u}$ and $\overline{u^3}$ are known, then (9.64) and (9.66) are simultaneous equations with two unknowns. Modern data collection and online analysis methods enable mean values to be continuously accumulated without storing individual records, so $\bar{u}$ and $\overline{u^3}$ are easily measured.

Another method for obtaining c and k is to measure $\bar{u}$, and the standard deviation of u about $\bar{u}$, to give $\overline{u^2} - \bar{u}^2$ and hence $\overline{u^2}$.

Example 9.2   Rayleigh distribution analysis

Show that for the Rayleigh distribution:

(1) $\Phi_{u>u'} = \exp\left[-\dfrac{\pi}{4}\left(\dfrac{u'}{\bar{u}}\right)^2\right]$

(2) $(\overline{u^3})^{1/3} = 1.24\bar{u}$
(3) $\Phi_u(\text{max})$ occurs at $u = (2/\pi)^{1/2}\bar{u} = 0.80\bar{u}$
(4) $(\Phi_n u^3)(\text{max})$ occurs at $u = 2(2/\pi)^{1/2} = 1.60\bar{u}$

Solution
In (9.62) with $k = 2$,

$$\bar{u} = c\Gamma(1 + 1/2) = c[(1/2)!] \tag{9.67}$$

where by definition

$$(1/2)! = \int_0^\infty u^{1/2}e^{-u}\,du$$

so by a standard integral

$$(1/2)! = \sqrt{\pi}/2$$

Hence in (9.67),

$$c = 2\bar{u}/\sqrt{\pi} \tag{9.68}$$

In (9.59), the Rayleigh distribution becomes

$$\Phi_u = \frac{\pi u}{2\bar{u}^2}\exp\left[-\frac{\pi}{4}\left(\frac{u}{\bar{u}}\right)^2\right] \tag{9.69}$$

and by (9.55)

$$\Phi_{u>u'} = \int_{u=u'}^\infty \Phi_u\,du = \exp\left[-\frac{\pi}{4}\left(\frac{u'}{\bar{u}}\right)^2\right] \tag{9.70}$$

Also

$$\overline{u^3} = \frac{\displaystyle\int_0^\infty \Phi_u u^3 \, du}{\displaystyle\int_0^\infty \Phi_u \, du} = \frac{1}{1}\left[\frac{\pi}{2\overline{u}^2} \int_0^\infty u^4 \exp - \frac{\pi}{4}\left(\frac{u}{\overline{u}}\right)^2\right] du \qquad (9.71)$$

By standard integrals of the gamma function this reduces to

$$\overline{u^3} = K(\overline{u})^3 \qquad (9.72)$$

where K is the energy pattern factor or more descriptively, in French, the *coefficient d'irregularité*. For the Rayleigh distribution of (9.71), $K = (6/\pi) = 1.91$.

A very useful relationship between mean wind speed and average annual power in the wind per unit area follows:

$$\frac{\overline{P_0}}{A} = \frac{1}{2}\rho\overline{u^3} \approx \rho(\overline{u})^3 \qquad (9.73)$$

Hence

$$(\overline{u^3})^{1/3} = 1.2\overline{u} \qquad (9.74)$$

By differentiation to obtain the values of u for maxima in Φ_u and $\Phi_u u^3$, and again using the standard integral relationships of the gamma function:

$$\Phi_u(\text{max}) \text{ occurs at } u = (2/\pi)^{1/2}\overline{u} = 0.80\overline{u} \qquad (9.75)$$

and

$$(\Phi_u u^3)(\text{max}) \text{ occurs at } u = 2(2/\pi)^{1/2}\overline{u} = 1.60\overline{u} \qquad (9.76)$$

Example 9.3 Rayleigh distribution fitted to measured data
Apply the results of Example 9.2 to the data from North Ronaldsay in Example 9.1.

Solution
For North Ronaldsay $\overline{u} = 8.2\,\text{m s}^{-1}$. Therefore by (9.75), $\Phi_u(\text{max})$ is at $(0.80)(8.2\,\text{m s}^{-1}) = 6.6\,\text{m s}^{-1}$. The measured value from Fig. 9.19 is $6.2\,\text{m s}^{-1}$.

By (9.76), $(\Phi_u u^3)(\text{max})$ is at $(1.60)(8.2\,\text{m s}^{-1}) = 13\,\text{m s}^{-1}$. The measured value from Fig. 9.19 is $12.5\,\text{m s}^{-1}$.

By (9.74), $(\overline{u^3})^{1/3} = (1.24)(8.2\,\text{m s}^{-1}) = 10.2\,\text{m s}^{-1}$. The measured value from Fig. 9.19 is $10.1\,\text{m s}^{-1}$.

Scaled and simplified distributions

For all distributions $u\Phi_u, \Phi_{u>u'}$ and $u/\bar{u}$ are dimensionless. For the Weibull distribution k is also dimensionless.

With the Rayleigh distribution ($k = 2 =$ constant) the dimensionless factors become scale factors appropriate for any range of wind speeds, and useful, for instance, if only the mean wind speed is known.

Weibull distribution fits to measured data often give values of k between about 1.6 and 3.0. In these cases $c \simeq 2\bar{u}/\sqrt{\pi}$ to within 1% as in the Rayleigh distribution, and so it may be shown that $\overline{u^3} = (\sqrt{\pi})(3c^3/k)/2$ (see Bowden *et al.*, 1983). Present investigations are exploring the hypothesis that k is related only to the form of the local topography and synoptic (general) wind pattern. If this is so, estimates of $\overline{u^3}$, and perhaps the extent and frequency of lulls, could be predicted by knowing only $\bar{u}$ from long term meteorological data.

9.7 Power extraction by a turbine

The fraction of power extracted from the wind by a turbine is, by (9.15), the power coefficient C_P as discussed in Section 9.3. C_P is most dependent on the tip speed ratio λ, unless the machine is controlled for other reasons (as happens below the cut-in wind speed and usually above the rated output). The strategy for matching a machine to a particular wind regime will range between the aims of (1) maximizing total energy production in the year (e.g. for fuel saving in a large electricity network), and (2) providing a minimum supply even in light winds (e.g. for water pumping to a cattle trough). In addition secondary equipment, such as generators or pumps, has to be coupled to the turbine, so its power matching response has to be linked to the turbine characteristic. The subject of power extraction is therefore complex, incorporating many factors, and in practice a range of strategies and types of system will be used according to different traditions and needs.

This section considers power extracted at the turbine, before this is used to generate electricity or rotate machinery. From Section 9.6.3, Φ_u is the probability per wind speed interval that the wind speed will be in the interval u to $(u + du)$.

Let E be the total energy extracted by the turbine in the period T, and let E_u be the energy extracted per interval of wind speed when the wind speed is u. Then

$$E = \int_{u=0}^{\infty} E_u \, du = \int_{u=0}^{\infty} [\tfrac{1}{2}\rho u^3 C_P(\Phi_u T)] \, du \tag{9.77}$$

The average power extracted if the air density is considered constant is

$$\frac{E}{T} = \bar{P}_T = \frac{\rho}{2} \int_{u=0}^{\infty} \Phi_u u^3 C_P \, du \tag{9.78}$$

This integral or summation cannot be evaluated until the dependence of C_P on the upstream wind speed u_0 is established. It is usually considered that there are four distinct wind speed regions to consider (e.g. Fig. 9.21):

(1) Below cut-in speed u_{ci}:

$$E_u = 0 \text{ for } u_0 < u_{ci} \tag{9.79}$$

(2) Above the rated speed u_R:

$$E_u = \Phi_{u>u_R} P_R T \tag{9.80}$$

where P_R is the rated power output.

(3) Above the cut-out speed u_{co}:

$$E_u = 0 \text{ for } u_0 > u_{co} \tag{9.81}$$

However in practice most machines do not cut out in high wind speeds, but continue to operate at greatly reduced efficiency yet at reasonably high power.

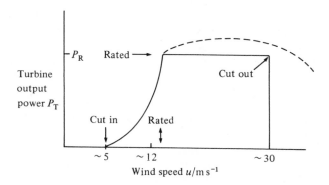

Fig. 9.21 Wind turbine operating regions:
————— standard characteristics
– – – – – actual operating characteristics of many machines

(4) Between u_{ci} and u_R.

The turbine power output P_T will depend on the operating conditions and type of machine. For many machines

$$P_T \approx au_0^3 - bP_R \tag{9.82}$$

where a and b are constants.

At cut-in, $P_T = 0$; so

$$u_{ci}^3 = bP_R/a$$

At rated power $P_T = P_R$; so

$$u_R^3 = (1+b)P_R/a$$

Thus

$$(u_{ci}/u_R)^3 = b/(1+b) \tag{9.83}$$

Hence a and b can be determined in terms of u_{ci}, u_R and P_R.

In practice turbines will often be operating in the region between cut-in and rated output, and it is wasteful of energy potential if the machine is unduly limited at high wind speeds. There are two extreme conditions of operation (see Fig. 9.22):

(1) *Constant tip speed ratio, hence constant C_P* In (9.78),

$$\bar{P}_T = \frac{\rho C_P}{2} \left(\int_{u=u_{ci}}^{u=u_R} \Phi_u u_0^3 \, du + \Phi_{u_R < u_0 < u_{co}} P_R \right) \tag{9.84}$$

so from Section 9.6, with a Rayleigh distribution of wind speed and u_{co} very large,

$$\bar{P}_T = \frac{\rho C_P}{2} \int_{u=u_{ci}}^{u=u_R} \frac{\pi u^4}{2\bar{u}^2} \exp\left[-\frac{\pi}{4}\left(\frac{u_0}{\bar{u}}\right)^2 \right] \, du + P_R \exp\left[-\frac{\pi}{4}\left(\frac{u_R}{\bar{u}}\right)^2 \right]$$

$$= \frac{\rho C_P}{2} \frac{6}{\pi} (\bar{u})^3 + P_R \exp\left[-\frac{\pi}{4}\left(\frac{u_R}{\bar{u}}\right)^2 \right] \tag{9.85}$$

(2) *Constant turbine rotational frequency, hence varying C_P* From Fig. 9.22 it can be seen that C_P can be obtained as a function of unperturbed wind speed u_0, and the turbine power calculated by numerical methods. By operating at constant frequency there is a loss of possible energy extraction. This may be particularly serious if there is a mismatch of optimum performance at higher wind speeds.

9.8 Electricity generation

9.8.1 *Basics*

Electricity is an excellent energy vector to transmit the high quality mechanical power of a wind turbine. Generation is usually $\sim95\%$ efficient, and transmission losses should be less than 10%. The frequency and voltage of transmission need

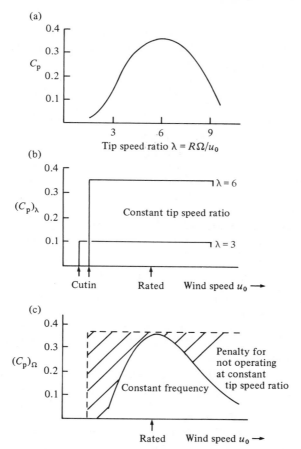

Fig. 9.22 Power coefficient C_p: (a) versus tip speed ratio (b) versus wind speed at constant tip speed ratio (c) versus wind speed at constant turbine frequency

not be standardized, since end-use requirements vary. Heating can accept wide variations in frequency and voltage, but grid connection to other generators or multiple uses may require careful standardization. There are already many designs of wind/electricity systems including a wide range of generators. Future development will produce radically new designs of generators and control systems as wind generated power becomes an electrical engineering speciality.

The distinctive features of wind/electricity generating systems are:

(1) Wind turbine efficiency is greatest if rotational frequency varies to maintain constant tip speed ratio, yet electricity generation is most efficient at constant or nearly constant frequency.

(2) Mechanical control of a turbine by flaps etc. to maintain constant frequency increases complexity and expense. An alternative method, usually cheaper and more efficient, is to vary the electrical load on the turbine to control the rotational frequency.

(3) The optimum rotational frequency of a turbine in a particular wind speed decreases with increase in radius in order to maintain constant tip speed ratio. Thus only small (~2 m radius) turbines can be coupled directly to generators. Larger machines require a gearbox to increase the generator drive frequency. Gearboxes are relatively expensive and heavy. They require maintenance and can be noisy. New types of generator, e.g. with large numbers of poles, are being produced to operate at lower frequency.

(4) The turbine can be decoupled from the generator to provide an indirect drive through a mechanical accumulator (e.g. a weight lifted by hydraulic pressure) or chemical storage (e.g. a battery). Thus generator control is independent of turbine operation. Even the provision of a 'soft coupling' using shock absorbers is useful to prevent electrical spikes or mechanical strain propagating from the turbine.

There are also some distinctive energy end-use features of wind power systems.

(5) Most wind power sites are in remote rural, island or marine areas. Energy requirements in such places are distinctive, and almost certainly will not require the high electrical power of large industrial complexes.

(6) Explicit end-use requirements for controlled electricity (e.g. 240 V/50 Hz or 110 V/60 Hz for lighting, machines and electronics) are likely to be only 5 to 10% of the total energy requirement for transport, cooking and heat. Thus a power system with mixed quality supplies can be a good match with total energy end-use, e.g. the supply of cheap variable voltage power for heating and expensive fixed voltage electricity for lights and motors.

(7) Rural grid systems are likely to be 'weak', since they carry relatively low voltage supplies (e.g. 33 kV) over relatively long distances with complicated inductive and resistive power loss problems. Interfacing a WECS in weak grids is difficult, and the safety of electric utility workers on the lines must be assured.

(8) There are always periods without wind. Thus WECS must be linked to energy storage or parallel generating systems if supplies are to be maintained.

It is obvious that wind/electricity systems will stimulate innovative improvements in electrical power systems, and so an adherence to conventional large system power engineering will handicap progress.

9.8.2 *Generators*

The basic operation of all generators is simple, but many complexities and variations are used to give particular properties and improvements in efficiency.

The principles of commercial generators used with wind machines are as follows. A magnetic field is arranged to cut a wire with a relative velocity, so inducing an electric current by the Faraday effect. Every generator has a stator and a rotor; one of these has a coil (winding) to produce the generated current, and the other has permanent magnets or other windings to produce the magnetic fields. If the magnetic fields are created by permanent magnets or DC currents, then the current will be induced at an AC frequency (f_1) in synchronization with the rotating shaft frequency (f_s) of the generator. With n pole pairs, $nf_s = f_1$ exactly. Such a generator is a *synchronous* AC generator with the output frequency locked to the shaft frequency.

A common variation is the *induction* AC generator. Here the currents that produce the magnetic field are in closed short circuited loops, say on the rotor. These currents are themselves induced from currents in the stator windings. The phase relationships are such that mechanical power can be transferred from the driven rotor shaft to electrical power in the stator circuit. If there are n windings on the rotor each acting as a magnet, if the induced rotor currents have a frequency f_2, and if the shaft frequency is f_s, then the AC output power has a frequency $f_1 = f_2 - nf_s$. The slip $s = (f_1 - f_s)/f_1$ is negative for a generator, but is positive when the same device acts as a motor unless power is dissipated by resistors in the rotor windings. Generator slip is usually less than 10%.

An induction generator can only generate when the induced closed loop currents have been initiated and maintained. There are generally two methods for this: (1) reactive power is drawn from the live grid to which the generator output is connected, or (2) autonomous self-excited generation is made possible by capacitors connected between the output and earth. In method 2 there has to be some residual magnetism in the framework or surroundings of the generator to provide the initial current, with the capacitors maintaining the correct phase relationships. It is also possible to maintain and control generation from an induction generator by running an idling synchronous generator in parallel as a 'synchronous compensator'. This system is attractive for small autonomous systems since the same synchronous generator can be used for, say, diesel generation at times of inadequate wind. The benefits of method (1) (grid linking) are: (a) the simplicity and cheapness of the system, (b) safety, since the generator should not generate if the grid power is off, and (c) the grid can be used to export power from the aerogenerator in windy conditions, and import power at times of low wind.

Direct current generators are essentially synchronous machines, with commutators (e.g. split rings) to insure the output current is unidirectional. This current, although DC, is not necessarily constant. Direct current can obviously also be produced by rectifying alternating current. Likewise, AC can be produced at controlled voltage and frequency by an inverter powered from a DC source such as a battery.

9.8.3 *Aerogenerator classification*

There are three classes of wind turbine electricity system, depending on the relative size of the aerogenerator and other electricity generators connected in parallel with it (Table 9.4). The aerogenerator rated capacity is P_T, and the other generators' capacity is P_G.

Table 9.4 Classes of wind turbine electricity systems

Class	A	B	C
P_T: aerogenerator P_G other generator capacity	$P_T \gg P_G$	$P_T \sim P_G$	$P_T \ll P_G$
Example	Autonomous	Wind/diesel	Grid slaved
Control modes	(a) Blade pitch (b) Load matching	(a) Wind or diesel separately (b) Wind and diesel together	(a) Direct induction generator (b) To DC then AC (c) Increased slip induction generator

Class A: aerogenerator capacity dominant, $P_T \geq 5 P_G$

Usually this will be a single autonomous stand-alone machine without any form of grid linking. Other generators would not be expected, but could for instance include smaller wind machines. For remote communication sites, household lighting, marine lights etc., $P_T \leq 5\,kW$. For full household supplies, including heat, $P_T \sim 20\,kW$. Aerogenerators of large capacity are likely to have standby generation of class B.

Control options have been discussed in Section 1.4.4 and are of extreme importance for efficient cost effective systems (Fig. 9.23). One choice is to have very little control so the output is of variable voltage (and, if AC, frequency) for use for heat or rectified power (Fig. 9.23(a)). There are many situations where such a supply will be useful. The relatively small amount of power that usually has to be controlled at, say, 240 V/50 Hz or 110 V/60 Hz can be obtained from batteries by inverters. Thus the high quality controlled electricity is obtained 'piggy-back' on the supply of less quality, and can be costed only against the marginal extras of battery and inverter.

However, it may be preferred to have the electricity at controlled frequency. There are two extreme options for this:

(1) *Mechanical control of the turbine blades*   As the wind changes speed, the pitch of the blades or blade tips is adjusted to control the frequency of turbine rotation (Fig. 9.23(b)). The disadvantage is that power in the wind is wasted (see Section 1.4.4) and the control method can be expensive and unreliable.

(2) *Load control*   As the wind changes speed the electrical load is changed by rapid switching, so the turbine frequency is controlled (Fig. 9.23(c)). This method makes greater use of the power in the wind because the blade pitch is kept at the optimum angles. Moreover local control by modern electronic methods is cheaper and more reliable than control of mechanical components exposed in adverse environments.

Generators for autonomous WECS can be of several types, as discussed earlier.

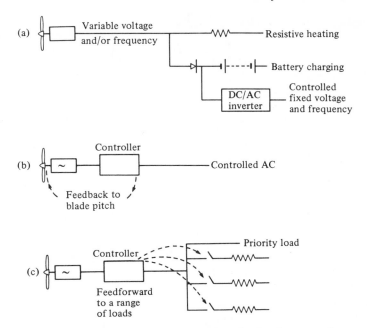

Fig. 9.23 Some supply options with the aerogenerator the dominant supply

Permanent magnet multipole generators are common for small machines. DC systems can be smoothed and the energy stored in batteries. AC systems may have synchronous generators producing uncontrolled output for heat, or controlled output by mechanical or load control. AC induction generators can be self-excited with a capacitor bank to earth, or may operate with an idling synchronous generator as a compensator.

Class B: aerogenerator capacity $\approx$ other generator capacity, $P_T \approx P_G$
This is a common feature of remote area, small grid systems. We shall first assume that the 'other generator' of capacity P_G is powered by a diesel engine. The principal purpose of the wind turbine is then to be a diesel oil saver. The diesel generator will be the only supply at windless periods, and will perhaps augment the wind turbine at periods of low wind. There are two extreme modes of operation:

(1) *Single mode electricity supply distribution* With a single set of distribution cables (usually a three-phase supply that takes single phase to domestic dwellings), the system must operate in a single mode at fixed voltage for 240 V or 110 V related use (Fig. 9.24(a)). A 24 h maintained supply without load management control will still depend heavily (at least 50% usually) on diesel generation since wind is often not available. The diesel is either left on continuously (frequently on light load even when the wind power is available) or switched off when an excessive wind is blowing.

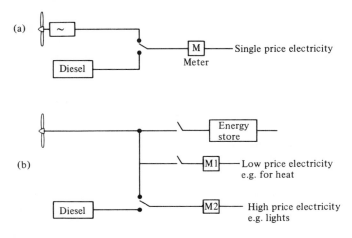

Fig. 9.24 Wind/diesel supply modes: (a) single mode (b) multiple mode

In practice a large amount (sometimes over 70%) of wind generated power has to be dumped into outside resistor banks owing to the mismatch of supply and demand in windy conditions.

(2) *Multiple mode distribution*   The essence of multiple mode operation is that every effort is made to use all wind generated power by offering cheap electricity for many uses in windy conditions (Fig. 9.24(b)). As the wind drops, the cheaper serviced loads are automatically switched off to decrease the demand. The same system can be used to control the rotation of the wind turbine. When no wind power is available, only the loads on the expensive supply are enabled for supply by the diesel generator. The economic advantage of successful multiple mode operation is that the full capital value of the wind machine is used at all times, and since the initial power in the wind is free the maximum benefit is obtained.

Class C: grid linked, aerogenerator slaved in a large system, $5 P_T \le P_G$
This is the most common arrangement for large (~ 3 MW), medium (~ 250 kW) and small (~ 50 kW) machines where a public utility or other large capacity grid is available (Fig. 9.25). The owner of the machine uses the wind power directly and sells (exports) any excess to the grid. Electricity is purchased (imported) from the grid at periods of low or no wind. The cheapest, and perhaps the safest, type of generator is an induction generator connected directly to the grid. The turbine has to operate at nearly constant frequency, within a maximum slip of about 10% ahead of the mains related frequency. In low wind there is an automatic cutout to prevent motoring.

The disadvantage of a directly coupled induction generator is that the turbine frequency cannot change sufficiently to maintain even approximate constant tip speed ratio. A common variation on small machines is to have two generators in

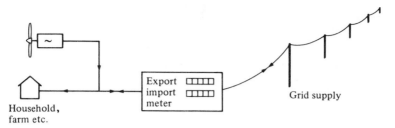

Fig. 9.25 Grid linked aerogenerator slaved in a large system

the same nacelle, say 5 kW and 22 kW, for automatic alternative use in light and strong winds. Other alternatives allowing variable turbine frequency are:

(1) Multiple combination windings in an induction generator to produce more poles in lighter winds for lower rotational frequency;
(2) Rectifying the immediate generator output to DC and then producing the prescribed AC mains frequency with an inverter; or
(3) Increasing the allowed slip on an induction generator with resistive loads on the windings.

9.9 Mechanical power

Historically the mechanical energy in the wind has been harnessed predominantly for transport with sailships, for milling grain and for pumping water. These uses still continue and can be expected to increase again in the future. This section briefly discusses those systems, bearing in mind that electricity can be an intermediate energy vector for such mechanical uses.

9.9.1 *Sea transport*

The old square rigged sailing ships operated by drag forces and were inefficient. Modern racing yachts, with a subsurface keel harnessing lift forces, are much more efficient and can sail faster than the wind. Large sail driven cargo ships requiring power of several megawatts are now being designed with automatic power operated sails that require little manual labor. Wind turbines can be placed on ships to drive propellers for controlled movement in harbor or near shore conditions. Such designs may be particularly useful for ferries operating short routes.

9.9.2 *Grain milling*

The traditional windmill (commonly described as a Dutch windmill) has been eclipsed by engine or electrical driven machines. It is unlikely that the intermittent nature of wind over land will ever be again suitable for commercial milling in direct mechanical systems.

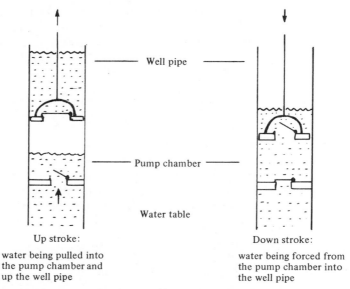

Up stroke:

water being pulled into
the pump chamber and
up the well pipe

Down stroke:

water being forced from
the pump chamber into
the well pipe

Fig. 9.26 Positive displacement water pump. The shaft would be connected to the rotating crankshaft of the wind turbine

9.9.3 *Water pumping*

Pumped water can be stored in tanks and reservoirs, or absorbed in the ground. This capacitor-like property gives smoothing to the intermittent wind source, and makes wind powered pumping economic. Farm scale pumps to about 10 kW maximum power are common in many countries, including the United States, Australia and Crete. The water is used mostly for cattle, irrigation or drainage. Continuity of supply is important, so high solidity, multiblade turbines are suitable, having high initial torque in light winds. The low rotational speed is not a handicap for direct mechanical action. The traditional cylinder pump with a fixed action (Fig. 9.26) is simple and reliable. At best, however, the delivered power is proportional to turbine rotational frequency ($P' \propto \Omega$), whereas at constant tip speed ratio the power at the turbine is proportional to Ω^3 ($P_T \propto \Omega^3$). Therefore the efficiency P'/P_T drops as $1/\Omega^2$. Improved pumps that maintain simplicity of operation will be important for more efficient water pumping. Since water is usually available at low locations, and wind increases with height, it is often sensible to have an electricity generating wind turbine placed on a hill operating an electric pump placed at the nearby water supply.

9.9.4 *Heat production*

The direct dissipation of wind harnessed mechanical power produces heat with 100% efficiency. Thus paddlewheel and other turbulent fluid systems have been used for producing heat directly at the wind turbine site. In practice such systems are noisy and difficult to control, and buildings tend to lessen the wind availability for the turbine. Likewise the mechanical shaft driven heat pump appears attractive. However, electrical generators are so common and efficient that it is likely that electricity will usually be favored as an intermediate energy vector.

9.10 Total systems

This section is a summary of a number of practical factors and proposals for WECS. Further detail can be found in the previous sections.

9.10.1 *Electricity generation for utility grids*

Government funded research and development in America and Europe has had the prime motive of generating electricity for high voltage regional grids. Aerogenerators have operated successfully for several years up to a rated capacity of about 200 kW. Other machines up to 3 and 4 MW have been constructed. These are expected to be accepted for regular commercial use producing electricity over a 20 year lifetime at costs less than from oil thermal stations. Early developments aimed to place machines on hilltop sites, but experience indicates that open plains may be more suitable, especially since numbers of machines arrayed on wind 'farms' make convenient and manageable units. This concept has been extended to deploy large arrays of aerogenerators at sea in shallow water, as planned for instance in the North Sea off the east coast of England.

9.10.2 *Private generation*

Machines of capacity between 10 kW and 100 kW are being accepted for household, farm and institutional use. Cost effective operation is most likely in regions where other energy supplies are expensive (e.g. oil for heat and electricity), or where excess electricity can be sold to a utility grid at a price of at least half the buying price. Commercial sales are most likely in countries where there is (1) a reliance on oil, (2) a high standard of living giving a large demand for energy, (3) an average wind speed $\geq 6 \, \mathrm{m\,s}^{-1}$, and (4) a sizeable rural population.

9.10.3 *New technology*

The principles of renewable energy supply, developed in Chapter 1, indicate that the new technology has to operate within quite different constraints than have fossil fuel and nuclear sources. Nowhere is this more clear than with wind energy. The dispersed and highly fluctuating nature of wind requires radically different approaches than those used for steady intensive sources. In particular, the end-uses of the wind generated power must change in response to the changing supply, and efficient, cheap methods of energy storage must be found.

Consider the following case study, especially as the lessons could be followed in more extensive systems.

Example 9.4 *Fair Isle multimode WECS*

Fair Isle is an isolated island in the North Sea between mainland Shetland and Orkney. The population of 70 is well established and progressive within the limits of the harsh yet beautiful environment. Until recently the people depended on coal and oil for heat, petroleum for vehicles, and diesel for electricity generation. The electricity cooperative has now installed a 50 kW rated capacity aerogenerator that operates in the persistent winds, of average speed over $8\,\mathrm{m\,s^{-1}}$. The control system was mentioned in Chapter 1 (see Fig. 1.6). High price electricity is available for lighting and electronic devices, and a low price controlled supply is available (wind permitting) for comfort heat and water heating (see Section 9.8.3B, multimode electricity generation). At the frequent periods of excessive wind power, further heat is to be available for growing food in a glasshouse or for a small swimming pool. An electric vehicle was charged from the system to include transport as an end-use. Despite the strong winds, the total generating capacity is small for the population served, and acceptable standards are only possible because the houses are well insulated, careful energy strategies are maintained, and sophisticated reliable control systems are incorporated.

Problems

9.1 From (9.16) the fraction of power extracted from the wind is the power coefficient $C_P = 4a(1-a)^2$. By differentiating with respect to a, show that the maximum value of C_P is 16/27 when $a = 1/3$.

9.2 The calculation of power coefficient C_P by linear momentum theory (Section 9.3) can proceed in terms of $b = u_2/u_0$. Show that (a) $C_P = (1-b^2)(1+b)/2$, (b) C_P is a maximum at 16/27 when $b = 1/3$, (c) $a = (1-b)/2$ where $a = (u_0 - u_1)/u_0$.

9.3 By considering the ratio of the areas A_0 and A_1 of Fig. 9.5, show that the optimum power extraction (according to linear momentum theory) per unit of area A_0 is 8/9 of the incident power in the wind.

9.4 For a wind speed pattern following a Rayleigh distribution, prove that:

(a) The most probable wind speed is $0.80\bar{u}$

(b) The most probable power in the wind occurs at a wind speed of $1.60\bar{u}$.

9.5 For a wind speed pattern following a Rayleigh distribution, prove

$$\overline{u^3} = \frac{6}{\pi}(\bar{u})^3$$

where $\overline{u^3}$ is the mean of u^3, and $\bar{u}$ is the mean of u and so $(\overline{u^3})^{1/3} = 1.24\bar{u}$

9.6 Compare the solutions for Problems 9.4 and 9.5 with the wind speed factors indicated on Figs 9.17 and 9.19 for North Ronaldsay, and comment on how well the relationships between these factors is explained by the wind having a Rayleigh distribution.

9.7 A number of designs of wind turbine pass the output wind from one set of blades immediately on to a second identical set (e.g. two contrary rotating blades set on the same horizontal axis). By considering two actuator disks in series, and using linear momentum theory, show that the combined maximum power coefficient C_P equals 0.64.

Note: this is only slightly higher than the maximum of 16/27 = 0.59 for a single pass of the wind through one set of blades. Thus in a tandem horizontal axis machine of identical blade sets, and indeed in a vertical axis turbine, little extra power is gained by the airstream passing the second set of blades.

9.8 (a) A WECS maintains a tip speed ratio of 8 at all wind speeds. At which wind speed will the blade tip exceed the speed of sound?

(b) A large WECS has a blade diameter of 100 m and rotates at constant frequency. At what frequency will the tip speed exceed the speed of sound?

Solutions

9.1 Simple algebra.

9.2 Simple algebraic substitution.

9.3 By conservation of mass, $\rho A_0 u_0 = \rho A_1 u_1$.
From (9.11) and with $a = 1/3$ for maximum power extraction (9.17), $2u_0 = 3u_1$; so $A_1 = 3A_0/2$.
At maximum power extraction:

$$\frac{\text{output power}}{\text{input power}} = \frac{(16/27)A_1 u_0^3}{A_0 u_0^3} = \left(\frac{16}{27}\right)\left(\frac{3}{2}\right) = \frac{8}{9}$$

9.4 (a) See (9.69):

$$\Phi(u) = au\exp[-bu^2], \text{ where } b = \pi/(4\bar{u}^2)$$

$$\frac{d\Phi(u)}{du} = 0 \text{ at } u^2 = 2\bar{u}^2/\pi, \quad u = 0.80\bar{u}$$

(b) $\Phi(u)u^3 = au^4\exp[-bu^2]$

$$\frac{d[\Phi(u)u^3]}{du} = 0 \text{ at } u^2 = 8\bar{u}^2/\pi, \quad u = 1.60\bar{u}$$

9.5 See (9.71)

$$\overline{u^3} = \frac{\pi}{2\bar{u}^2}\int_0^\infty u^4\exp\left[-\frac{\pi}{4}\left(\frac{u}{\bar{u}_1}\right)^2\right]du$$

Let $bu^2 = v$, where $b = \pi/4\bar{u}^2$. Hence

$$u = (v/b)^{1/2}$$

and

$$du = \frac{1}{2}\left(\frac{1}{bv}\right)^{1/2} dv$$

Hence

$$\overline{u^3} = \frac{\pi}{2\overline{u}^2} \int_0^\infty \frac{v^2}{\alpha^2} \exp[-v] \frac{1}{2(bv)^{1/2}} \, dv$$

$$= \frac{\pi}{4\overline{u}^2 b^{5/2}} \int_0^\infty v^{3/2} e^{-v} \, dv$$

$$= \frac{\pi}{4\overline{u}^2} \left[\frac{4\overline{u}^2}{\pi}\right]^{5/2} \left[\frac{3}{2}\right]!$$

$$= \left[\frac{4\overline{u}^2}{\pi}\right]^{3/2} \frac{3}{2}\left(\frac{1}{2}!\right)$$

$$= \frac{8(\overline{u})^3}{\pi^{3/2}} \frac{3}{2} \frac{\sqrt{\pi}}{2}$$

$$= \frac{6}{\pi}(\overline{u})^3 = 1.91(\overline{u})^3$$

Hence

$$[\overline{u^3}]^{1/3} = 1.24\overline{u}$$

9.6

	$\dfrac{u[\Phi(u)_{max}]}{\overline{u}}$	$\dfrac{u[\Phi(u)\,u^3]_{max}}{\overline{u}}$	$\dfrac{[\overline{u^3}]^{1/3}}{\overline{u}}$	$\dfrac{u[\Phi(u)\,u^3]_{max}}{u[\Phi(u)]_{max}}$
Rayleigh distribution	0.80	1.60	1.24	2.00
North Ronaldsay data	0.76	1.53	1.24	2.00

9.7 Using prime notation for the first set of blades, and double prime for the second, the overall power extraction is

$$P = \tfrac{1}{2}\rho A u_0^3 C_P$$

But

$$P = P' + P'' = \tfrac{1}{2}\rho A[u_0^3 C_P' + (u_2')^3 C_P'']$$

$$= \tfrac{1}{2}\rho A u_0^3 \left[C_P' + \left(\frac{u_2'}{u_0}\right)^3 C_P''\right]$$

So

$$C_P = C_P' + \left(\frac{u_2'}{u_0}\right)^3 C_P''$$

But

$$a' = \frac{u_0 - u_2}{2u_0}$$

So

$$\frac{u_2'}{u_0} = 1 - 2a'$$

and

$$C_P' = 4a'(1 - a')^2$$
$$C_P = 4a'(1 - a')^2 + (1 - 2a')^3 C_P''$$

C_P'' is independent of a', so C_P' is maximum when C_P'' is maximum at 16/27.
Thus

$$C_P = 4a'(1 - a')^2 + (1 - 2a')^3 (16/27)$$

C_P is a maximum when $a' = 0.2$:

$$C_P = 0.8^3 + (0.6^3) \left(\frac{16}{27}\right) = 0.640$$

9.8 (a) $v_{tip} = \lambda u_0$

    If $v_{tip} = 330 \, \mathrm{m \, s^{-1}}$,

$$u_0 = \frac{330 \, \mathrm{m \, s^{-1}}}{8} = 41 \, \mathrm{m \, s^{-1}}$$

(b) $\Omega = v_{tip}/R = (330/50) \, \mathrm{s^{-1}} = 6.6 \, \mathrm{rad^{-1}}$

    $f = \Omega/2\pi = 1.1 \, \mathrm{Hz}$

Bibliography

Barbour, D. (1984) *Energy study of the island of North Ronaldsay, Orkney*, MSc thesis, University of Strathclyde.

Bowden, G. J., Barker, P. R., Shestopal, V. O. and Twidell, J. W. (1983a) 'The Weibull distribution function and wind power statistics', *Wind Engineering*, **7**, 85–98.

Bowden, G. J., Barker, P. R., Robbins, B., Williams, J., Outhred, H. and Twidell, J. W. (1983b) *Wind Energy Study of Lord Howe Island*.

Golding, E. W. (1976) *The Generation of Electricity by Wind Power*, reprinted with additional material by E. and F. N. Spon, London.
A classic text that has become a guide for much modern work.

Hunt, V. D. (1981) *Wind Power*, Van Nostrand, New York.

Jeffreys, H. and Jeffreys, B. (1966) *Methods of Mathematical Physics*, Cambridge University Press.

Petersen, E. L. (1975) *On the Kinetic Energy Spectrum of Atmospheric Motions in the Planetary Boundary Layer*, Report 285 of the Wind Test Site, Riso, Denmark.

Shepherd, D. G. (1978) 'Wind power', in Amer, P. (ed.) *Advances in Energy Systems and Technology*, vol. 1, Academic Press, New York.
An excellent summary of theory and application at research level. Includes summaries of extension of linear momentum theory and of aerodynamic analysis.

Sisterson, D. L., Hicks, B. B., Coulter, R. L. and Wesley, M. L. (1982) 'Difficulties of using power laws for wind energy assessment', *Solar Energy*, **31**, 201–4.

Sørensen, B. (1979) *Renewable Energy*, Academic Press, London.
The various sections on wind properties and machines are outstanding and probably the most rigorous treatment of any text.

World Meteorological Organization (1981) *Meteorological Aspects of the Utilization of Wind as an Energy Source*, Technical Note no. 175, Geneva, Switzerland.

10 *The photosynthetic process*

10.1 Introduction

Photosynthesis is the making (synthesis) of organic structures and chemical energy stores by the action of solar radiation (photo). It is by far the most important renewable energy process, because living organisms are made from material fixed by photosynthesis, and our activities rely on oxygen in which the solar energy is mostly stored. The photosynthetic energy flux on the earth is about 0.9×10^{14} W (250 kW per person; 100 000 large nuclear power stations). The energy supply opportunities that this represents are discussed in the next chapter, on biofuels.

Solar radiation incident on green plants and other photosynthetic organisms must relate to two dominant functions, (1) temperature control for chemical reactions to proceed, and (2) photo excitation of electrons for the production of oxygen and carbon structural material. For an isolated plant it is so important to maintain leaf temperature in the correct range, that solar radiation might be reflected or transmitted rather than absorbed to increase photosynthesis. The organic material produced is mainly carbohydrate, with carbon in a medium position of oxidation and reduction (e.g. glucose $C_6H_{12}O_6$). If this (dry) material is burnt in oxygen, the heat released is about 16 MJ kg^{-1} (4.8 eV per carbon atom, 470 kJ per mole of carbon). The fixation of one carbon atom from atmospheric CO_2 to carbohydrate proceeds by a series of stages in green plants, including algae:

(1) Reactions in light, in which photons produce O_2 from H_2O, and electrons are excited in two stages to produce strong reducing chemicals.
(2) Reactions not requiring light (called dark reactions), in which these reducing chemicals reduce CO_2 to carbohydrates, proteins and fats.

Combining both the light and dark reactions gives an overall reaction, neglecting many intermediate steps:

$$CO_2 + 2H_2\dot{O} \xrightarrow{\text{light}} \dot{O}_2 + [CH_2O] + H_2O \qquad (10.1)$$

where the products are about 5 eV per C atom higher in energy than the initial material, from the absorption of at least eight photons. [CH$_2$O] represents a basic unit of carbohydrate, so the reaction for glucose production is

$$6CO_2 + 12H_2\dot{O} \xrightarrow{\text{light}} 6\dot{O}_2 + C_6H_{12}O_6 + 6H_2O \tag{10.2}$$

In these equations, the oxygen atoms initially in CO_2 and $H_2\dot{O}$ are distinguished, the latter being shown with a dot over the O.

Most studies of photosynthesis depend on biochemical analysis considering the many complex chemical processes involved. This chapter however will emphasize the physical processes, and will relate to the branch of spectroscopy called photophysics. There will also be interesting similarities and comparisons with photovoltaic devices (Chapter 7). We shall proceed in three stages:

(1) The trophic level (Fig. 10.1)
(2) The plant level (Fig. 10.2)
(3) The molecular level (Fig. 10.3): this is a complex system, which will be studied in Section 10.6.

There is extensive variety in all aspects of photosynthesis, from the scale of plants down to molecular level. It must not be assumed that any one system is as straightforward as described in this chapter, which considers only the general principles.

10.2 Trophic level photosynthesis

Animals exist by obtaining energy and materials directly or indirectly from plants. This is called the trophic (feeding) system. Fig. 10.1 is an extremely simplified diagram to emphasize the essential processes.

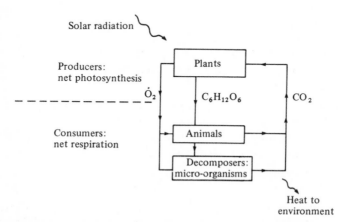

Fig. 10.1 Trophic level global photosynthesis. Fluxes: energy, 10^{14} W; carbon, 10^{11} t y^{-1}; oxygen, 3×10^{11} t y^{-1}. Concentrations: oxygen, 20%; CO_2, 0.03% (by volume)

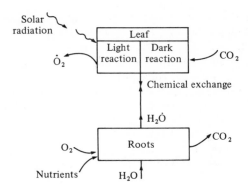

Fig. 10.2 Plant level photosynthesis

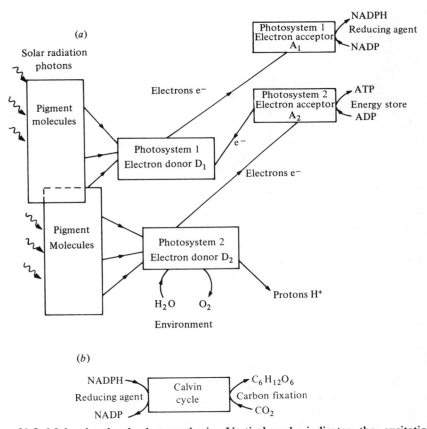

Fig. 10.3 Molecular level photosynthesis. Vertical scale indicates the excitation energy of the electron. (a) Light reaction, indicating the flow of energy and materials in the two interacting photosystems of green plants. (b) Dark reaction, using the reducing agents produced from the light reaction of photosystem 1

During photosynthesis CO_2 and H_2O are absorbed to form carbohydrates, proteins, and fats. The generalized symbol $[CH_2O]$ is used to indicate the basic building block for these materials. CO_2 is released during respiration of both plants and animals, and by the combustion of biological material. This simplified explanation is satisfactory for energy studies, but neglects the essential roles of nitrogen, nutrients and environmental parameters in the processes. The net energy absorbed from solar radiation during photosynthesis can be measured from combustion, since

$$\Delta H + CO_2 + 2H_2O \underset{\text{combustion}}{\overset{\text{photosynthesis}}{\rightleftharpoons}} [CH_2O] + O_2 + H_2O \tag{10.3}$$

$\Delta H = 470\,J$ per mole $C = 5.8\,eV$ per atom C

   $\approx 16\,MJ\,kg^{-1}$ of dry carbohydrate material

Here ΔH is the enthalpy change of the combustion process, equal to the energy absorbed from the photons of solar radiation in photosynthesis, less the energy of respiration during growth. Note that combustion requires temperatures of $\sim 400\,°C$, whereas respiration proceeds by catalytic enzyme reactions at $\sim 20\,°C$. The uptake of CO_2 by a plant leaf is a function of many factors, especially temperature, CO_2 concentration and the intensity and wavelength distributions of light (Fig. 10.4).

Photosynthesis can occur by reducing CO_2 in reactions with compounds other than water. In general these reactions are of the form

$$CO_2 + 2H_2X \rightarrow [CH_2O] + X_2 + H_2O$$

For example, $X = S$. Certain photosynthetic bacteria can grow in the absence of oxygen by such mechanisms.

The efficiency of photosynthesis η is defined for a wide range of circumstances. It is the ratio of the net enthalpy gain of the biomass per unit area (H/A) to the incident solar energy per unit area (E/A), during the particular biomass growth over some specified period:

$$\eta = \frac{H/A}{E/A} \tag{10.4}$$

Here A may range from the surface area of the earth (including deserts) to the land area of a forest, the area of a field of grain, and the exposed or total surface area of a leaf. Periods range from several years to minutes, and conditions may be natural or laboratory controlled. It is particularly important with crops to determine whether quoted growth refers to just the growing season or a whole year. Table 10.1 gives values of η for different conditions.

The quantities involved in a trophic level description of photosynthesis can be appreciated from the following example. Healthy green leaves in sunlight

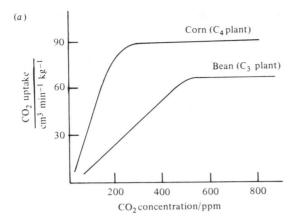

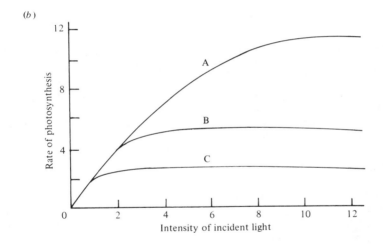

Fig. 10.4 (a) CO_2 uptake of fresh leaf as a function of CO_2 concentration. After Keeton (1980). (b) Effect of external factors on rate of photosynthesis. Effect of light intensity at (A) 25 °C and 0.4% CO_2 (B) 15 °C and 0.4% CO_2 (C) 25 °C and 0.01% CO_2. All units arbitrary. From Hall and Rao (1978)

produce about 3 litres of O_2 per hour per kg of leaf (wet basis). This is an energy flow of 16 W, and would be obtained from an exposed leaf area of about 1 m². A person metabolizes at about 100 W (resting), 200 W (active). Thus each person obtains metabolic energy for 24 hours from reaction with oxygen derived from about 15 to 30 m² of leaf area. Thus in temperate regions one person's annual bodily oxygen intake is provided by approximately one large tree. In the tropics a tree would provide for about three people. Industrial, transport and domestic fuel consumption require far more oxygen per person.

Table 10.1 Approximate photosynthetic efficiency for a range of circumstances. Reported data vary widely for many different circumstances

Conditions	Photosynthetic efficiency (%)
Whole plant (net photosynthesis)	
Whole earth: 1 year average (radiation incident beneath the atmosphere on to all land and sea)	0.1
Forest: annual general average	1
Grassland: annual (tropical, average; temperate, well managed)	2
Cereal crop: closely planted, good farming, growing season only; temperate or tropical crops	3
Continuing crop: e.g. cassava	2
Laboratory conditions: enhanced CO_2, temperature and lighting optimized, ample water and nutrients	10
Initial photosynthetic process	
Filtered light, controlled conditions, plant respiration *not* included, theoretical maximum for the primary photosynthetic process	30

Consideration of the bond energies in photosynthesis shows that the absorbed solar energy is predominantly stored in the oxygen molecules and not the carbon compounds (see Morowitz, 1969). The oxygen gas is able to move freely in the atmosphere, and eventually disperses this store of energy evenly over the earth's surface. In a similar manner carbon, in CO_2, is also dispersed. The coupled reaction of oxygen with carbon compounds has to occur before the energy can be released. Thus the usual, but erroneous, implication that food and biomass contain the stored energy is acceptable.

10.3 Photosynthesis at the plant level

10.3.1 *Absorption of light*

Solar radiation incident on a leaf is reflected, absorbed and transmitted. Part of the absorbed radiation (<5%) provides the energy stored in photosynthesis and produces oxygen; the remainder is absorbed as sensible heat producing a temperature increase, or as latent heat for water evaporation (kinetic and potential energy changes are negligible). Oxygen production is a function of the wavelength of the radiation and may be plotted as an *action spectrum*.

Fig. 10.5 shows typical absorption and action spectra for a leaf. Note that photosynthesis, as measured by oxygen production, occurs across nearly the whole visible spectrum. Absorption is usually most marked in the blue and the red – hence the green colour of most leaves. Spectroscopic techniques of great precision and variety are used to investigate photosynthesis, but two basic experiments are fundamental:

(1) The absorption spectrum from the pigments of a living leaf is different from the sum of the individual absorption spectra of the same pigments separated by chemical methods. Thus within the cooperative structure of an assembly

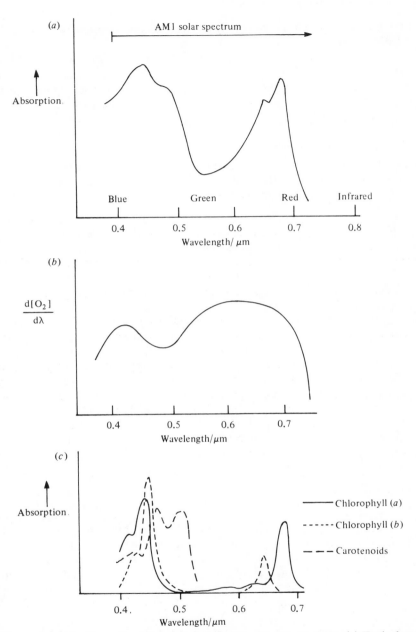

Fig. 10.5 Absorption and action spectra of plant leaves and pigments. (a) Typical spectral absorption spectrum of a green leaf *in vivo*. (b) Action spectrum of a typical green plant. $d[O_2]/d\lambda$ is the spectral distribution of the rate of oxygen production per unit area per unit radiation intensity. (c) Absorption spectra of important pigments when separated in laboratory conditions *in vitro*. Note that no combination of spectra could produce the *in vivo* spectrum of (a), showing that the pigments are changed when part of a cooperative structure

of pigments *in vivo*, the absorption spectrum of any one pigment molecule is changed because it is no longer isolated as an individual.

(2) For green plants photosynthesis increases if there is absorption in one part of the spectrum, e.g. 700 nm, and also at another frequency, e.g. 650 nm. This is a further cooperative effect, called the *Emmerson effect*, giving substantial evidence that photosynthesis in green plants occurs in two systems that act together in series.

10.3.2 *Structure of plant leaves*

A general structure and scale of green leaves is given in Fig. 10.6. In practice there is extensive variety and complexity in all these aspects down to molecular scale, as described in texts on plant physiology. In outline:

(1) Photosynthesis occurs in plant material, usually green leaves and algae which we consider here. It also occurs in some simple organisms (e.g. purple bacteria) without the associated O_2 producing system.

(2) The active cells of green plants, e.g. palisade and spongy mesophyll cells $\sim$0.5 mm in length, have membranes permeable to gases and water (Fig. 10.6(a)). These cells contain distinct intracellular bodies, organelles, also with membranes.

(3) The photosynthetically active ellipsoidal shaped organelles are called chloroplasts. These are $\sim$10 μm long, and contain liquid, the stroma, and membrane structure, the lamellae (Fig. 10.6(b)).

(4) The lamellae have a layered structure that is in general either stacked or open (unstacked) (Fig. 10.6(c)). The stacks are called grana, and the unstacked membranes are stroma lamellae. The open structure is linked to the stacks and maintains the enclosed pockets within the chloroplast.

(5) The lamellae are made of thylakoids. These are like flattened balloons having a double membrane structure with four different surfaces (outer top, inner top, inner lower, outer lower) (Fig. 10.6(d)). The structure divides the chloroplast so the fluid may be different each side of the thylakoid lamellae. This division is not easily discerned in two-dimensional sections of chloroplasts.

(6) The thylakoid membrane contains the components of the photosynthetic light reaction (Fig. 10.6(e)). These include the pigment molecules, mostly chlorophyll, that absorb photons in a structural array like a telecommunication antenna. This array is called the *antenna*. The pigment molecules act cooperatively to channel the absorbed energy 'packet' (called an exciton) to central reaction centers. The light trapping and energy channeling system is called the *light harvesting* (LH) system. There are about 300 pigment molecules associated with each reaction center.

(7) The reaction centers contain the final pigment molecules of the LH systems in chemical contact with large molecular weight enzyme molecules. At the reaction center the energy from the LH system enables oxidation/reduction reactions to occur in complex catalytic sequences.

(8) There are two types of reaction center in green plants that may partly, but

not entirely, share LH systems. The centers are of photosystem 2 and photosystem 1 (PS1, PS2).

(9) At reaction centers of PS2 oxygen gas is produced from water and some excess energy is used to form energy storage molecules ATP (adenosine triphosphate).

(10) At reaction centers of PS1 a strong reducing agent NADPH (reduced nicotinamide adenine dinucleotide phosphate) is produced.

(11) NADPH is able to initiate reactions to fix CO_2 outside the thylakoid membrane in the outer liquid of the stroma. These reactions can occur in the dark or light, since the production of NADPH has separated the CO_2 uptake from the immediate absorption of light. The reactions are called the *Calvin cycle dark reaction*.

(12) Protons, formed at oxygen and ATP production, are held by the thylakoid membrane within the inner regions of the lamellae. During ATP formation the protons are 'pumped' through the membrane to maintain the cycle of elements in the entire process.

(13) In the Calvin cycle, CO_2 from solution is fixed into carbohydrate structures, and also protein and fat formation can be initiated.

(14) *Photosynthetic bacteria* are prokariotic cells (without internal nuclear membranes) and the photosynthetically active pigments are located in the membrane of the cell itself. Only one photosystem, PS1, operates and oxygen is not produced.

10.4 Thermodynamic considerations

In this section we shall first consider photosynthesis as an aspect of thermodynamics. The implications are important to guide strategy for renewable energy research and to give basic understanding.

Consider an ideal Carnot heat engine operated from solar energy (see Section 6.9) and producing work at efficiency η with a heat sink at ambient temperature, say 27 °C (300 K). The heat source is derived from solar radiation. With a flat plate absorbing collector, the maximum source temperature is about 200 °C (473 K) and the maximum efficiency is $(473 - 300)/473 = 37\%$. For a power tower with radiation concentration on to the collector, the maximum efficiency might be $(773 - 300)/773 = 61\%$. If the sun's outer temperature could be used as the source, the maximum efficiency would be $(5900 - 300)/5900 = 95\%$. Thus from an engineering point of view there is much interest in seeking to link processes to the highest temperature available to us, namely the sun's temperature.

The only connection between the earth and the sun is via solar radiation. From the previous paragraph it might seem sensible to seek for a thermal process to raise the temperature of the collector. This would, however, be similar to the thermodynamic aims of nuclear fusion, and would suffer similar difficulties. For if the bulk temperature of a material is raised it will tend to be structurally weakened and have a short working life. However, it is possible to absorb the radiation by a photon process into the electron states of materials, without

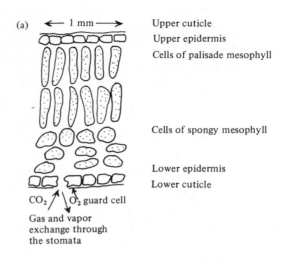

(a) ← 1 mm →

Upper cuticle
Upper epidermis
Cells of palisade mesophyll

Cells of spongy mesophyll

Lower epidermis
Lower cuticle

CO_2 / \ O_2 guard cell
Gas and vapor
exchange through
the stomata

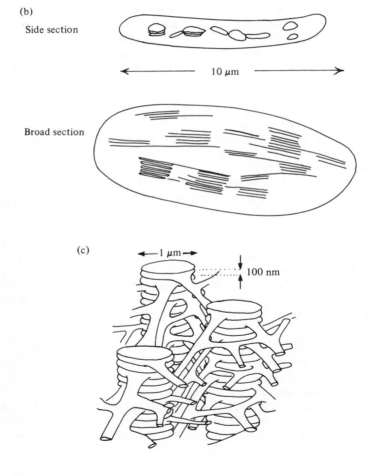

(b)

Side section

← 10 μm →

Broad section

(c) ← 1 μm →
100 nm

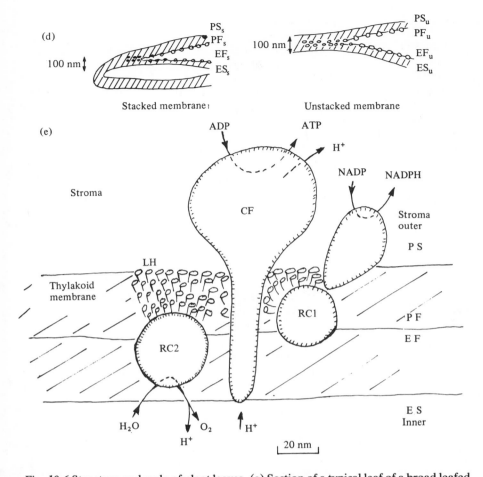

Fig. 10.6 Structure and scale of plant leaves. (a) Section of a typical leaf of a broad leafed plant. Photosynthetically active green cells shown dotted with chloroplast organelles. Approximate scale shown. Actual cells press together, i.e. do not have the gaps shown in the figure for clarity. (b) Section through chloroplast organelle. The thylakoid internal membrance are shown in the liquid stroma. Certain regions have stacked thylakoid membrane (the grana) which are connected by unstacked stroma lamellae membrane. (c) Perspective of the stacked and unstacked thylakoid membrane structure. Stacked grana linked by bridges of the stroma lamellae, all within the liquid stroma of the chloroplast organelle. Approximate scale shown. (d) Thylakoid membrane shown fractured by freezing techniques along natural lines of weakness. Four distinct faces and surfaces appear: outer, PS (protoplasmic surface); inner of outer, PF (protoplasmic face); inner of inner, EF (ectoplasmic face); outer of inner, ES (ectoplasmic surface). Stacked (subscript s) and unstacked (u) regions are identified. The inner surfaces show distinct bumps with electron microscopy. These bumps seem to be associated with ATP and NADPH production. Loose protein structures associated with CO_2 assimilation are on the outer PS surface. Structures associated with O_2 and proton production are on the outer inner surface ES_s. (e) Conceptual diagram of thylakoid membrane of the grana stacks, with the liquid stroma beyond the outer surface. LH: light harvesting system of pigment molecules about 5 nm in length. About 200 to 400 molecules per reaction center. RC: reaction centers, about 20 nm diameter containing protein molecule complexes of about 50 000 molecular weight, 1 and 2 indicate the photosystems. CF: coupling factor. Very large complexes producing ATP and allowing protons to 'pump' through the membrane

immediately increasing the bulk temperature. Such a process occurs in photo-voltaic power generation (see Chapter 7). Let us compare the two processes, namely thermal and photon excitation.

Fig. 10.7 represents a material that can exist in two electronic states, normal and excited. The difference between these states is due solely to a different electronic configuration, and the core or 'lattice' of the material remains unaffected.

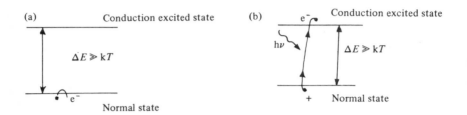

Fig. 10.7 Electron excitation by (a) heat and (b) photon absorption. The vertical scale indicates the excitation energy of the electron

In Fig. 10.7(a) the excited state can only be reached by heating the whole material, and the proportion of excited states N_e to normal states N_n is calculated as for intrinsic semiconductors:

$$N_e/N_n = \exp(-\Delta E/kT) \tag{10.5}$$

We shall be considering pigment molecules where $\Delta E \sim 2\,\text{eV}$, and $T < 373\,\text{K} = 100\,°\text{C}$, since the cellular material is water based. Thus $N_e/N_n \sim 10^{-14}$. Even at the sun's temperature of $5900\,\text{K}$, $N_e/N_n = 0.02$. It is concluded that thermal excitation does not produce many excited states.

However, in Fig. 10.7(b) the excited electronic state is formed by electro-magnetic absorption of a photon of energy $h\nu \geq \Delta E$. This process does not immediately add energy to the remaining 'lattice', which remains at the same temperature. The population of the excited state depends on the rate of absorp-tion of photons and coupling of the excited electronic states to the 'lattice'. According to our ideal model the population limit is $N_e = N_n$, when the radiation is transforming equal numbers of states back and forth, and the electronic temperature is effectively infinite. This ideal would not be reached in practice, but the model does show how 10^{10} more electronic excited states can be formed by the absorption of photons directly into electron states, rather than into the entire material as heat.

The thermodynamic analysis is not complete until the energy has performed a function. In photosynthesis the photon energy is transformed into excited states by electromagnetic absorption, and the energy is eventually stored in chemical products. There is no production of 'work' in the normal mechanical engineering sense, but work has been done by the rearrangement of electronic states producing organic structures and chemical stores of energy.

The chemical changes occurring in photosynthesis are in some ways similar to energy state changes in semiconductor physics. In chemistry the changes occur by reduction and oxidation. The *reduction level* is the number of oxygen

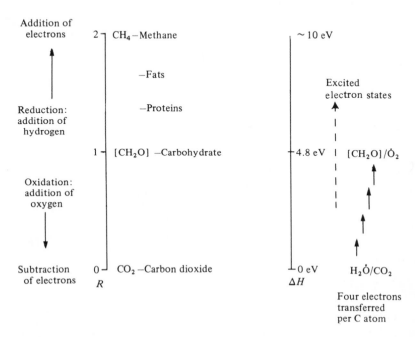

Fig. 10.8 Reduction level R of carbon compounds. Enthalpy change per carbon atom, ΔH, of chemical couples referred to CO_2/H_2O

molecules per carbon atom needed to transform the material to CO_2 and H_2O. For carbon compounds of the general form $C_cH_hO_o$, the reduction level is

$$R = \frac{c + 0.25h - 0.5o}{c} \qquad (10.6)$$

The energy to form these compounds from CO_2 and H_2O is about 460 kJ (mole carbon)$^{-1}$ per unit of reduction level R.

The relationship of reduction level to the energy states involved in photosynthesis is shown in Fig. 10.8. Photosynthesis is essentially the reduction of CO_2 in the presence of H_2O to carbohydrate and oxygen. In the process four electrons have to be removed from four molecules of water (Fig. 10.9). The full process will be explained in Section 10.6.

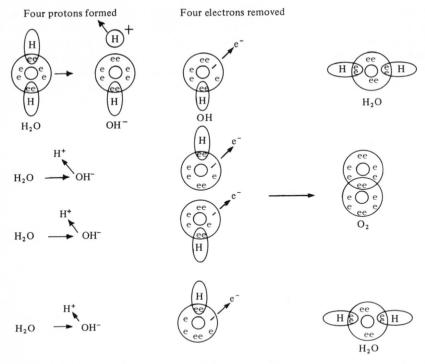

Fig. 10.9 Reduction of water to oxygen and protons at reaction center of photosystem 2 as four electrons are removed:

$$4H_2O \underset{h\nu}{\overset{\rightarrow}{\rightleftharpoons}} 2H_2O + O_2 + 4H^+ + 4e^-$$

10.5 Photophysics

The physical aspects of photosynthesis involve the absorption of photons of light by electrons within pigment molecules. These molecules absorb the energy to form excited states. When the molecules are isolated, the energy is usually re-emitted as fluorescent radiation. However, when the pigments are bound in chloroplast structures the majority of the energy is transferred cooperatively to reaction centers for chemical reductions, and there is little fluorescence.

The isolated properties are explained by the Franck-Condon diagram (see Fig. 10.10). This portrays the ground and excited energy states of the molecule as a function of the relative position of its atoms. This relative position is measured by some spatial coordinate, such as the distance x between two particular neighboring atoms. Note that the minima in energy occur at different values of x owing to molecular changes in size or position after excitation. A photon traveling at $3 \times 10^8 \, \mathrm{m\,s^{-1}}$ passes the molecule, of dimension $\sim 10^{-9}\mathrm{m}$, in time $\sim 10^{-18}\mathrm{s}$. During this time electromagnetic interaction with the electronic states can occur, and the photon energy of $\sim 2\,\mathrm{eV}$ is absorbed. However, vibrational and rotational motions are occurring in the molecule, with thermal energy $kT \sim 0.03\,\text{eV}$ and period $\sim 10^{-13}\,\text{s}$. These states are indicated by horizontal lines on the diagram as the molecule oscillates about its minimum energy positions. Absorption (A) takes place too fast for the molecule structure to adjust, and so the excited state is formed away from the minimum. If the excited electron was originally paired with another electron (as will be probable), the excited state will be expected to be a singlet state (spin $= \frac{1}{2} - \frac{1}{2} = 0$) with lifetime $\sim 10^{-8}\,\text{s}$.

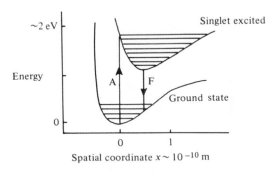

Fig. 10.10 Franck-Condon diagram illustrating Stokes shift in energy between the absorbed photon A and the fluorescent photon F. The spatial coordinate x indicates the change in position or size between the excited system and its ground state

During this time there are $\sim 10^6$ molecular vibrations, and two main processes occur. Either

(1) The molecule is close to other similar molecules, and the absorbed energy (called an exciton) is passed to these by resonant transfer linked with the thermal motion. This is the dominant process for pigment molecules *in vivo*. Or

(2) The excited state relaxes to the minimum position. After $\sim 10^{-8}\,\text{s}$ fluorescent emission (F) may occur as the molecule returns to the ground state. The wavelength of fluorescence is longer than the absorbed light, as described by the *Stokes shift*. Alternatively the electron may change orientation in the excited state, by magnetic interaction with the nucleus, to form a triplet state (spin $= \frac{1}{2} + \frac{1}{2} = 1$).

The lifetime of triplet states is long ($\sim 10^{-3}\,\text{s}$) and again loss of energy occurs, by phosphorescence (P: Fig. 10.11) or by resonant transfer.

Resonant transfer can occur between molecules when they are close ($\sim 5 \times 10^{-10}\,\text{m}$), and when the fluorescence radiation of the transferring molecule overlaps with the absorption band of the neighbor (Fig. 10.12). In these conditions the excited electronic state energy (the exciton) may transfer without radiation to the next molecule. This may be described by separate energy level diagrams of the form of Fig. 10.13(a), or, when molecules are very close, by a graded band gap diagram like Fig. 10.13(b). In either description there is a

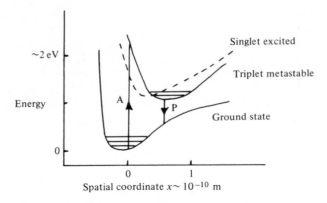

Fig. 10.11 Triplet state and phosphorescent photon P. However, states may overlap, so return to ground state may occur without emitted radiation

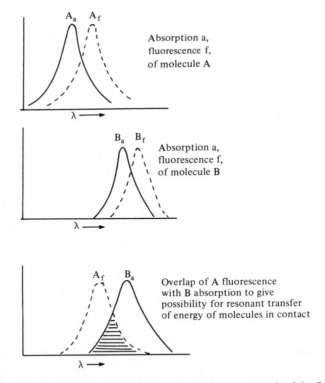

Fig. 10.12 Resonant transfer of energy. The abscissa is the wavelength of the fluorescence and absorption bands of separated molecules. The ordinate indicates the intensity of the bands

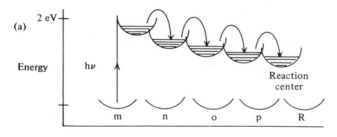

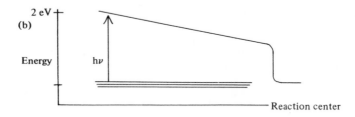

Fig. 10.13 Transfer of energy by pigment molecules of the light harvesting system to the particular reaction center. (a) Spatial position of light harvesting pigment molecules (m, n, o, p) transferring energy to a reaction center R. (b) Graded band gap model: continuous electronic structure of light harvesting pigment molecules acting as a continuous 'super molecule'

spatial transfer of energy down a potential gradient through the assembly of molecules. The process is similar to conduction band electron movement in graded gap photovoltaic cells (Section 7.8.1). However, in photosynthesis, energy is transferred as whole molecules adjust their position and structure during electronic excitation and relaxation, and not just by the transport of a free electron.

There is nevertheless a most significant difference between electron transport in photovoltaic semiconductors and energy transport in pigment molecules. In photovoltaics the structural material is manufactured with graduated dopant properties across the cell. Each element of material has a precise dopant level and must remain at the suitable location. If the photovoltaic cell is broken up, each piece keeps its distinguishing characteristic. However, in the photosynthetic light harvesting system, it is the *cooperative structure* of all the pigments that gives each pigment the necessary electronic structure required for its precise location. It does not matter where a pigment molecule finds itself; it will always be given the correct properties to fit into the light harvesting array, suitable for its position. So when the array is broken up, each pigment reverts to its isolated properties. This accounts for the difference between *in vivo* and *in vitro* properties of pigment molecules, during absorption and fluorescence.

10.6 Molecular level photosynthesis

The previous sections have outlined the whole photosynthetic process. In this section we shall consider some of the physical details, although we stress that many of the conclusions are still speculative. Note that we purposely do not include biological and chemical details, since these may detract from physical understanding and also such detail is well described elsewhere (e.g. Clayton, 1980; Govinjee, 1975).

10.6.1 *Light harvesting and the reaction center*

The photosynthetically active light is absorbed in pigment molecules, of which the most common is chlorophyll type (a). Other pigments occur of similar molecular structure and shape. A chlorophyll molecule consists of a hydrophobic (water fearing) tail and hydrophilic (water loving) ring. The tail is a long chain hydrocarbon (phytyl residue $C_{20}H_{39}$), about four times longer than the 'diameter' of the ring. The tail is presumed to be fixed in the thylakoid membrane outer surface, with the ring in the water based stroma of the chloroplast (Fig. 10.6(e)).

Light photons may be absorbed by the molecules, which are close enough together to (1) cooperatively interact to produce wideband *in vivo* absorption, and (2) to channel (harvest) excited state energy excitons down a potential gradient to the reaction center by resonant transfer. Assemblies of about 200 to 400 pigment molecules occur with 'cores' each containing one reaction center. Assemblies with less or more pigment molecules per reaction center occur in some species of plant, algae, or photosynthetic bacteria. In general, systems adapted to weak light have more pigment molecules per core.

Photons may be absorbed at the outer or inner light harvesting molecules, and the resulting energy exciton passes to the reaction center along paths of least resistance. This consists of the final pigment molecule and some very large specialized molecules (molecular weight ~50 000) in which the 'chemistry' occurs. Most evidence for this general structure of parallel light harvesting paths to a central center has come from optical absorption, fluorescence and action spectra linked with light flashes of varying interval and intensity.

It seems that, in general, two types of core exist, one for each of photosystems 2 and 1. It is possible for different cores to share the same light harvesting arrays.

10.6.2 *The light reactions*

The physical processes of photosystems 2 and 1 are similar (Fig. 10.14). The light harvesting (LH) system channels energy to the reaction center. The last pigment chlorophyll (a) molecule in the system is used to identify the particular center. This is called P680 for PS2, and P700 for PS1, since the respective molecules absorb distinctively at $680 \pm 10\,nm$ (1.82 eV) and $700 \pm 10\,nm$ (1.77 eV) respectively. Note that both these molecules absorb at the red end of the spectrum, indicating their position at the lower potential energy end of the LH system.

In the reaction center there are both donor and acceptor molecules. (These names are used in the sense of semiconductor physics. However, beware; the

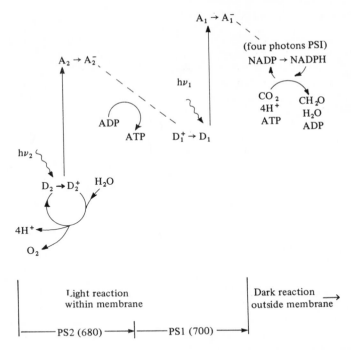

Fig. 10.14 The Z scheme of photosynthesis: eight photons per C fixed. The vertical scale indicates the excitation energy of the electron

letter p may be used for pigment donor molecules, in the *opposite* sense of p type semiconductors.) The channeled exciton from each absorbed photon lifts an electron from the particular donor (D) to the acceptor (A). Chemical reactions then occur with the D^+ and A^- radicals so formed. These reactions are cyclic, so D and A are reformed by the transfer of another electron to D, and the excited electron from A. Two sets of chemical reactions occur, one set as D^+ returns to D, and the other set as A^- returns to A. These chemical reactions are complex, and not yet fully determined for the many variations of detail that occur. However, the physical processes are clear. For green plants there are five distinct steps of the Z scheme (Fig. 10.14). See also Fig. 10.3:

(1) $D_2^+ \rightarrow D_2$ D_2^+ is formed by electron excitation of PS2. It is a strong oxidant (attracting electrons) with a redox potential of $+0.9\,\mathrm{V}$. There is evidence that D_2^+ is a chlorophyll (a) molecule, or molecular pair, in the nonaqueous environment of the thylakoid membrane. The oxidizing action of D_2^+ is to extract an electron from H_2O, with the ultimate result that one molecule of O_2 is formed by four D_2^+ reactions (see Section 10.4 and Fig. 10.9). The O_2 molecules diffuse through the various membranes and out of the plant cells. The protons, H^+, remain in the thylakoid membrane double layer.

(2) *Photon absorption in PS2*   The donor D_2 releases an electron that is lifted to acceptor A_2, forming A_2^-. A_2 has been given the label Q by research workers and may be a plastoquinone. A_2^- is a mild reductant of redox potential about $-0.1\,V$, i.e. the photon excitation from D_2^+ is 1.0 to 1.1 eV. D_2 absorbs at $680 \pm 10\,nm$.

(3) *Link between PS2 and PS1; ATP production*   After A_2^- formation, a series of reactions now occur to transfer the electron down an energy potential to D_1^+ at redox potential $+0.4\,V$. The energy difference is used to form the energy storage molecule ATP (adenosine triphosphate) from ADP (adenosine diphosphate) at 0.34 eV per molecule. Minimum ATP production seems to be three ATP per four operations of PS2 and PS1, as required for the Calvin cycle (see later):

$$ADP + P_i \text{ (inorganic)} \rightarrow ATP + H_2O \tag{10.7}$$

(4) *Photon absorption in PS1*   Further photon absorption through the LH system to the reaction center of PS1 results in the formation of D_1^+ and A_1^-. D_1 appears to be a further chlorophyll (a) monomer or pair (absorbing at $700 \pm 10\,nm$). A_1 is a ferredoxin protein.

(5) *NADPH production*   The electron raised to A_1^- then passes to form NADPH from $NADP^+$. NADP is nicotinamide adenine dinucleotide phosphate. Its reduced form is NADPH, and its oxidized form is $NADP^+$, sometimes written NADP.

$$NADP^+ + 2e^- + H^+ \rightarrow NADPH \tag{10.8}$$

Thus each reduction of $NADP^+$ requires two electrons, from two operations of PS1.

10.6.3 *Dark reactions (Calvin cycle)*

The formation of NADPH by PS1 supplies the main reducing agent required to fix CO_2 into carbohydrate. Subsequent reactions occur by ordinary 'thermal' chemistry, not photochemistry, and so may occur in the dark *after* a period of light absorption, as well as *during* periods of light absorption. The reactions – the Calvin cycle – occur in the stroma outside the thylakoid structures, but within the chloroplast. The reactions may be followed using the radioisotope ^{14}C. In photosynthetic bacteria PS1 is the only photosystem, and CO_2 fixation occurs with no O_2 production (since there is no PS2). In green plants, chemical products from all three sets of reactions associated with PS2 and PS1 (D_2^+, A_2^- and D_1^+, and A_1^-) are used in the dark reaction. Thus the main initial inputs for the Calvin cycle are NADPH, ATP and H^+ from the photosystems, and CO_2 and H_2O from the environment. A complex of intermediate chemicals are required within the cycle.

The overall Calvin cycle reaction is

$$6CO_2 + 18ATP + 12NADPH + 12H^+ + 12H_2O$$
$$\rightarrow \text{glucose} + 18P_i + 18ADP + 12NADPH^+ \tag{10.9}$$

Thus three ATP, two H^+ and two NADPH are required per C fixed from reduced CO_2. Each $NADPH^+$ requires two electrons from two operations of PS1 to be reduced again to NADPH. Thus the Calvin cycle is powered by four photon absorptions in PS1. The protons are made available from within the thylakoid membrane by various 'pumping' mechanisms. P_i indicates inorganic phosphate.

The first product of the Calvin cycle is a three-carbon (C_3) compound in most plants. Certain tropical plants (e.g. sugarcane, maize and sorghum) have a preliminary chemical cycle involving a C_4 compound before the Calvin cycle. These C_4 plants have two different types of photosynthetic cells that function cooperatively in the plant. In high light levels ($\sim 0.5\,\text{kW m}^{-2}$) and high temperatures ($\sim 40\,°C$) the C fixation and hence biomass production of C_4 plants may be twice that of C_3 plants.

10.6.4 *Number of photons per carbon fixed*
The main requirement for light absorption is that individual photons can be absorbed, and the energy stored for sufficient time to be used in later chemical reactions or further photon excitation. Thus each photosystem is triggered by single photons. If the molecules of one system are in operation (saturated) then it seems that the exciton can be passed to others. A minimum of four operations of PS2 are needed to produce one molecule of O_2 (i.e. four electrons have to be lifted off H_2O; Fig. 10.9). Four other photons are needed to produce the NADPH for CO_2 reduction. So in green plants with coupled PS2 and PS1, at least eight photons are needed to fix one C atom as carbohydrate. In practice it seems that more photons are needed, either because an effective chemical saturation or loss occurs, or because further ATP is required. Thus most plants probably operate at about ten photons per C fixed in optimum conditions.

10.6.5 *Energy states*
In the simplified reaction (10.1) the energy difference of C in $[CH_2O]$ and CO_2 is 4.8 eV. This averages 1.2 eV each cycle, per four cycles of PS2 and PS1 together. The 1.2 eV may be accounted for at the reaction centers by the sum of:

$-0.2\,\text{eV}$ O_2 formation by D_2^+
$+1.1\,\text{eV}$ absorption PS2
$-0.5\,\text{eV}$ link of PS2 and PS1, ATP formation
$+1.0\,\text{eV}$ absorption PS1
$-0.2\,\text{eV}$ NADPH formation

The signs of these energies are used in the sense of electron excitation of physics. The chemical redox potentials have opposite sign, but the essential meaning is the same. The zero energy is with separated CO_2 and H_2O before solutions are formed and the reactions begin.

10.6.6 *Efficiency of photosynthesis*

This may be defined in various ways. The minimum photon energy input at the outside antenna pigment molecules (i.e. *not* at the reaction center) may be given as four photons of $1.77\,eV$ (PS2 absorption for D_2 at 700 nm) and four of $1.82\,eV$ (PS1 absorption for D_1 at 680 nm), totalling $14.4\,eV$. The actual excitations D_2 to A_2, and D_1 to A_1, are about $1.1\,eV$ each. So four operations of each require $8.8\,eV$. The outputs may be considered to be four electrons lifted from H_2O to NADP over redox potential $1.15\,eV$ ($4.60\,eV$), plus three ATP molecules at $0.34\,eV$ each ($1.02\,eV$), to give a total output of $5.6\,eV$. The output may also be considered as one O_2 molecule, and one C atom fixed in carbohydrate, requiring $4.8\,eV$.

A reasonable maximum efficiency from light absorption to final product can be taken as $4.8/14.4 = 33\%$. The higher figures $5.6/14.4$, $5.6/8.8$ and $4.8/8.8$ may sometimes be considered.

In discussing photon interactions, the unit of einstein is often used. One einstein is Avogadro's number of photons of the same frequency, i.e. one mole of identical or similar photons.

However the solar spectrum consists of many photons of too low energy to be photosynthetically active ($\lambda > 700$ nm, $h\nu < 1.8\,eV$), and photons of greater energy than the minimum necessary ($h\nu > 1.8\,eV$). The situation is very similar to that with photovoltaic cells (see Fig. 7.12), so that only about 50% of the energy of the photons photosynthetically absorbed is used to operate PS2 and PS1. This effect would reduce the maximum efficiency to about 16%. However, leaves are not black and there is considerable reflection and transmission, reducing the maximum efficiency to about 12%. Efficiencies near to this have been obtained in controlled laboratory conditions, but are not reached in even the best agriculture, or in natural conditions (Table 10.1).

10.7 Synthetic photosynthesis

Technology continually advances from fundamental studies in science. It is likely that the same process will follow the eventual full understanding of photosynthesis in its many varied details. This section speculates about future applications.

10.7.1 *Plant physiology and biomass*

Plants will be selected and bred to be better suited to their environment. Examples have already resulted from the understanding of why C_4 plants are more productive in high light levels than C_3 plants (see Section 10.6). As biomass energy becomes more important (Chapter 11), plants will be developed to optimize fuel supplies rather than just for their fruit, grain or similar part product. It is even conceivable that artificial carbon based structures, materials and food could be manufactured from synthetically controlled forms of photosynthesis.

10.7.2 *Hydrogen production*

The free protons formed during the operation of the photosynthetic cycles may, in some instances, be emitted as hydrogen gas. The H^+ ions are reduced (have an electron added) from the acceptor A_1^- of PS1. This reaction is aided by certain natural enzymes (hydrogenases) and does occur naturally, e.g. with the bacteria *bacteriorhodopsin* and in human metabolism associated with digestion deficiencies. In general, however, hydrogenases are inoperative in the presence of O_2 and, if H_2 is emitted, the concentration is extremely low. Nevertheless, the prospect of producing considerable amounts of hydrogen from chemical reactions activated by sunlight is so important commercially that vigorous research is essential.

10.7.3 *Photochemical electricity production*

The driving function of photosynthesis is the photon induced molecular excitations of PS2 and PS1. These involve electronic excitations within molecular structure, and are not exactly comparable with electron–hole separation across the band gap photovoltaic devices. Nevertheless these molecular excitons have sufficient energy to drive an external electric circuit, and eventually a method may be found to link the processes. The advantage over conventional photovoltaics might be the manufacture of the base material by liquid chemical methods in bulk and the possible link with electrolytic battery storage. Research and development of photochemistry is now of prime interest.

Problem

10.1 Calculate very approximately how many trees are necessary to produce the oxygen used to maintain the per caput total fuel consumption of your country. Compare this with the number of trees per caput in your country.

Bibliography

Clayton, R. K. (1980) *Photosynthesis: Physical Mechanisms and Chemical Patterns*, Cambridge University Press.
  An excellent explanation of physical and chemical processes. Reviews current research progress and strategy. Strongly recommended.
Govinjee. (1975) *Bioenergetics of Photosynthesis*, Academic Press, New York.
  A set of papers useful in their own right, and as citation sources for ongoing literature searches.
Hall, D. O. and Rao, K. K. (1978) *Photosynthesis*, Edward Arnold, London.
  A short and stimulating introduction with more physical bias than many others.
Keeton, W. T. (1980) *Biological Sciences*, Norton, New York.
Monteith, J. (1973) *Principles of Environmental Physics*, Edward Arnold, London.
  Considers the physical interaction of plant and animal life with the environment. Chemical aspects are not considered. Of background relevance to photosynthesis.

Morowitz, H. J. (1969) *Energy Flow in Biology: Biological Organization as a Problem in Thermal Physics*, Academic Press, New York.

A stimulating book considering physical processes and constraints.

Ward, R. L. (1980) 'Biological energy conversion – the photosynthetic process', in Dickinson, W. C. and Cheremisinoff, P. N. (eds) *Solar Energy Technology Handbook*, Marcel Dekker, New York.

11 *Biofuels*

11.1 Introduction

The material of plants and animals is called *biomass*. It is organic carbon based material that reacts with oxygen in combustion and natural metabolic processes to release heat. The initial material may be transformed by chemical and biological processes to produce intermediate *biofuels* such as methane gas, ethanol liquid or charcoal solid. The initial energy of the biomass–oxygen system is captured from solar radiation in photosynthesis, as described in Chapter 10. When released in combustion the biofuel energy is dissipated, but the elements of the material should be available for recycling in natural ecological or agricultural processes, as described in Chapter 1 and Fig. 11.1. Thus the use of industrial biofuels, when linked carefully to natural ecological cycles, may be nonpolluting and sustainable. Such systems are called *agro-industries*, of which the most established are the sugarcane and forest products industries.

The dry matter mass of biological material cycling in the biosphere is about $250 \times 10^9 \, \text{t} \, \text{y}^{-1}$, incorporating about $100 \times 10^9 \, \text{t} \, \text{y}^{-1}$ of carbon. The associated energy bound in photosynthesis is $2 \times 10^{21} \, \text{J} \, \text{y}^{-1}$ ($0.7 \times 10^{14} \, \text{W}$). Of this, about 0.5% by weight is biomass used for human food. Biomass production varies with local conditions, and is about twice as great per unit surface area on land than at

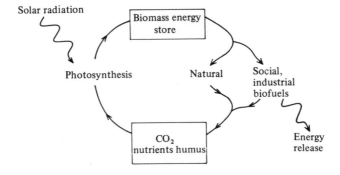

Fig. 11.1 Natural and managed biomass systems

sea. The industrial use of biomass energy may be large, e.g. about 40% of national commercial supplies in some sugarcane producing countries, where crop residues (bagasse) are burnt for process heat. The domestic use of biofuel in wood, dung and plant residues for cooking is of prime importance for about 50% of the world's population, the supply totalling about 300 GW. But if biomass is to be considered renewable, growth must at least keep pace with use. It is disastrous that forest and firewood consumption is significantly outpacing growth in ever increasing areas of the world.

The *energy storage* aspect of biomass and biofuels is of fundamental importance. All of the many processes to be described in this chapter have the aim of producing convenient fuels, at economical prices, for a full range of end uses, including liquid fuel for transport. The net energy density available in combustion ranges from about 10MJ kg^{-1} (green wood) to about 40MJ kg^{-1} (fats and oils), and 55MJ kg^{-1} for methane. Biomass is however mostly carbohydrate material with a heat of combustion of about 20MJ kg^{-1} dry matter; refer to Table B.6, Appendix B for detail.

The success of biomass systems is regulated by principles that are often not appreciated:

(1) Every biomass activity produces a wide range of products and services. For instance where sugar is made from cane, many commercial products can be obtained from the otherwise waste molasses and fiber. If the fiber is burnt, then any excess process heat can be used to generate electricity. Washings and ash can be returned to the soil as fertilizer.

(2) Some fuel products may require more energy to manufacture than they produce, e.g. ethanol from starch crops. However, such an energy deficiency need not be an economic handicap, since process energy can be available at very low cost by consuming otherwise waste material, e.g. straw, crop fiber, forest trimmings.

(3) The full economic benefit of agro-industries is likely to be widespread and difficult to assess. The terms of reference have to be carefully established. For example distinctions have to be made between:

  (a) The national benefits of increased employment in the rural sector, self-sufficiency, reduced imports etc., and

  (b) The profitability of local rural businesses having to operate for their own interests without national planning.

(4) Biofuel production is only likely to be economic if the production process uses materials *already concentrated*, probably as a byproduct and so available at low cost. Thus there has to be a supply of biomass already passing near the proposed place of production, just as hydro-power depends on a natural flow of water already concentrated by a catchment. Examples are the wastes from animal enclosures, offcuts and trimmings from sawmills, municipal sewage, husks and shells from coconuts, and straw from cereal grains. It is extremely important to identify and quantify these flows of biomass in a national or local economy *before* specifying likely biomass

developments. If no such concentrated biomass already exists as a previously established system, then the cost of biomass collection is usually too great and too complex for economic development.

(5) The main dangers of extensive biomass fuel use are deforestation, soil erosion and the displacement of food crops by fuel crops.

(6) Biofuels are organic materials, so there is always the alternative of using these materials as *chemical feedstock* or *structural materials*. For instance palm oil is an important component of soaps; plastics and pharmaceutical goods can be made from natural products; and building board can be made from plant fibers constructed as composite materials.

11.2 Biofuel classification

Biomass is largely composed of organic material and water. However, significant quantities of soil, shell or other extraneous material may be present in commercial supplies. It is essential that biomass is clearly assessed as wet or dry matter mass, and the exact moisture content should be given.

If m is the total mass of the material as it is, and m_0 is the mass when completely dried, the dry basis moisture content is $w = (m - m_0)/m_0$ and the wet basis moisture content is $w' = (m - m_0)/m$. The moisture content is in the form of extracellular and intracellular water, and so drying may be necessary (see Section 6.3). When harvested the wet basis moisture content of plants is commonly 50%, and may be as high as 90% in aquatic algae including seaweed (kelps). The material is considered 'dry' when it reaches long term equilibrium with the environment, usually at about 10% to 15% water content by mass.

Carbon based fuels may be classified by their reduction level (Section 10.4). When biomass is converted to CO_2 and H_2O, the energy made available is about 450 kJ per mole of carbon (38 MJ per kg carbon), per unit of reduction level R. This is not an exact figure because of other energy changes. Thus sugars ($R = 1$) have a heat of combustion of about 450 kJ per 12 g of carbon content. Fully reduced material, e.g. methane CH_4 ($R = 2$), has a heat of combustion of about 900 kJ per 12 g of C (i.e. per 16 g of methane).

The presence of moisture in biomass fuel often leads to a significant loss in useful thermal output, since the evaporation of water requires 2.3 MJ per kg of water.

The density of biomass, and the bulk density of stacked fibrous biomass, is important. In general three to four times the volume of dry biological material has to be accumulated to provide the same energy as coal. Thus transport and fuel handling become difficult and expensive, especially if the biofuels are not utilized at source.

We have classified seven general types of biomass energy process for full discussion in the later sections. These are as follows (see Fig. 11.2).

11.2.1 *Thermochemical*

(1) Direct *combustion* for immediate heat. Dry homogeneous input is preferred.

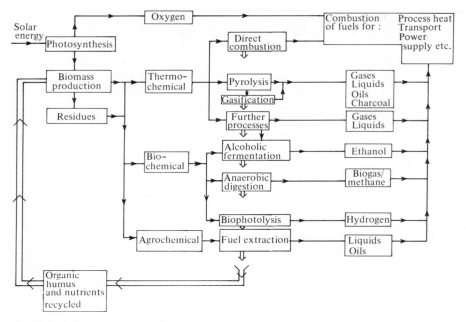

Fig. 11.2 Biofuel production processes

(2) *Pyrolysis*   Biomass heated either in the absence of air, or by the partial combustion of some of the biomass in a restricted air or oxygen supply. The products are extremely varied, consisting of gases, vapors, liquids and oils, and solid char. The output depends on temperature, type of input material and treatment process. In some processes the presence of water is necessary and therefore the material need not be dry. If output of combustible gas is the main product, the process is called *gasification*.

(3) *Other thermochemical processes*   A wide range of pretreatment and process operations are possible. These normally involve sophisticated chemical control and industrial scale of manufacture. Of particular importance are those processes that break down cellulose and starches into sugars, for subsequent fermentation.

11.2.2 *Biochemical*

(4) *Alcoholic fermentation*   Ethanol is a volatile liquid fuel that may be used in place of refined petroleum. It is manufactured by the action of micro-organisms and is therefore a fermentation process. Conventional fermentation has sugars as feedstock.

(5) *Anaerobic digestion*   In the absence of free oxygen, certain micro-organisms can obtain their own energy supply by reacting with carbon compounds of medium reduction level (see Section 10.4) to produce CO_2 and the fully reduced carbon fuel methane, CH_4. The process is also a

fermentation, but is called digestion because of the similar process that occurs in the digestive tracts of ruminant animals. The evolved mix of CO_2, CH_4 and trace gases is biogas fuel.

(6) *Biophotolysis*   Photolysis is the splitting of water into hydrogen and oxygen by the action of light. Recombination occurs when hydrogen is burnt or exploded as a fuel in air. Certain biological organisms produce, or can be made to produce, hydrogen in biophotolysis. Similar results can be obtained chemically, without living organisms, under laboratory conditions. Commercial exploitation of these effects has not yet occurred (see Section 10.7).

11.2.3 *Agrochemical*

(7) *Fuel extraction*   Occasionally liquid or solid fuels may be obtained directly from living or freshly cut plants. The materials are called exudates and are obtained by cutting into (tapping) the stems or trunks of the living plant, or by crushing freshly harvested material. A well-known similar process is the production of natural rubber latex. Related plants to the rubber plant *Herea*, such as species of *Euphorbia*, produce hydrocarbons of lower molecular weight than rubber which may be used as petroleum substitutes.

11.3 Biomass production for energy farming

This section links the discussion of photosynthesis (Chapter 10) to the production of crops. Of particular importance are the efficiencies of photosynthesis and biomass production (Sections 10.2 and 10.6).

11.3.1 *Energy farming*

We use this term in the very broadest sense to mean the production of fuels or energy as a main or subsidiary product of agriculture (fields), silviculture (forests), aquaculture (fresh and sea water), and also of industrial or social activities that produce organic waste residues (e.g. food processing, urban refuse). Table 11.1 gives some examples from an almost endless range of opportunity. The main purpose of the activity may be to produce energy (as with wood lots), but more commonly it is found best to integrate the energy and biofuel production with crop or other biomass material products.

An outstanding and established example of energy farming is the sugarcane industry (Fig. 11.3). The process depends upon the combustion of the crushed cane residue (bagasse) for powering the mill and factory operations. With efficient machinery there should be excess energy for the production and sale of byproducts (e.g. molasses, chemicals, animal feed, ethanol, fiber board, electricity). Note that these include ethanol and electricity, suitable for energy supplies to the farming and transport operations.

The variety of opportunity for energy farming has distinct advantages and disadvantages (Table 11.2). An important danger is that energy crops will substitute for food production. For example, the grain farms of the United States grow about 10% of the world's cereal crops, and export over one-third of this to

Table 11.1 Biomass supply and conversion examples

Biomass source or fuel	Biofuel produced	Conversion technology	Approx. conversion efficiency (%)	Energy required in conversion n: necessary o: optional	Approx. range of energy from biofuels MJ
Forest logging (a)	(Heat)	Combustion	70	Drying (o)	16–20 (kg wood)$^{-1}$
Timber mill residues (a)	(Heat)	Combustion	70	Drying (o)	16–20 (kg wood)$^{-1}$
Wood lot cropping (b)	Gas	Pyrolysis	85	Drying (o)	†40 (kg gas)$^{-1}$
	Oil				40 (kg oil)$^{-1}$
	Char				20 (kg char)$^{-1}$
Grain crops	Straw	Combustion	70	Drying (o)	14–16 (kg dry straw)$^{-1}$
Sugarcane juice	Ethanol	Fermentation	80	Heat (n) Electricity (o)	3–6 (kg fresh cane)$^{-1}$
Sugarcane residue	Bagasse	Combustion	65	Drying (o)	5–8 (kg fresh cane)$^{-1}$
Sugarcane total	–	–	–	–	8–14 (kg fresh cane)$^{-1}$
Animal wastes (tropical)	Methane	Anaerobic digestion	50	–	4–8 (kg dry input)$^{-1}$
Animal wastes (temperate)	Methane	Anaerobic digestion	50	Heat (n)	*2–4 (kg dry input)$^{-1}$
Municipal sewage	Methane	Anaerobic digestion	50	Heat	2–4 (kg dry input)$^{-1}$
Urban refuse	(Heat)	Combustion	50	–	5–16 (kg dry input)$^{-1}$

* This value is net, having deducted the biogas fed back to heat the boiler
† Nitrogen removed

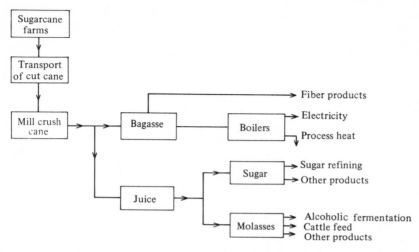

Fig. 11.3 Sugar cane agro-industry. Bagasse is plant fiber residue: molasses is sugar rich residue

grain deficient countries. A change to producing biofuels (e.g. ethanol from corn) on a large scale will therefore seriously affect world food supplies. A second major danger is that intensive energy farming will be a further pressure towards soil infertility and erosion. The obvious strategy to avoid these excesses is to always grow plants that can supply both human food (e.g. grain) and energy (e.g. straw), and to decrease dramatically the feeding of animals from crops.

Table 11.2 Advantages and dangers of energy farming

Advantages	Dangers and difficulties
Large potential supply	May lead to soil infertility and erosion
Variety of crops	May compete with food production
Variety of supplies (including transport fuel and electricity)	May encourage genetic engineering of uncontrollable organisms
Link with established agriculture and forestry	Large scale agro-industry may be too complex for efficient operation
Encourages integrated farming practice	Bulky biomass material handicaps transport to the processing factory
Efficient use of byproducts, residues, wastes	
Environmental improvement by utilizing wastes	Poorly designed, and incompletely integrated systems may produce water and air pollution
Fully integrated and efficient systems need have little air and water pollution (e.g. sulfur content low)	
Encourages rural development	
Diversifies the economy with respect to product, location and employee skill	
Greatest potential is in tropical countries, frequently of the Third World	
Establishes agro-industry that may include full range of technical tasks and processes, including the need for skilled and trained personnel	

Table 11.3 Estimated potential for biomass energy: $10^{15}\,\mathrm{J\,y^{-1}}$ ($10^{15}\,\mathrm{J\,y^{-1}} = 320\,\mathrm{MW}$). Estimated total potential biofuel resources harvested per year for various countries (after Vergara and Pimental, 1978)

Source	Sudan	Brazil	India	Sweden	USA
Animal manure	93	640	890	18	110
Sugarcane	660	1 000	430	–	420
Fuelwood	290	3 200	420	160	510
Urban refuse	5	94	320	23	170
Municipal sewage	2	11	66	1	5
Other					630
Total potential (two significant figures)	1 000	4 800	2 100	200	1 800
Present national energy consumption	180	2 700	5 800	1 500	72 000
Ratio potential to consumption	5.5	1.8	0.4	0.13	0.03

11.3.2 *Geographical distribution*

Clearly the greatest potential for energy farming occurs in tropical countries, especially those with adequate rainfall and soil condition. Table 11.3 lists estimated biomass potential for a range of countries. The challenge for development in certain Third World countries is obvious.

11.3.3 *Crop yield*

It is not possible to predict crop yields without detailed knowledge of meteorological conditions, soil type, farming practice, fertilizer use, irrigation etc. Comparison between different crops is made even more difficult by differences in growing seasons and harvesting methods. Some crops are planted annually (e.g. cereal grains) and may be cropped more than once (e.g. grasses); others are planted every few years and harvested annually (e.g. sugarcane); others may grow over long periods (e.g. more than 10 months for some varieties of cassava). Trees may grow for many years and be totally harvested (timber logging); other tree crops may grow from the continuing roots and be harvested as coppice every few years (e.g. some eucalyptus). Table 11.4 is a summary of data to provide an estimate of the maximum biofuel potential of crops in terms of heat of combustion and continuing energy supply. The data for aquatic crops assume abundant nutrients. Grasses are assumed to have frequent harvesting in the growing season. We emphasize the great unreliability of such data and the rule that such generalizations are almost always unrealistic for particular situations.

11.3.4 *Energy analysis*

Crop growth requires two forms of energy: (1) solar irradiance, and (2) energy expended in labor, fuel for tractors, and manufacturing machines and fertilizer etc. The total of the second form of energy is the gross energy requirement

Table 11.4 Maximum practical biomass yields. Total plant mass, not just the grain etc. The data are from a variety of sources and summarized by the authors. Accuracy of no more than ±25% is claimed. The majority of plants and crops yield much less than these maxima

| Crop (Assume one crop per year unless indicated otherwise) | (*) | Biomass yield $t\,ha^{-1}\,y^{-1}$ | | Energy density | Power from yield |
		Wet basis	Dry basis	$MJ\,(kg\,dry)^{-1}$	$GJ\,ha^{-1}\,y^{-1}$
Natural					
Grassland		7	3		
Forest, temperate	C_3	14	7	18	130
Forest, tropical	C_3	22	11	18	200
Forage					
Sorghum (3 crops)	R, C_4	200	50	17	850
Sudan grass (6 crops)	R, C_4	160	40	15	600
Alfalfa	C_3	40	25		
Rye grass, temperate	C_3	30	20		
Napier grass	C_4	120	80		
Food					
Cassava (60% tubers)		50	25		
Maize (corn) (50% grain)	C_4	30	25		
Wheat (35% grain)	C_3	30	22		
Rice (60% grain)	C_3		20		
Sugarbeet	C_3	45			
Sugarcane	R, C_4	100	50		
Plantation					
Oil palm	R, C_3	50	40		
Combustion energy					
Eucalyptus	R, C_3	55	20	19	380
Sycamore	R, C_3	20	10	19	190
Populus	R, C_3	18	29	19	380
Water hyacinth	C_3	300	36	19	680
Kelp (macroalgae)	C_3	250	54	21	1100
Algae (microalgae)	C_3	230	45	23	1000
Tree exudates					
Good output		1	1	40	40

*C_3, C_4: photosynthesis type (see Section 10.6). R: harvested above the root (coppiced).

(GER), and is the total of all forms of energy other than incident solar energy sequestered (used up) in producing the crop. It is best to explain the technique of energy analysis by an example.

Table 11.5 is taken from Slesser and Lewis (1979), where further references on energy analysis are given. The table lists the energy sequestered for all the inputs to make ethanol from various crop substrates. The total GER is given per kg of ethanol (row 7 of the table). Note that in most instances the energy obtainable from the final product (ethanol at $30\,MJ\,kg^{-1}$) is less than the GER, which gives the impression of a useless fuel manufacturing process, with negative net energy production (row 8).

However, the greatest amount of energy in production is associated with process heat and factory machines (rows 3, 4 and 5). Frequently all, or a large

Table 11.5 Energy analysis of ethanol production from various crop substrates. Data refer to the gross energy requirement for the crop input and each component of manufacture: unit MJ per kg of anhydrous ethanol produced. Rows 1 to 7 are from Slesser and Lewis (1979). The heat of combustion of the output ethanol is 30 MJ kg^{-1}

	Sugarcane	Cassava	Timber (enzyme hydrolysis)	Timber (acid hydrolysis)	Straw
(1) Substrate	7.3	19.2	12.7	20.0	4.4
(2) Chemicals	0.6	0.9	4.7	6.4	4.7
(3) Water pumping	0.3	0.4	0.8	0.3	0.8
(4) Electricity	7.0	10.5	176	7.8	167
(5) Fuel oil	8.0	29	42	62	42
(6) Machinery and buildings	0.5	1.2	3.3	0.6	3.3
(7) Total (1) to (6) (MJ kg^{-1})	24	61	239	98	222
(8) Net energy: (30 MJ kg^{-1}) − (7)	+8	−31	−209	−68	−192
(9) Total (1) + (2) + (6)	8.4	21	21	27	12
(10) Net energy: (30 MJ kg^{-1}) − (9)	+21	+9	+9	+3	+18

part, of this energy is available at very low cost within the factory from the combustion or pyrolysis of otherwise waste material (bagasse from sugarcane, trimmings and other waste from timber, part of the straw from cereal crops). Thus such energy supplies can be treated as low cost gains rather like solar radiation. Row 10 gives the net energy gain without the components of rows 3, 4 and 5, showing the dramatic change produced if otherwise waste material is used as a free energy gain.

In practice the energy analysis and subsequent economic analysis of biomass agro-industries is much more complicated than this simplistic approach. However, the crucial factor remains that use of low cost biomass residues for process heat and electricity production can be of overriding importance.

Energy analysis is a useful tool in assessing energy consuming and producing systems, since it emphasizes the technical aspects and choices of the processes. For instance from Table 11.5 it would seem obvious that ethanol production from sugarcane is most reasonable. However, the final choice must involve economic factors, such as the need for independent fuel supplies and the value of the alternative products.

11.4 Direct combustion for heat

Biomass is burnt to provide heat for cooking, comfort heat (space heat), crop drying, factory processes, and raising steam for electricity production and transport. In most developing countries, biomass combustion provides the largest component of total national fuel use. This is due to the extensive use of firewood for cooking and occasional heavy industrial use of biomass for sugar-cane milling, tea or copra drying, oil palm processing and paper making. By contrast, in industrialized countries, there is an entirely dominant use of fossil fuels and perhaps nuclear electricity.

Table B.6 gives the heat of combustion for a range of energy crops, residues, derivative fuels and organic products, assuming dry material. Such data are important for the industrial use of biomass fuel.

11.4.1 *Domestic cooking and heating*

About half of the world's population depends on fuelwood or other biomass for *cooking* and other domestic uses. Average daily consumption of fuel is about 0.5 to 1 kg of dry biomass per person, i.e. $10–20 \, \text{MJ} \, \text{d}^{-1} \approx 150 \, \text{W}$. Multiplied by 2×10^9 people, this represents energy usage at the very substantial rate of 300 GW.

An average consumption of 150 W 'continuous', solely for cooking, may seem surprisingly large. Such a high consumption arises from the widespread use of inefficient cooking methods, the most common of which is still an open fire. This 'device' has a thermal efficiency of only about 5%. That is, only about 5% of the heat that could be released by complete combustion of the wood reaches the interior of the cooking pot. The rest is lost by incomplete combustion of the wood, by wind and light breezes carrying heat away from the fire, and by radiation losses etc. resulting from the mismatch of fire and pot size. Consider-able energy is also wasted in evaporation from uncovered pots and from wet fuel. Smoke (i.e. unburnt carbon and tars) from a fire is evidence of incomplete combustion. Completely burnt wood – in which these tars burn in a secondary reaction – yields only CO_2 and H_2O. Moreover, the smoke is a health hazard to the cook, and there is little control over the rate at which wood is burnt.

Cooking efficiency and facilities can be improved by

(1) Introducing alternative foods and cooking methods, e.g. steam cookers
(2) Decreasing heat losses using enclosed burners or stoves, and well-fitting pots with lids
(3) Encouraging the secondary combustion of unburnt flue gases
(4) Introducing stove controls that are robust and easy to use.

With these improvements, the best cooking stoves using fuelwood and natural air circulation can place more than 20% of the combustion energy into the cooking pots. Designs using forced and actively controlled ventilation, say with an electric fan, can produce ~50% efficiency. There are now many scientifically based programs to improve cooking stoves, yet full market acceptability is seldom reached.

A parallel method for reducing fuelwood demand is to encourage alternative supplies such as biogas (methane) (see Section 11.8); fuel from crop wastes; solar cookers (refer to Section 6.8); and small scale hydro-power (Section 8.5). The need for such improvements is overwhelming when forests are dwindling and deserts increasing.

Fig. 11.4 shows two types of wood burning stoves, designed to make better use

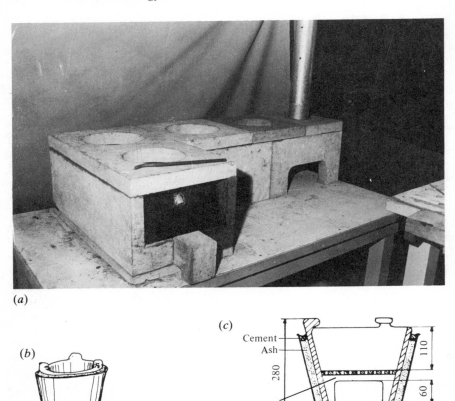

Fig. 11.4 Improved efficiency cooking stoves. (a) A large stove designed by the Fiji Ministry of Energy. It is a modification of the Indian (Hyderabad) *chula*, and is constructed mainly from concrete mouldings. Its operation is described in the text (photo: J. Richolson). (b) The 'Thai bucket' stove (sketch). (c) Vertical section through same (units millimeters). (d) Thai bucket stoves in use, in this case using charcoal as a fuel (photo: J. Richolson)

of wood as a cooking fuel. Both designs are cheap enough to allow widespread use in developing countries.

In the stove of Fig. 11.4(a) the fire is completely inclosed in the firebox at the left. The iron (dark colored) door is removed only when fuel is inserted. Air enters through a hole of adjustable size beneath the door (fully shut in the photo). Thus the rate of combustion can be closely controlled, to match the type of cooking being done. Hot gases from the fire are led through a narrow channel underneath the cooking pots (which sit in the holes on top). At this stage air could enter

(*d*)

through further channels for secondary combustion. The fully burnt gases and vapors pass to the outside environment through the chimney (on the right of the picture). This prevents pollution in the cooking place and encourages air flow.

The stove of Fig. 11.4(b) is simpler and cheaper, but has less control and less flexibility. Nevertheless its low mass means that little heat is used in heating the stove as distinct from the pot, which is an advantage for quick cooking. Air reaches the fuel from below, through a grate. Since the fire is contained, and the heat is all channeled towards the pot, the efficiency is high. This stove is well suited for use with charcoal as a fuel, since charcoal burns cleanly, without smoke.

Many of these remarks on cooking apply also to the use of biofuels for comfort (space) heat in buildings. It is important to have a controlled fire with good secondary combustion. In some systems air for combustion is introduced directly to the stove from outside the building. This decreases air circulation and heat losses in the room.

11.4.2 *Crop drying*

The drying of crops (e.g. copra, cocoa, coffee, tea, fruit) for storage is commonly accomplished by burning wood and the crop residues, or by using the waste heat from electricity generation. The material to be dried may be placed directly in the flue exhaust gases, but there is a danger of fire and contamination of food products. More commonly air is heated in a gas/air heat exchanger before passing through the crop. Drying theory is discussed in Section 6.3.

Residue combustion is a rational use of biofuel, since the fuel is close to where it is needed. Combustion in an efficient furnace yields a stream of hot clean exhaust gas ($CO_2 + H_2O$ + excess air) at about 1000°C, which can be diluted with cold air to the required temperature. In almost all cases, the improved utilization of biomass means that the flow of biomass exceeds that required for crop drying, leaving an excess for other purposes, such as industrial steam raising.

11.4.3 *Process heat and electricity*

Steam process heat is commonly obtained for factories by burning wood or other biomass residues in boilers, perhaps operating with fluidized beds. It is physically sensible to use the higher temperature to generate electricity before the heat degrades to the lower useful temperature. Thus electricity is produced and the process heat is retained. Note however that steam boilers are generally expensive. The efficiency of electricity generation is low, so little process energy is lost and a useful final temperature is maintained. Frequently the optimum operation of such processes treats electricity as a byproduct of process heat generation, with excess electricity being sold to the local electricity supply agency.

11.4.4 *Wood resource*

We emphasize again that wood is a renewable energy resource only if it is grown faster than it is consumed. Regrowth may occur in natural forest, or in manmade plantations (which usually grow faster, and are to be encouraged). The world's wood resource is consumed not just for firewood, but for sawn timber, paper making and other industrial uses. In addition, much forest is cleared for agriculture with its timber just burnt as 'waste'.

In many countries (e.g. Sudan, Kenya, Nepal) firewood consumption exceeds forest growth, so that fuelwood is a depleting resource. Moreover, the populations of firewood using countries are increasing at 2–3% p.a., thus further increasing the demand for cheap cooking fuels. To alleviate these problems requires both intensive reafforestation and a switch to more efficient cooking methods.

11.5 Pyrolysis (destructive distillation)

We use 'pyrolysis' as a general term for all processes whereby organic material is heated or partially combusted to produce secondary fuels and chemical products. The input may be wood, biomass residues, municipal waste or indeed coal. The products are gases, condensed vapors as liquids, tars and oils, and solid residue as char (charcoal) and ash. Traditional *charcoal making* is pyrolysis with the vapors and gases not collected. *Gasification* is pyrolysis adapted to produce a maximum amount of secondary fuel gases. Various pyrolysis units are shown in Fig. 11.5. Vertical top-loading devices are usually considered the best. The fuel

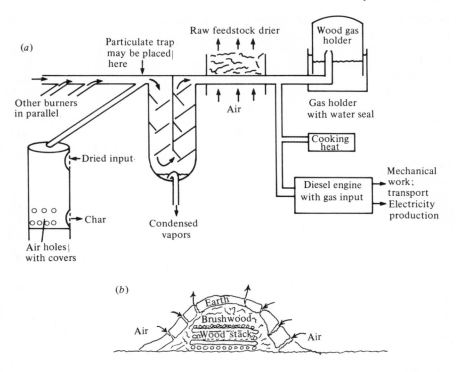

Fig. 11.5 Pyrolysis systems: (a) small scale pyrolysis unit (b) traditional charcoal kiln

products are more convenient, clean or transportable than the original biomass. The chemical products are important as chemical feedstock for further processes, or as directly marketable goods. Partial combustion devices, which are designed to maximize the amount of combustible gas rather than char or volatiles, are usually called *gasifiers*. The process is essentially pyrolysis, but may not be described as such.

Efficiency is measured as the heat of combustion of the secondary fuels produced, divided by the heat of combustion of the input biomass as used. Very high efficiencies of 80 to 90% can be reached. For instance gasifiers from wood can produce 80% of the initial energy in the form of combustible gas (largely H_2 and CO), suitable for operation in converted petroleum fueled engines. In this way the overall efficiency of electricity generation (say $80\% \times 30\% = 24\%$) could be higher than that obtained with a steam boiler. Such gasifiers are potentially useful for small scale power generation ($\lesssim 150\,kW$).

The chemical processes in pyrolysis are much related to similar distillations of coal to produce synthetic gases, tars, oils, and coke. For instance the large scale use of piped town gas ($H_2 + CO$) in Europe, before the recent change to natural gas (mainly CH_4), was possible from the reaction of water on heated coal with reduced air supply:

$$H_2O + C \rightarrow H_2 + CO$$

$$C + O_2 \rightarrow CO_2; \qquad CO_2 + C \rightarrow 2CO \tag{11.1}$$

The following is given as a summary of the wide range of conditions and products of pyrolysis. The input material needs to be graded to remove excessive non-combustible material (e.g. soil, metal), dried if necessary (usually completely dry material is avoided with gasifiers), chopped or shredded, and then stored for use. The air/fuel ratio during combustion is a critical parameter affecting both the temperature and the type of product. Pyrolysis units are most easily operated below 600 °C. Higher temperatures of 600 to 1000 °C, need more sophistication, but more hydrogen will be produced in the gas. Below 600 °C there are generally four stages in the distillation process:

(1) 100 to 120 °C   The input material dries with moisture passing up through the bed.
(2) 275 °C   The output gases are mainly N_2, CO and CO_2; acetic acid and methanol distill off.
(3) 280 to 350 °C   Exothermic reactions occur, driving off complex mixtures of chemicals (ketones, aldehydes, phenols, esters), CO_2, CO, CH_4, C_2H_6 and H_2. Certain catalysts, e.g. ZnCl, enable these reactions to occur at lower temperature.
(4) Above 350 °C   All volatiles are driven off, a higher proportion of H_2 is formed with CO, and carbon remains as charcoal with ash residues.

The condensed liquids, called tars and pyroligneous acid, may be separated and treated to give identifiable chemical products (e.g. methanol, CH_3OH, a liquid fuel). Table 11.6 gives examples and further detail.

The secondary fuels from pyrolysis have less total energy of combustion than the original biomass, but are far more convenient to use. Some of the products have significantly greater energy density (e.g. CH_4 at $55\,MJ\,kg^{-1}$) than the

Table 11.6 Pyrolysis yields from dry wood

Yields per 1000 kg dry wood	
Charcoal	300 kg
Gas (combustion $10\,465\,kJ\,m^{-3}$)	$140\,m^3$ (NTP)
Methyl alcohol	14 l
Acetic acid	53 l
Esters	8 l
Acetone	3 l
Wood oil and light tar	76 l
Creosote oil	12 l
Pitch	30 kg

average input. Convenience includes better control of combustion, easier handling and transport, greater variety of end-use devices, and less air pollution. The following general comments apply:

11.5.1 *Solid char (mass yield 25–35% maximum)*

Modern charcoal retorts operating at about 600 °C produce 25 to 35% of the dry matter biomass as charcoal. Traditional earthen kilns usually give yields closer to 10% since there is less control. Charcoal is 75 to 85% carbon, unless great care is taken to improve quality (as for chemical grade charcoal), and the heat of combustion is about $30 \, MJ \, kg^{-1}$. Thus if charcoal alone is produced from wood, between 15 and 50% of the original chemical energy of combustion remains. Charcoal is useful as a clean controllable fuel. Chemical grade charcoal has many uses in laboratory and industrial chemical processes. Charcoal is superior to coal products for making high quality steel.

11.5.2 *Liquids (condensed vapors, mass yield ~30% maximum)*

These divide between (1) a sticky phenolic tar (creosote) and (2) an aqueous liquid, pyroligneous acid, of mainly acetic acid, methanol (maximum 2%) and acetone. The liquids may be either separated or used together as a crude fuel with a heat of combustion of about $22 \, MJ \, kg^{-1}$. The maximum yield corresponds to about 400 liters of combustible liquid per tonne of dry biomass.

11.5.3 *Gases (mass yield ~80% maximum in gasifiers)*

The mixed gas output with nitrogen is known as wood gas, synthesis gas, producer gas or water gas, and has a heat of combustion in air of 5 to $10 \, MJ \, kg^{-1}$ (4 to $8 \, MJ \, m^{-3}$ at STP). It may be used directly in diesel cycle or spark ignition engines with adjustment of the fuel injector or carburetor, but extreme care has to be taken to avoid intake of ash and condensable vapors. The gas is mainly N_2, H_2 and CO, with perhaps small amounts of CH_4 and CO_2. The gas may be stored in gas holders near atmospheric pressure, but is not conveniently compressed. A much cleaner and more uniform gas may be obtained by gasification of wet charcoal rather than wood, since the majority of the tars from the original wood have been removed.

11.6 Further thermochemical processes

In the previous sections biomass has been used directly after preliminary sorting and cutting for combustion or pyrolysis. However the biomass may be treated chemically, (1) to produce material suitable for alcoholic fermentation (Section 11.7), or (2) to produce secondary or improved fuels.

Consider the following few important examples from the great number of possibilities.

11.6.1 *Hydrogen reduction*

Dispersed, shredded or digested biomass, e.g. manure, is heated in hydrogen to about 600 °C under pressure of about 50 atmospheres. Combustible gases, mostly methane and ethane, are produced that may be burnt to give about 6 MJ per kg of initial dry material.

11.6.2 *Hydrogenation with CO and steam*

As above, but heated with CO and steam to about 400 °C and 50 atmospheres. A synthetic oil is extracted from the resulting products that may be used as a fuel. A catalyst is needed to produce reactions of the following form:

$$CO + H_2O \rightarrow CO_2 + H_2$$

$$C_n(H_2O)_n + (n+1) H_2 \rightarrow nH_2O + H(CH_2)_nH \tag{11.2}$$

where the latter reaction implies the conversion of carbohydrate material to hydrocarbon oils. The energy conversion efficiency is about 65%.

11.6.3 *Acid and enzyme hydrolysis*

Cellulose is the major constituent (30% to 50%) of plant dry biomass and is very resistant to hydrolysis and hence fermentation by micro-organisms (see later sections). Conversion to sugars, which can be fermented, is possible by heating in sulfuric acid or by the action of enzymes (cellulases) of certain micro-organisms. The products may also be used as cattle feed.

11.6.4 *Conversion of coconut oils to esters as fuel*

This is an example of many chemical processes becoming available to improve biofuels. About 50% of the white 'meat' of coconuts, copra, is oil that is extracted by squeezing and rolling. Substantial trade exists in this oil, e.g. the Philippines' annual production of coconut oil is $10^6 t y^{-1}$. The oil may be used directly in diesel engines with adjustment to the injection system, but toxic fumes are produced, deposits occur and the fuel solidifies below 23 °C. Reaction with 20% by volume of ethanol or methanol produces volatile esters and glycerol. The esters are better diesel engine fuels than the oil, and the glycerol is a valuable byproduct. The ethanol can be made by fermentation from natural sugars, and the methanol via gasification from wood wastes. The energy density of the esters is about $38 MJ kg^{-1}$, which is higher than the oil and near to petroleum at about $46 MJ kg^{-1}$. Other vegetable oils can be used similarly in diesel cycle engines.

11.6.5 *Methanol liquid fuel*

Methanol, a toxic liquid, is made from the catalytic reaction of H_2 and CO at 330 °C and at 150 atmosphere pressure:

$$2H_2 + CO \rightarrow CH_3OH \tag{11.3}$$

The gases are the components of synthesis gas (Section 11.5) and may be obtained from gasification of biomass. Methanol may be used as a liquid fuel in petroleum engines with an energy density of $23 MJ kg^{-1}$.

11.7 Alcoholic fermentation

11.7.1 *Alcohol production methods*

Ethanol, C_2H_5OH, is produced naturally by certain micro-organisms from sugars under acidic conditions, pH 4 to 5. This alcoholic fermentation process is used world wide to produce alcoholic drinks. The most common micro-organism, the yeast *Saccharomyces cerevisiae*, is poisoned by C_2H_5OH concentration greater than 10%, and so higher concentrations up to 95% are produced by distilling and fractionating (Fig. 11.6). When distilled, the remaining constant

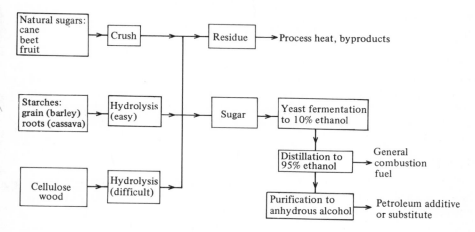

Fig. 11.6 Ethanol production

boiling point mixture is 95% ethanol, 5% water. Anhydrous ethanol is produced commercially with azeotropic removal of water by codistillation with solvents such as benzene. Only about 0.5% of the energy potential of the sugars is lost during fermentation, but significant amounts of process heat are required for the concentration and separation processes. This process heat may be provided by the combustion or gasification of otherwise waste biomass.

The sugars may be obtained by the following routes, listed in order of increasing difficulty.

(1) *Directly from sugarcane* Usually commercial sucrose is removed from the cane juices, and the remaining molasses used for the alcohol production process (Fig. 11.3). These molasses themselves have about 55% sugar content. If the molasses have low commercial value, then ethanol production from molasses has favorable commercial possibilities, especially if the cane residue (bagasse) is available to provide process heat.

In this case the major reaction is sucrose to ethanol:

$$C_{12}H_{22}O_{11} + H_2O \overset{\text{yeast}}{\rightarrow} 4C_2H_5OH + 4CO_2 \tag{11.4}$$

In practice the yield is limited by other reactions and the increase in mass of yeast. Commercial yields are about 80% of those predicted by (11.4). The fermentation reactions for other sugars (e.g. glucose, $C_6H_{12}O_6$) are very similar.

(2) *Sugar beet* provides sugars for fermentation, but there is little opportunity to obtain process heat from the crop residue, so ethanol production is expensive.

(3) *Starch crops*, e.g. grain and cassava, can be hydrolyzed to sugars. Starch is the main energy storage carbohydrate of plants, and is composed of two large molecular weight components, amylose and amylopectin. These large molecules are essentially linear and branched chains of glucose molecules, linked by distinctive carbon bonds. These links can be broken by enzymes from malts associated with specific crops, e.g. barley or corn, or enzymes from certain moulds (fungi). Such methods are common in whisky distilleries, corn syrup manufacture, and in obtaining ethanol from cassava roots. The links can also be broken by acid treatment at pH 1.5 and 2 atmospheres pressure, but yields are low and the process more expensive than enzyme alternatives. An important byproduct of the enzyme process is the residue used for cattle feed or soil conditioning.

(4) *Cellulose* comprises about 40% of all biomass dry mass and is potentially a vast renewable energy source. It has a polymer structure of linked glucose molecules, and so forms the main mechanical structure component of the woody parts of plants. The links are considerably more resistant to breakdown under hydrolysis than in starch. Cellulose is found in close association with lignin in plants, which discourages hydrolysis to sugars. Acid hydrolysis is possible as with starch, but the process is expensive and energy intensive. Hydrolysis is less expensive, and less energy input occurs, with the enzymes of natural wood rotting fungi, but the process is slow. Prototype commercial processes use pulped wood or, more preferably, old newspaper as input. The physical breakdown of woody material is the most difficult and expensive stage, usually requiring much electricity for the rolling and hammering machines.

11.7.2 *Ethanol fuel use*

Liquid fuels are of extreme importance because of the ease of handling and the controlled combustion in engines. Ethanol may be a direct substitute or additive for petrol (gasoline) in two ways: (1) 95% (hydrous) ethanol can be used directly in modified engines, and (2) 100% (anhydrous) ethanol can be mixed with dry petrol to produce gasohol, usually a 10% anhydrous ethanol mix with 90%

petrol. Note that water does not mix with petrol, and so is often present as a sludge in petroleum storage tanks without causing difficulty. This cannot be tolerated for gasohol storage, however.

Anhydrous ethanol is a liquid between $-117\,°C$ and $+78\,°C$, flash point $130\,°C$, ignition temperature $423\,°C$. However, the low vapor pressure of the anhydrous ethanol requires special carburation for internal combustion engines. For this reason it is more common to mix up to 20% by volume of anhydrous ethanol with petrol and to use this mixture, gasohol, in regular (preferably retuned) petrol engines. Gasohol is now standard in Brazil (ethanol mostly from sugarcane, but some from cassava) and common in the USA (ethanol from corn grain). The ethanol additive has antiknock properties and is preferable to the more common tetraethyl lead, which produces serious air pollution. The excellent combustion properties of ethanol enable an engine to produce up to 20% more power with ethanol than with petroleum. The mass density and calorific value of ethanol are both less than those of petroleum, so the energy per unit volume of ethanol $(24\,GJ\,m^{-3})$ is 40% less than for petroleum $(39\,GJ\,m^{-3})$ (see Table B.6). However, the better combustion properties of ethanol compensate, so that in practice some trials have shown that a car could use about the same volume of ethanol, gasohol or petrol.

Production costs of ethanol fuels depend greatly on local conditions and the prices paid for alternative products. Government policy and tax concessions are extremely important. In general, in favorable circumstances, ethanol fuel prices may be competitive with petroleum (1984 prices).

11.7.3 *Ethanol production from crops*

Table 11.7 summarizes some outline data of ethanol production and crop yield. Commercial operations depend on many other factors, including energy analysis (see Section 11.3.4 and Table 11.5) and economic analysis. We emphasize that the use of otherwise waste biomass residues for electricity production and factory process heat will be crucial in these analyses.

Table 11.7 Ethanol yields of various crops, based on average yields in Brazil (except for corn, which is based on US yields). Two crops a year are possible in some areas. Cassava yields could be boosted to $3600\ \text{liter}\,ha^{-1}\,y^{-1}$ by improved cultivation technologies

	Liters of ethanol per tonne of crop	Liters of ethanol per hectare per year
Sugarcane	70	3500
Cassava	180	2160
Sweet sorghum	86	3010
Sweet potato	125	1875
Corn (maize)	370	2220
Wood	160	3200

11.8 Anaerobic digestion for biogas

11.8.1 *Introduction*

Decaying biomass and animal wastes are broken down to elementary nutrients and soil humus by decomposer organisms, fungi and bacteria. The processes are favored by wet, warm and dark conditions. The final stages are accomplished by many different species of bacteria classified as either aerobic or anaerobic. Aerobic bacteria are favored in the presence of oxygen with the biomass carbon being fully oxidized to CO_2. In closed conditions with no oxygen available from the environment, anaerobic bacteria are able to exist by breaking down general carbohydrate material. The carbon may be ultimately divided between fully oxidized CO_2 and fully reduced CH_4 (see equation 10.6). Nutrients such as

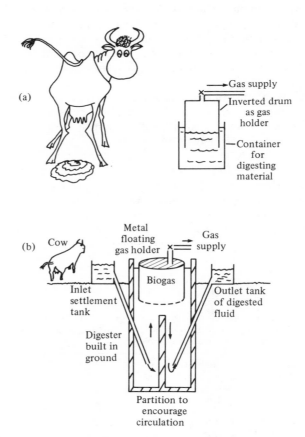

Fig. 11.7 Biogass digesters, four basic designs: (a) oil drum batch digester (b) typical Indian gobar gas digester (c) Chinese pit design (from Van Buren, 1979) (d) high rate farm digester for use in mid-latitude climates where heating is necessary (from Meynell, 1976)

soluble nitrogen compounds remain available in solution, so providing excellent fertilizer and humus. Being accomplished by micro-organisms, the reactions are all classed as fermentations, but in anaerobic conditions leading to methane the term 'digestion' is often preferred.

Biogas is the CH_4/CO_2 gaseous mix evolved from purpose built digesters constructed and controlled to favor methane production (Fig. 11.7). The energy available from the combustion of biogas is between 60 and 90% of the dry matter heat of combustion of the input material. However, the gas is obtained from slurries of 95% water, so in practice the energy is often available where none would otherwise have been obtained. Another, perhaps dominant, benefit is that the digested effluent forms significantly less of a health hazard than the input material. Note however that *not* all parasites and pathogens are destroyed in the digestion.

The economics and general benefit of biogas are always most favorable when

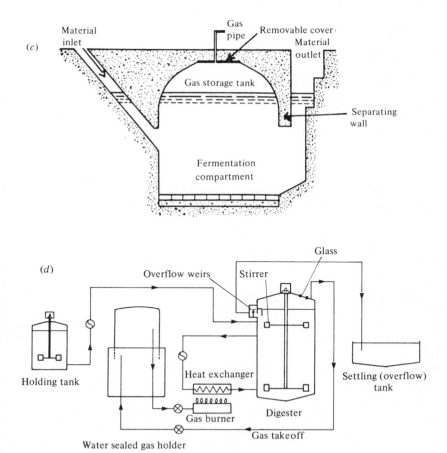

the digester is placed in a flow of waste material already present. Examples are sewage systems, piggery washings, cattle shed slurries, abattoir wastes, food processing residues and municipal refuse landfill dumps. The economic benefits are that input material does not have to be specially collected, administrative supervision is present, waste disposal is improved, and uses are likely to be available for the biogas and nutrient rich effluent.

Biogas generation is suitable for small to large scale operation. It is particularly attractive for *integrated farming*, where the aim is to emulate a full ecological cycle on the single farm. Thus plant and animal wastes are digested, with the effluent passing for further aerobic digestion in open tanks. The biogas is used for lighting, machines, vehicles, generators, and domestic and process heat. Algae may be grown on the open air tanks and removed for cattle feed. From the aerobic digestion, the fully treated effluent passes to fish tanks and duck ponds before finally being passed to the fields as fertilizer. Success for such schemes depends ultimately on successful total integrated design, high standards of construction, and regular maintenance.

11.8.2 *Basic processes and energetics*
The general equation for anaerobic digestion is

$$C_xH_yO_z + (x - y/4 - z/2)H_2O$$
$$\rightarrow (x/2 - y/8 + z/4)CO_2 + (x/2 + y/8 - z/4)CH_4 \tag{11.5}$$

For cellulose this becomes

$$(C_6H_{10}O_5)_n + nH_2O \rightarrow 3nCO_2 + 3nCH_4 \tag{11.6}$$

Some organic material (e.g. lignin) and all inorganic inclusions do not digest. These add to the bulk of the material, form a scum and can easily clog the system. In general 95% of the mass of the material is water.

The reactions are slightly exothermic, with typical heats of reaction being about 1.5 MJ per kg dry digestible material, equal to about 250 kJ per mole of $C_6H_{10}O_5$. This is not sufficient to significantly affect the temperature of the bulk material.

If the input material had been dried and burnt, the heat of combustion would have been about 16 MJ kg^{-1}. Only about 10% of the potential heat of combustion need be lost in the digestion process. This is 90% conversion efficiency. Moreover very wet input has been processed to give a highly convenient and controllable gaseous fuel, whereas drying of 95% aqueous input would have taken a further 40 MJ per kg of solid input. In practice digestion is seldom left to go to completion because of the long time involved, and 60% conversion is common. Gas yield is about 0.2 to 0.4 m^3 per kg dry digestible input at STP, with throughput of about 5 kg dry digestible solid per m^3 of liquid.

It is generally considered that three ranges of temperature favor particular types of bacteria. Digestion at higher temperature proceeds more rapidly than at lower temperature, with gas yield rates doubling at about every 5°C increase. The temperature ranges are (1) psicrophilic, about 20°C, (2) mesophilic, about 35°C, and (3) thermophilic, about 55°C. In tropical countries unheated digesters are likely to be at average ground temperature between 20 and 30°C. Consequently the digestion is psicrophilic, with retention times being at least 14 days. In colder climates the digesters have to be heated, probably by using part of the biogas output, and a temperature of about 35°C is likely to be chosen. Few digesters operate at 55°C unless the purpose is to digest material rather than produce excess biogas.

The *biochemical processes* occur in three stages, each facilitated by distinct sets of anaerobic bacteria:

(1) Insoluble biodegradable materials, e.g. cellulose, polysaccharides and fats, are broken down to soluble carbohydrates and fatty acids. This occurs in about a day at 25°C in an active digester.
(2) Acid forming bacteria produce mainly acetic and propionic acid. This stage likewise takes about one day at 25°C.
(3) Methane forming bacteria slowly, in about 14 days at 25°C, complete the digestion to $\sim$70% CH_4, $\sim$30% CO_2 with trace amounts of H_2 and perhaps H_2S. H_2 may play an essential role, and indeed some bacteria (e.g. *Clostridium*) are distinctive in producing H_2 as the final product.

The methane forming bacteria are sensitive to pH, and conditions should be mildly acidic (pH 6.6 to 7.0) and certainly not below pH 6.2. Nitrogen should be present at 10% by mass of dry input, and phosphorus at 2%. A golden rule for successful digester operation is to *maintain constant conditions* of temperature and suitable input material. As a result a suitable population of bacteria is able to become established to suit these conditions.

11.8.3 *Digester sizing*

The energy available from a biogas digester is given by:

$$E = \eta H_b V_b \tag{11.7}$$

η is the combustion efficiency of burners, boilers etc. ($\sim$60%). NB: some of the heat of combustion of the methane goes to heating the CO_2 of the biogas, and is therefore unavailable for other purposes. The net effect is to decrease the efficiency. H_b is the heat of combustion per unit volume biogas ($20\,MJ\,m^{-3}$ at 10 cm water gauge pressure, 0.01 atmosphere) and V_b is the volume of biogas. Alternatively

$$E = \eta H_m f_m V_b \tag{11.8}$$

H_m is the heat of combustion of methane ($56\,MJ\,kg^{-1}$, $28\,MJ\,m^{-3}$ at STP) and f_m is the fraction of methane in the biogas ($\sim$0.7).

The volume of biogas is given by

$$V_b = cm_0 \tag{11.9}$$

c is the biogas yield per unit dry mass of whole input (0.2 to $0.4\,\mathrm{m^3kg^{-1}}$) and m_0 is the mass of dry input (e.g. $2\,\mathrm{kg\,d^{-1}}$ per cow).

The volume of fluid in the digester is given by

$$V_f = m_0/\rho_m \tag{11.10}$$

where ρ_m is the density of dry matter in the fluid ($\sim 50\,\mathrm{kg\,m^{-3}}$).

The volume of the digester is given by

$$V_d = \dot{V}_f t_r \tag{11.11}$$

where $\dot{V}_f$ is the flow rate of the digester fluid and t_r is the retention time in the digester (~ 8 to 20 days).

Example 11.1
Calculate (1) the volume of a biogas digester suitable for the output of four cows, (2) the power available from the digester. Retention time is 14 days, temperature 30°C, dry matter consumed $2\,\mathrm{kg\,d^{-1}}$, biogas yield $0.24\,\mathrm{m^3kg^{-1}}$, burner efficiency 0.6, methane proportion 0.8.

Solution

$$m_0 = (2\,\mathrm{kg\,d^{-1}})(4) = 8\,\mathrm{kg\,d^{-1}} \tag{11.12}$$

From (11.10), fluid volume is

$$V_f = \frac{(8\,\mathrm{kg\,d^{-1}})}{(50\,\mathrm{kg\,m^{-3}})} = 0.16\,\mathrm{m^3\,d^{-1}} \tag{11.13}$$

In (11.11), digester volume is

$$V_d = (0.16\,\mathrm{m^3\,d^{-1}})(20\,\mathrm{d}) = 3.2\,\mathrm{m^3} \tag{11.14}$$

From (11.9),

$$V_b = (0.24\,\mathrm{m^3kg^{-1}})(8\,\mathrm{kg\,d^{-1}}) \tag{11.15}$$

In (11.8),

$$E = (0.6)(28\,\mathrm{MJ\,m^{-3}})(0.8)(1.9\,\mathrm{m^3\,d^{-1}})$$

$$= 26\,\mathrm{MJ\,d^{-1}} = 7.1\,\mathrm{kWh\,d^{-1}} \tag{11.16}$$

$$= 300\,\mathrm{W}\ (\text{continuous, thermal})$$

11.8.4 *Working digesters*

Fig. 11.7 shows designs for four basic biogas digesters as examples.

(1) *Household unit for the tropics* This is the most simple design, comprising an upturned metal cylinder in another larger tank, e.g. a 200 liter oil drum

with the top removed (Fig. 11.7(a)). It is suitable for batch use, being filled each occasion with fresh animal manure (seeded if possible with anaerobic bacteria from a previous active digester). Systems like this are messy and usually do not last for more than a brief period of enthusiasm. The biogas is trapped in the top cylinder to be piped to the household for cooking and lighting.

Single batch treatment does not give a constant yield, however, so a continuous process is preferable.

(2) *Indian gobar gas system*   *Gobar* means cow dung, and is the word used for the sun dried cow pats used in tropical countries and previously in Europe for cooking fuel. Since about 1939 there has been experience of household and village digesters in India, and there are now about 100000 installations. Fig. 11.7(b) gives a standard design. Material is placed in the inlet settlement tank to separate out nondigestible straw and inclusions. The flow moves slowly through the buried brick tank in about 14 to 30 days to the outlet, from which nutrient rich fertilizer is obtained. Gas pressure of ~10cm water column is maintained by the heavy metal gas holder, which is the most expensive item of the digester. The holder is lifted regularly (~six-monthly intervals) so that any thick scum at the top of the fluid can be removed. Daily inspection of pipes etc. and regular maintenance is essential. Lack of maintenance is the predominant reason for the failure of biogas digesters generally. (See Khadi and Village Industries Commission, 1975.)

(3) *Chinese digester*   Fig. 11.7(c) is a recommended design in the Republic of China for households and village communes. The main feature of the design is the permanent concrete cap, which is far cheaper than a heavy metal floating gas holder. As the gas evolves, its volume replaces digester fluid and the pressure increases.

(4) *Industrial design*   Fig. 11.7(d) shows a design for commercial operation under fully controlled conditions. The digester tank is usually heated to at least 35°C.

11.9 Agrochemical fuel extraction

We use this title to describe the production of fuels from plants while the plant usually remains alive and unharmed. It is also possible for the whole plant to be harvested for fuel extraction. We do not include the taking of plant branches or stems for combustion while leaving the roots intact (cropping), which is included in Section 11.4. The fuels are mostly natural oils and solvents, which it is hoped will require a minimum of secondary processing.

Categories of suitable materials, with examples, are:

(1) Seeds: sunflower with 50% oil.
(2) Nuts: oil palm; coconut copra to 50% by mass of oil.
(3) Fruits: olive.
(4) Leaves: eucalyptus with 25% oil.
(5) Stems, roots remain: see Section 11.4.

(6) Tapped exudates: rubber latex; jojoba, *Simmondsia chinensis* tree oil.
(7) Harvested plants: oils and solvents to 15% of the tree dry mass, e.g. turpentine, rosin, oleoresins from pine trees; oil from *Euphorbia*.

Agrochemical fuel extraction has been strongly favored by Nobel laureate Melvin Calvin (of the Calvin cycle, Section 10.6.3). The subject is well surveyed by Bungay (1981). The viability of agrochemical fuel farms is hotly debated, and frequently the material is more valuable for its chemical properties than as a fuel. It is likely that fuels suitable for vehicles will be the most successful.

Research station production has given encouraging results, (e.g. *Euphorbia* oil at $10\,t\,ha^{-1}y^{-1}$; the Brazilian tree *Compaifera langsdorffia*, producing $45\ 1y^{-1}$ of tapped oil), but production on marginal land is much less.

The advantages and disadvantages may be summarized as follows:

Disadvantages
(1) Small yield, average $2\,t\,ha^{-1}y^{-1}$ to a maximum of $10\,t\,ha^{-1}y^{-1}$ from a standing crop of $\sim 40\,t\,ha^{-1}y^{-1}$.
(2) Displaces food production unless on marginal land.
(3) Much labor required (employer's viewpoint).

Advantages
(1) Good chemical feedstock for high quality fuels (also has chemical product value).
(2) Ecologically sound, since the plants remain; heavy machines are not used, land may be improved.
(3) Intercropping or cattle farming may be integrated.
(4) Much labor required (employee's viewpoint).

Problems

11.1 A farmer with 50 pigs proposed to use biogas generated from their wastes to power his motor car.
  (a) Discuss the feasibility of doing this. You should calculate both the energy content of gas and the energy used in compressing the gas to a usable volume, and compare these with the energy required to run the car.
  (b) Briefly comment on what other benefits (if any) he might gain by installing a digester.
  You may assume that a 100 kg pig excretes about 0.5 kg of volatile solids (VS) per day (plus 6 kg water), and that 1 kg of VS yields $0.4\,m^3$ of biogas at STP.

11.2 Recent studies show that the major energy consumption in Fijian villages is wood which is used for cooking on open fires. Typical consumption of wood is $1\,kg\,person^{-1}day^{-1}$.

  (a) Estimate the heat energy required to boil a 2 liter pot full of water. Assuming this to be the cooking requirement of each person, compare this with the heat content of the wood, and thus estimate the thermal efficiency of the open fire.

(b) How much timber has to be felled each year to cook for a village of 200 people?

Assuming systematic replanting, what area of crop must the village therefore set aside for fuel use if it is not to make a net deforestation? *Hint:* refer to Table 11.4.

(c) Comment on the realism of the assumptions made, and revise your estimates accordingly.

11.3 (a) A bag of total volume $3\,m^3$ is used as a biogas digester. Each day it is fed an input of $0.2\,m^3$ of slurry, of which $4\,kg$ is volatile solids, and a corresponding volume of digested slurry is removed. (This input corresponds roughly to the waste from 20 pigs.)

Assuming that a typical reaction in the digestion process is

$$C_{12}H_{22}O_{11} + H_2O \xrightarrow{\text{bacteria}} 6CH_4 + 6CO_2$$

and that the reaction takes 7 days to complete, calculate (i) the volume of gas (ii) the heat obtainable by combustion of this gas for each day of operation of the digester. Finally, (iii) how much kerosene would have the same calorific value as one day's biogas?

(b) The reaction rate in the digester can be nearly doubled by raising the temperature of the slurry from 28 °C (ambient) to 35 °C. (i) What would be the advantage of doing this? (ii) How much heat per day would be needed to achieve this? (iii) What proportion of this could be contributed by the heat evolved in the digestion reaction?

11.4 (a) Write down a *balanced* chemical equation for the conversion of sucrose $(C_{12}H_{22}O_{11})$ to ethanol (C_2H_5OH). Use this to calculate how much ethanol could be produced in theory from one tonne of sugar. What do you think would be a realistic yield?

(b) Fiji is a small country in the South Pacific, whose main export crop is sugar. Fiji produces $300\,000\,t\,y^{-1}$ of sugar, and imports $300\,000\,t$ of petroleum fuel. If all this sugar were converted to ethanol, what proportion of petroleum imports could it replace?

Solutions

11.1 (a) Gas yield $200\,MJ\,day^{-1}$. Car requires $4\,l$ petrol $day^{-1} = 160\,MJ\,day^{-1}$. Compressor work $\bar{p}V \sim 30\,MJ$.

(b) See text.

11.2 (a) $mc\Delta T \approx 0.6\,MJ$ (heat losses from pot imply actual requirement is higher). $\eta \approx 3\%$.

(b) 70 tonnes; 7 ha.

11.3 (a) (i) $3\,m^3$ gas (ii) $63\,MJ$ (iii) from $1.7\,l$ kerosene.
  (b) (i) Smaller tank, smaller cost. (ii) Heat required $6\,MJ\,day^{-1}$. (iii) Heat
     evolved $0.3\,MJ\,(mole\ sucrose)^{-1} = 3.6\,MJ\,day^{-1}$.

11.4 (a) $680\,l$ at 100% yield.
  (b) About 90%, if suitable machinery available.

Bibliography

Anderson, R. E. (1979) *Biological Paths to Self Reliance*, Van Nostrand Reinhold, New
  York.
  Important book, also includes the wider implications of biomass development (e.g.
  social, ecological), with a strong awareness of the need and opportunities in developing
  countries. Includes small scale applications, basic science, details on particular crops.
Bungay, H. R. (1981) *Energy, the Biomass Options*, John Wiley and Sons, New York.
  Comprehensive account with emphasis on commercial utilization of favorable pro-
  cesses. Stronger on the biological and chemical aspects than the physical. Reviews
  economics of processes, which are mostly large scale as applicable to the USA.
Hall, D. D., Barnard, G. W. and Moss, P. A. (1982). *Biomass for Energy in Developing
  Countries*, Pergamon Press, Oxford.
  Comprehensive but wordy, without much theoretical detail.
Kovarik, B. (1981) *Fuel Alcohol*, International Institute for Environment and Develop-
  ment, London.
Lewis, C. W. (1983) *Biological Fuels*, Edward Arnold, London.
  Excellent basic text at a low price.
San Pietro, A. (ed.) (1980) *Biochemical and Photosynthetic Aspects of Energy Pro-
  duction*, Academic Press, London.
  Useful articles by different authors, good basic scientific data for mechanized processes
  especially of biomass potential (article by D. O. Hall), anaerobic fermentation,
  seaweed aquaculture.
Slesser, M. and Lewis, C. (1979). *Biological Energy Resources*, E. and F. N. Spon,
  London.
  Recommended for microbiological information and energy analysis of processes.
Vergara, W. and Pimental, D. (1978) 'Fuels from biomass', in Auer, P. (ed.), *Advances in
  Energy Systems and Technology*, vol. 1, Academic Press, New York, pp. 125–73.
  Reviews biomass application, particularly for a range of countries including tropical,
  less developed countries. Useful references and some case studies.

Direct combustion, especially wood fuel (Section 11.4)
De Lepeleire, G., Prasad, K. K., Verhaart, P. and Visser, P. (1981) *A Woodstove
  Compendium*, Eindhoven University, Holland.
  Gives principles of woodburning, and technical descriptions of many stoves designed
  for domestic cooking in developing countries.
Earl, D. E. (1975) *Forest Energy and Economic Development*, Oxford University Press.
  Concentrates on the resource and its potential, especially in developing countries.
Openshaw, K. (1978) *Woodfuel: A Time for Reassessment, Natural Resources Forum*,
  vol. 3, United Nations.
  Describes pressure on world forest resources.

Shelton, J. and Shapiro, A. B. (1976) *The Woodburner's Encyclopedia*, Vermont Crossroads Press.
  Excellent account of physical principles. Applications emphasis is on heating in cold climates.

Biogas (Section 11.8)

Khadi and Village Industries Commission (1975) *Gobar Gas – why and how*, Bombay.
  A short manual written from practical experience.
Meynell, P. J. (1976) *Methane – Planning a Digester*, Prism Press, Dorchester, UK.
  A practical book with basic technical and biochemical explanations.
Pyle, D. L., Imperial College, London. Various excellent publications, e.g. (1979) Paper at the Conference on Small Scale Energy for Developing Countries, Reading University and Intermediate Technology Development Group.
Van Brakel (1980) *The Ignis Fatuus of Biogas*, Delft University Press, Delft.
  A careful and critical review of pre-1970 literature, so drawing on much practical experience.
Van Buren, A. (1979) *A Chinese Biogas Manual*, Intermediate Technology Publications, London.
  A stimulating and useful handbook, based on the considerable experience of small scale digesters in rural China.

12 *Wave energy*

12.1 Introduction

Very large energy fluxes can occur in deep water sea waves. The power in the wave is proportional to the square of the amplitude and to the period of the motion. Therefore the long period (~ 10s), large amplitude (~ 2m) waves have considerable interest for power generation, with energy fluxes commonly averaging between 50 and 70 kW per meter width of oncoming wave.

The possibility of generating electrical power from these deep water waves has been realized for many years, and there are countless ideas for machines to extract the power. In recent years interest has revived, particularly in Japan, Britain and Scandinavia, so research and small scale development has progressed to the stage of commercial construction for meaningful power extraction. As with all renewable energy supplies the scale of operation has to be determined, and present trends support moderate power generation levels at about 1 MW from modular devices about 50 m wide across the wavefront. Such devices should be economic to replace diesel generated electricity, especially on islands.

It is important to appreciate the many difficulties facing wave power developments. These will be analyzed in later sections, but may be summarized here:

(1) Wave patterns are irregular in amplitude, phase and direction. It is difficult to design devices to extract power efficiently over the wide range of variables.

(2) There is always a probability of extreme gales or hurricanes producing waves of freak intensity. The structure of the power devices must be able to withstand this. Commonly the 50 year peak wave is 10 times the height of the average wave. Thus the structures have to withstand ~ 100 times the power intensity to which they are normally matched. Allowing for this is expensive and will probably reduce normal efficiency of power extraction.

(3) Peak power is generally available in deep water waves from open sea swells, e.g. beyond the Western Islands of Scotland in one of the most tempestuous areas of the North Atlantic. The difficulties of constructing power devices for these types of wave regimes, of maintaining and fixing or mooring them in position, and of transmitting power to land, are fearsome.

(4) Wave periods are commonly ~5 to 10 s (frequency ~0.1 Hz). It is extremely difficult to couple this irregular slow motion to electrical generators requiring ~500 times greater frequency.

(5) So many types of device may be suggested for wave power extraction that the task of selecting a particular method is made complicated and somewhat arbitrary.

(6) The large power requirement of industrial areas makes it tempting to seek for equivalent wave energy supplies. As a consequence power plans are scaled up so only mammoth schemes are contemplated in the most demanding wave regimes. Smaller sites of far less power potential, but more reasonable economics, tend to be ignored.

The distinctive advantages of wave power are the large energy fluxes available and the predictability of wave conditions over periods of days. Waves are created by wind, and effectively store the energy for transmission over great distances. For instance large waves appearing off Europe will have been initiated in stormy weather in the mid-Atlantic or as far as the Caribbean.

The following sections aim to give a general basis for understanding wave energy devices. Firstly we outline the theory of deep water waves and calculate the energy fluxes available in single frequency waves. Then we review the patterns of sea waves that actually occur. Finally we describe attempts being made to construct devices that efficiently match variable natural conditions. With the complex theory of water waves we have sacrificed mathematical rigor for, we hope, physical clarity, since satisfactory theoretical treatments exist elsewhere.

12.2 Wave motion

Most wave energy devices are designed to extract energy from deep water waves. This is the most common form of wave, found when the mean depth of the sea bed D is more than about half the wavelength λ. For example an average sea wave for power generation may be expected to have a wavelength of ~100 m and an amplitude of ~3 m, and to behave as a deep water wave at depths of sea bed greater than ~30 m. Fig. 12.1(a) illustrates the motion of water particles in a

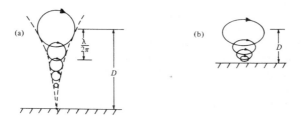

Fig. 12.1 Particle motion in water waves: (a) deep water, circular motion of water particles (b) shallow water, elliptical motion of water particles

deep water wave. The circular particle motion has an amplitude that decreases exponentially with depth and becomes negligible for $D > \lambda/2$. In shallower water (Fig. 12.1(b)) the motion becomes elliptical and water movement occurs against the sea bottom, producing energy dissipation.

The properties of deep water waves are distinctive, and may be summarized as follows:

(1) The surface waves are sets of unbroken sine waves of irregular wavelength, phase and direction.

(2) The motion of any particle of water is circular. Whereas the surface form of the wave shows a definite progression, the water particles themselves have no net progression.

(3) Water on the surface remains on the surface.

(4) The amplitudes of the water particle motions decrease exponentially with depth. At a depth of $\lambda/2\pi$ below the mean surface position, the amplitude is reduced to $1/e$ of the surface amplitude ($e = 2.72$, base of natural logarithms). At depths of $\lambda/2$ the motion is negligible, being less than 5% of the surface motion.

(5) The amplitude a of the surface wave is essentially independent of the wavelength λ, velocity c or period T of the wave, and depends on the history of the wind regimes above the surface. It is rare for the amplitude to exceed one-tenth of the wavelength, however.

(6) A wave will break into white water when the slope of the surface is about 1 in 7, and hence dissipate energy potential.

The formal analysis of water waves is difficult, but is well covered in specialist texts (see Barber, 1969, for an introduction; and Coulson and Jeffrey, 1977, for standard theory). We shall draw out from the theory those results that are important for understanding the most common wave energy devices.

For deep water waves, frictional, surface tension and inertial forces are small compared with the two dominant forces of gravity and circular motion. As a result, the water surface always takes up a shape so that its tangent lies perpendicular to the resultant of these two forces (Fig. 12.2).

It is of the greatest importance to realize that there is no net motion of water in deep water waves. Objects suspended in the water show the motions of Fig. 12.1, where deep water waves are contrasted with the kinds of motion occurring in shallower water.

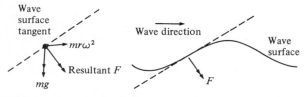

Fig. 12.2 Water surface perpendicular to resultant of gravitational and centrifugal forces acting on an element of water, mass m

A particle of water in the surface has a circular motion of radius *a* equal to the amplitude of the wave (Fig. 12.3). The wave height *H* from the top of a crest to the bottom of a trough is twice the amplitude: $H = 2a$. The angular velocity of the water particles is ω (radian per second). The wave surface has a shape that progresses as a moving wave, although the water itself does not progress. Along the direction of the wave motion the moving shape results from the phase differences in the motion of successive particles of water. As one particle in the

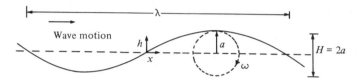

Fig. 12.3 Wave characteristics

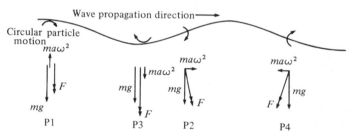

Fig. 12.4 Resultant forces on surface particles

crest drops to a lower position, another particle in a forward position circles up to continue the crest shape and the forward motion of the wave.

The resultant forces *F* on water surface particles of mass *m* are indicated in Fig. 12.4. The water surface takes up the position produced by this resultant, so that the tangent to the surface is perpendicular to *F*.

A particle at the top of a crest, position P1, is thrown upwards by the centrifugal force $ma\omega^2$. A moment later the particle is dropping, and the position in the crest is taken by a neighboring particle rotating with a delayed phase. At P2 a particle is at the average water level, and the surface orientates perpendicular to the resultant force *F*. At the trough, P3, the downward force is maximum. At P4 the particle has almost completed a full cycle of its motion.

The accelerations of a surface particle are drawn in Fig. 12.5(b). Initially $t = 0$, the particle is at the average water level, and subsequently:

$$\Phi = \frac{\pi}{2} - \omega t \tag{12.1}$$

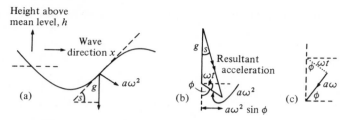

Fig. 12.5 Accelerations and velocities of a surface water particle (a) water surface (b) particle acceleration, general derivation (c) particle velocity

and

$$\tan s = \frac{a\omega^2 \sin \Phi}{g + a\omega^2 \cos \Phi} \approx \frac{a\omega^2 \sin \Phi}{g} \tag{12.2}$$

since in practice $g \gg a\omega^2$ for nonbreaking waves (e.g. $a = 2\,\mathrm{m}$, T (period) $= 8\,\mathrm{s}$, $a\omega^2 = 1.2\,\mathrm{m\,s^{-2}}$ and $g = 9.8\,\mathrm{m\,s^{-2}}$). Let h be the height of the surface above the mean level. The slope of the tangent to the surface is given by

$$\frac{dh}{dx} = \tan s \tag{12.3}$$

From (12.1), (12.2) and (12.3),

$$\frac{dh}{dx} = \frac{a\omega^2}{g} \sin \Phi = \frac{a\omega^2}{g} \cos \left(\frac{\pi}{2} - \Phi \right) = \frac{a\omega^2}{g} \cos \omega t \tag{12.4}$$

From Fig. 12.5(c), the vertical particle velocity is

$$\frac{dh}{dt} = a\omega \sin \Phi = a\omega \cos \omega t \tag{12.5}$$

The solution of (12.4) and (12.5) is

$$h = a \sin \left(\frac{\omega^2 x}{g} - \omega t \right) \tag{12.6}$$

Comparing this with the general traveling wave equation of wavelength λ and velocity c, we obtain

$$h = a \sin \frac{2\pi}{\lambda} (x - ct)$$

$$= a \sin \left(\frac{2\pi}{\lambda} x - \omega t \right) = a \sin(kx - \omega t) \tag{12.7}$$

where $k = 2\pi/\lambda$ is called the wave number.

It is apparent that the surface motion is that of a traveling wave, where

$$\lambda = \frac{2\pi g}{\omega^2} \tag{12.8}$$

This equation is important; it gives the relationship between the frequency and the wavelength of deep water surface waves.

The period of the motion is $T = 2\pi/\omega = 2\pi/(2\pi g/\lambda)^{1/2}$. So

$$T = \sqrt{\left(\frac{2\pi\lambda}{g}\right)} \tag{12.9}$$

The velocity of a particle at the crest of the wave is

$$v = a\omega = a\sqrt{\left(\frac{2\pi g}{\lambda}\right)} \tag{12.10}$$

The wave surface velocity in the x direction, from (12.7), is

$$c = \frac{\omega\lambda}{2\pi} = \frac{g}{\omega} = g\sqrt{\left(\frac{\lambda}{2\pi g}\right)}$$

$$c = \frac{g\lambda}{2\pi} \tag{12.11}$$

The velocity c is called the phase velocity of the traveling wave made by the surface motion. Note that the phase velocity c does not depend on the amplitude a, and is not obviously related to the particle velocity v.

Example 12.1
What is the period and phase velocity of a deep water wave of 100 m wavelength?

Solution
From (12.8),

$$\omega^2 = \frac{2\pi g}{\lambda} = \frac{(2\pi)(10\,\mathrm{m\,s^{-2}})}{100\,\mathrm{m}}, \omega = 0.8\,\mathrm{s^{-1}}$$

and so $T = 2\pi/\omega = 8\,\mathrm{s}$.
From (12.11),

$$c = \sqrt{\left[\frac{(10\,\mathrm{m\,s^{-2}})(100\,\mathrm{m})}{2\pi}\right]} = 13\,\mathrm{m\,s^{-1}}$$

So

$$\lambda = 100\,\mathrm{m}, T = 8\,\mathrm{s}, c = 13\,\mathrm{m\,s^{-1}} \tag{12.12}$$

12.3 Wave energy and power

12.3.1 *Basics*

The elementary theory of deep water waves begins by considering a single regular wave. The particles of water near the surface will move in circular orbits, at varying phase, in the direction of propagation x. In a vertical column the amplitude equals half the crest to trough height at the surface, and decreases exponentially with depth.

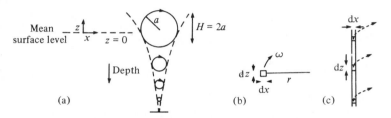

Fig. 12.6 Elemental motion of water, drawn to show the exponential decrease of amplitude with depth

The particle motion remains circular if the sea bed depth $D > 0.5\lambda$, when the amplitude becomes negligible at the sea bottom. For these conditions (Fig. 12.6(a)) it is shown in standard texts that a water particle whose mean position below the surface is z moves in a circle of radius given by

$$r = a\,e^{kz} \tag{12.13}$$

Here k is the wave number, $2\pi/\lambda$, and z is the mean depth below the surface (a negative quantity).

We consider elemental 'strips' of water across unit width of wavefront, at position (x, z). The volume per unit width of wavefront of this strip of density ρ is

$$dV = dx\,dz \tag{12.14}$$

and the mass is

$$dm = \rho\,dV = \rho\,dx\,dz \tag{12.15}$$

Let E_K be the kinetic energy of the total wave motion to the sea bottom, per unit length along the x direction, per unit width of the wavefront. The total kinetic energy of a length δx of wave is $E_K\,\delta x$. Each element of water of height δz, length δx and unit width is in circular motion at constant angular velocity ω, radius of circular orbit r, and velocity $v = r\omega$ (Fig. 12.6(b)). The contribution of this element to the kinetic energy in a vertical column from the sea bed to the surface is $\delta E_K\,\delta x$, where

$$\delta E_K\,\delta x = \tfrac{1}{2}mv^2 = \tfrac{1}{2}(\rho\,\delta z\,\delta x)\,r^2\omega^2 \tag{12.16}$$

Hence

$$\delta E_K = \tfrac{1}{2}\rho r^2 \omega^2 \delta z \tag{12.17}$$

It is easiest to consider a moment in time when the element is at its mean position, and all other elements in the column are moving vertically at the same phase in the z direction (Fig. 12.6(c)).

From (12.13) the radius of the circular orbits is given by

$$r = a\,e^{kz} \tag{12.18}$$

where z is negative below the surface.

Hence from (12.17),

$$\delta E_K = \tfrac{1}{2}\rho(a^2 e^{2kz})\omega^2 \delta z \tag{12.19}$$

and the total kinetic energy in the column is

$$E_K \delta x = \int_{z=-\infty}^{z=0} \frac{\rho\omega^2 a^2}{2} e^{2kz}\,\delta z\,\delta x = \frac{1}{4}\rho\frac{\omega^2 a^2}{k}\,\delta x \tag{12.20}$$

Since $k = 2\pi/\lambda$, and from (12.8) $\omega^2 = 2\pi g/\lambda$,

$$E_K = \frac{1}{4}\rho a^2 \frac{2\pi g}{\lambda}\frac{\lambda}{2\pi} = \frac{1}{4}\rho a^2 g$$

The kinetic energy per unit width of wavefront per unit length of wave is

$$E_K = \frac{1}{4}\rho\frac{a^2\omega^2}{k} = \frac{1}{4}\rho a^2 g \tag{12.21}$$

In Problem 12.1 it is shown that the potential energy per unit width of wave per unit length is

$$E_P = \rho a^2 g/4 \tag{12.22}$$

Thus, as would be expected for harmonic motions, the average kinetic and potential contributions are equal.

The total energy per unit width per unit length of wavefront, i.e. total energy per unit area of surface, is

total = kinetic + potential

$$E = E_K + E_P = \tfrac{1}{2}\rho a^2 g \tag{12.23}$$

Note that the root mean square amplitude is $\sqrt{(a^2/2)}$, so

$$E = \rho g\,(\text{root mean square amplitude})^2 \tag{12.24}$$

The energy per unit wavelength in the direction of the wave, per unit width of wavefront, is

$$E_\lambda = E\lambda = \tfrac{1}{2}\rho a^2 g\lambda \qquad (12.25)$$

From (12.20) $\lambda = 2\pi g/\omega^2$, so

$$E_\lambda = \pi\rho a^2 g^2/\omega^2 \qquad (12.26)$$

Or, since $T = 2\pi/\omega$,

$$E_\lambda = \frac{1}{4\pi}\rho a^2 g^2 T^2 \qquad (12.27)$$

It is necessary to show the kinetic, potential and total energies in these various forms, since all are variously used in the literature.

12.3.2 *Power extraction from waves*

So far we have calculated the total excess energy (kinetic plus potential) in a dynamic sea due to continuous wave motion in deep water. The energy is associated with water that remains at the same location when averaged over time. However, these calculations have told us nothing about the transport of energy (the power) across vertical sections of the water.

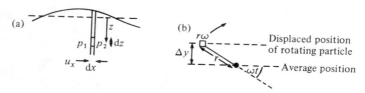

Fig. 12.7 Local pressure fluctuations in the wave (a) pressures in the wave (b) local displacement of water particle

Standard texts (e.g. Coulson and Jeffrey, 1977) calculate this power from first principles by considering the pressures in the water and the resulting displacements. The applied mathematics required is rigorous and comprehensive, and of fundamental importance in fluid wave theory. We can extract the essence of the full analysis, which is simplified for deep water waves.

Consider an element or particle of water below the mean surface level (Fig. 12.7). For a surface wave of amplitude a and wave number k, the radius of particle motion below the surface is

$$r = a e^{kz} \qquad (12.28)$$

The vertical displacement Δy (Fig. 12.7(b)) from the average position is

$$\Delta y = r \sin \omega t = a e^{kz} \sin \omega t \tag{12.29}$$

The horizontal component of velocity u_x is given by

$$u_x = r \omega \sin \omega t = \omega a e^{kz} \sin \omega t \tag{12.30}$$

From Fig. 12.7(a) the power carried in the wave at x, per unit width of wavefront at any instant, is given by

$$P' = \int_{z=-\infty}^{z=0} (p_1 - p_2) u_x \, dz \tag{12.31}$$

where p_1 and p_2 are the local pressures experienced across the element of height dz and unit width across the wavefront. Thus $(p_1 - p_2)$ is the pressure difference experienced by the element in a horizontal direction. The only contribution to the energy flow that does not average to zero at a particular average depth in the water is associated with the change in potential energy of particles rotating in the circular paths (see Coulson and Jeffrey, 1977).

$$p_1 - p_2 = \rho g \Delta y \tag{12.32}$$

Substituting for Δy from (12.29),

$$p_1 - p_2 = \rho g a e^{kz} \sin \omega t \tag{12.33}$$

In (12.31), and with (12.30) and (12.33),

$$\begin{aligned} P' &= \int_{z=-\infty}^{z=0} (\rho g a e^{kz} \sin \omega t)(\omega a e^{kz} \sin \omega t) \, dz \\ &= \rho g a^2 \omega \int_{z=-\infty}^{z=0} e^{2kz} \sin^2 \omega t \, dz \end{aligned} \tag{12.34}$$

The time average over many periods of $\sin^2 \omega t$ equals $1/2$, so

$$P' = \frac{\rho g a^2 \omega}{2} \int_{z=-\infty}^{z=0} e^{2kz} \, dz = \frac{\rho g a^2 \omega}{2} \frac{1}{2k} \tag{12.35}$$

The phase velocity of the wave is, from (12.7),

$$c = \frac{\omega}{k} = \frac{\lambda}{T} \tag{12.36}$$

So the power carried forward in the wave per unit width across the wavefront becomes

$$P' = \frac{\rho g a^2}{2} \frac{c}{2} = \frac{\rho g a^2 \lambda}{4T} \tag{12.37}$$

From (12.23) and (12.37) the power P' equals the total energy (kinetic plus potential) E in the wave per unit area of surface, times $c/2$. $c/2$ is called the group velocity of the deep water wave, i.e. the velocity at which the energy in the group of waves is carried forward. Thus, with the group velocity $u = c/2$,

$$P' = Eu = Ec/2 \tag{12.38}$$

where $E = \rho g a^2/2$.
    From (12.8),

$$k = \omega^2/g \tag{12.39}$$

and so the phase velocity is

$$c = \frac{\omega}{k} = \frac{g}{\omega} = \frac{g}{(2\pi/T)} \tag{12.40}$$

This difference between the group velocity and the wave (phase) velocity is common to all waves where the velocity depends on the wavelength. Such waves are called dispersive waves and are well described in the literature, both descriptively (e.g. Barber, 1969) and analytically (e.g. Lighthill, 1978).
    In (12.37),

$$P' = \frac{\rho g a^2}{2} \frac{1}{2} \left(\frac{gT}{2\pi} \right)$$

So

$$P' = \frac{\rho g^2 a^2 T}{8\pi} \tag{12.41}$$

Therefore the power in the wave increases directly as the square of the wave amplitude and directly as the period. The attraction of long period, large amplitude ocean swells to wave power engineers is apparent. This relationship is perhaps not obvious, and may be written in terms of wavelength using (12.9).

Example 12.2
What is the power in a deep water wave of wavelength 100 m and amplitude 1.5 m?

Solution

From (12.12) for Example 12.1, $c = 13\,\text{m s}^{-1}$. With (12.38),

$$u = c/2 = 6.5\,\text{m s}^{-1}$$

where u is the group velocity of the energy and c is the phase velocity.

The sea water waves have an amplitude $a = 1.5\,\text{m}$ ($H = 3\,\text{m}$), not unrealistic for Atlantic waves, so in (12.37)

$$P' = \tfrac{1}{2}(1025\,\text{kg m}^{-3})(9.8\,\text{m s}^{-2})(1.5\,\text{m})^2(6.5\,\text{m s}^{-1}) = 73\,\text{kW m}^{-1}$$

From this realistic example, we can appreciate that there can be extremely large power densities available in the deep water waves of ocean swells.

12.4 Wave patterns

Wave systems are not in practice the single sine wave patterns idealized in the previous sections. Very occasionally natural or contrived wave diffraction patterns, or channeled waves, approach this condition, but normally a sea will be an irregular pattern of waves of varying period, direction and amplitude. Under the stimulus of a prevailing wind the wave trains may show a preferred direction (e.g. the south west to north east direction of Atlantic waves off the British Isles), and produce a significant long period sea 'swell'. More erratic winds produce irregular water motion typical of shorter periods, called a 'sea'. At sea bottom depths, ~30 m or less, significant focusing and directional effects can occur, however, possibly producing more regular or enhanced power waves at local sites. Wave power devices must therefore match a broad band of natural conditions, and be designed to extract the maximum power averaged over a considerable time for each particular deployment position. In designing these devices it will be first necessary to understand the wave patterns of the particular site that may arise over a 50 year period.

The height of waves at one position was traditionally monitored on a wave height analog recorder. Separate measurements and analysis are needed to obtain the direction of the waves. Fig. 12.8 gives a simulated trace of such a recorder. A crest occurs whenever the vertical motion changes from upwards to downwards, and a trough vice versa. Modern recorders use digital methods for computer based analysis of large quantities of data. If H is the height difference

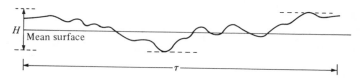

Fig. 12.8 Simulated wave height record at one position (with an exaggerated set of crests to explain terminology)

between a crest and its succeeding trough, there are various methods of deriving representative values, as defined in the following.

The basic variables measured over long intervals of time are:

(1) N_c, the number of crests; in Fig. 12.8 there are 10 crests.
(2) $H_{1/3}$, the 'one-third' significant wave height. This is the average height of the highest one-third of waves as measured between a crest and subsequent trough. Thus $H_{1/3}$ is the average of the $N_c/3$ highest values of H.
(3) H_s, the 'true' significant wave height. H_s is defined as

$$H_s = 4a_{rms} = 4\left[\left(\sum_{i=1}^{n} h^2\right)/n\right]^{1/2} \tag{12.42}$$

where a_{rms} is the root mean square displacement of the water surface from the mean position, as calculated from n measurements at equal time intervals. Care has to be taken to avoid sampling errors, by recording at a frequency at least twice that of the highest wave frequency present.
(4) H_{max} is the measured or most probable maximum height of a wave. Over 50 years H_{max} may equal 50 times H_s, and so this necessitates considerable overdesign for structures in the sea.
(5) T_z, the mean zero crossing period, is the duration of the record divided by the number of upward crossings of the mean water level. In Fig. 12.8, $T_z = \tau/3$.
(6) T_c, the mean crest period, is the duration of the record divided by the number N of crests. In Fig. 12.8, $T_c = \tau/10$; in practice N is very large, so reducing the error in T_c.
(7) The spectral width parameter ϵ gives a measure of the variation in wave pattern:

$$\epsilon^2 = 1 - (T_c/T_z)^2 \tag{12.43}$$

For a uniform single frequency motion, $T_c = T_z$, so $\epsilon = 0$. In our example $\epsilon = \sqrt{[1 - (0.3)^2]} = 0.9$, implying a mix of many frequencies. The full information is displayed by Fourier transformation to a frequency spectrum (e.g. Fig. 12.9).

From (12.41) the power per unit width of wavefront in a pure sinusoidal deep water wave is

$$P' = \frac{\rho g^2 a^2 T}{8\pi} = \frac{\rho g^2 H^2 T}{32\pi} \tag{12.44}$$

where the trough to crest height is $H = 2a$. The root mean square (rms) wave displacement for a pure sinusoidal wave is $a_{rms} = a/\sqrt{2}$, so in (12.44)

$$P' = \frac{\rho g^2 a_{rms}^2 T}{4\pi} \tag{12.45}$$

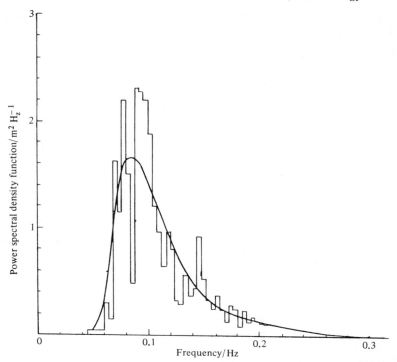

Fig. 12.9 Distribution of power per frequency interval in a typical Atlantic deep water wave pattern. From Shaw (1982). The smoothed spectrum is used to find T_e, the energy period

In practice sea waves are certainly not continuous single frequency sine waves. The power per unit width of wavefront is therefore written in the form of (12.45) as

$$P' = \frac{\rho g^2 H_s^2 T_e}{64\pi} \tag{12.46}$$

Here H_s is the significant wave height defined (12.42), and T_e, called the energy period, is the period of the dominant power oscillations given by the peak in the power spectrum (see Fig. 12.9). For many seas

$$T_e \approx 1.12 T_z \tag{12.47}$$

Until modern developments in wave power an approximate value of P could be obtained from analog recording wave meters such that

$$P' \approx \frac{\rho g^2 H_{1/3}^2 T_z}{64\pi} \approx (490\,\mathrm{W\,m^{-1}m^{-2}s^{-1}})\,H_{1/3}^2 T_z \tag{12.48}$$

However, with modern equipment and computer analysis, more sophisticated methods can be used to calculate (1) a_{rms} and hence H_s, and (2), T_z or T_e. Thus

$$P' = (490\,\mathrm{W\,m^{-1}m^{-2}s^{-1}})H_s^2 T_e$$
$$= (550\,\mathrm{W\,m^{-3}s^{-1}})H_s^2 T_z \qquad (12.49)$$

Since a wave pattern is not usually composed of waves all progressing in the same direction, the power received by a directional device will be significantly reduced.

Wave pattern data are recorded and tabulated in detail from standard meteorological sea stations. Perhaps the most important graph for any site is the wave scatter diagram over a year (e.g. Fig. 12.10). This records the number of occurrences of wave measurements of particular ranges of significant wave height and zero crossing period. Assuming the period is related to the wavelength by (12.9) it is possible to also plot on the diagram lines of constant wave height to wavelength. Contours of equal number of occurrences per year are also drawn.

From the wave data, it is possible to calculate the maximum, mean, minimum etc. power in the waves. This can be plotted on maps, e.g. for averages through the year. See Fig. 12.11 and Fig. 12.12 for average world and north west European power intensities.

12.5 Devices

As a wave passes a stationary position the surface changes height, water near the surface moves as it changes kinetic and potential energy, and the pressure under the surface changes. A great variety of devices have been suggested for extracting energy using one or more of these variations as input to the device. Included are devices that catch water at the crest of the waves, and allow it to run back into the mean level or troughs after extracting potential energy. In addition, various natural or artificial constructions can be arranged to produce wave diffraction or channeled effects that increase the power intensity at the site of the device.

It is not possible here to review all the types of wave power extraction devices that have been researched or developed (see NEL, 1976 for an excellent summary), so we shall describe only some of the most important.

12.5.1 *Wave profile devices*

Stephen Salter of Edinburgh University designed his 'ducks' to have a shape for optimum power extraction (Fig. 12.13). A wave entering from the left sets the beak of the duck into oscillation. The back of the duck has a circular shape, so that in oscillating about the axis at O no wave is propagated to the right. Power can be extracted via the axis of oscillation so that a minimum of reflected energy occurs. With very little energy transmitted or reflected, there is a very high conversion of energy from the incident wave into the power available from the axis, over a broad frequency band to match the conditions (Fig. 12.14).

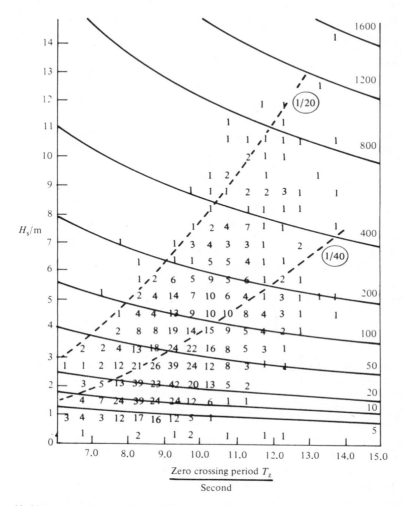

Fig. 12.10 Scatter diagram of significant wave height H_s against zero crossing period T_z. The numbers on the graph denote the average number of occurrences of each H_s, T_z in each 1000 measurements made over one year. The most frequent occurrences are at $H_s \sim 3\,\text{m}$, $T_z \sim 9\,\text{s}$, but note that maximum likely *power* occurs at longer periods.

_ _ _ _ _ _ these waves have equal maximum gradient or slope
⊂⊃ the maximum gradient of such waves
_____ lines of constant wave power, kW m^{-1}

Data for 58°N 19°W in the mid-Atlantic. After Glendenning (1977)

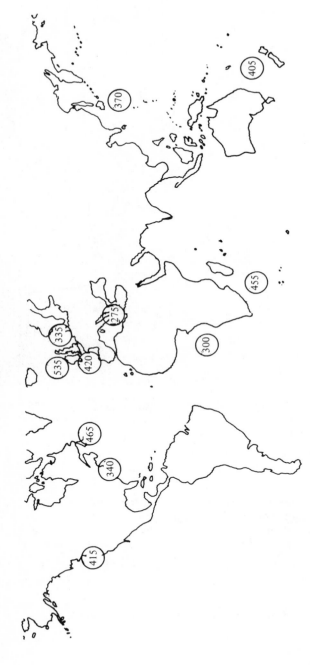

Fig. 12.11 Average annual wave energy (MWh per m) in certain sea areas of the world. After NEL (1976)

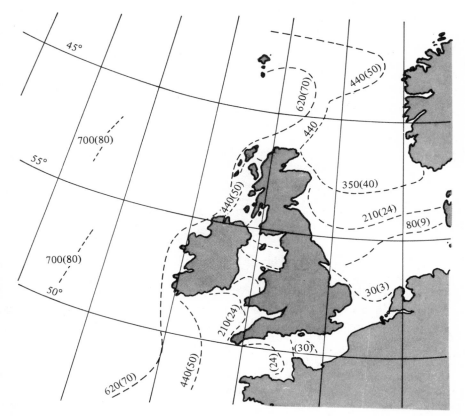

Fig. 12.12 Contours of average wave energy off north west Europe. Numbers indicate annual energy in the unit of MWh, and power intensity (bracketed) in the unit of kW m^{-1}. NB Local effects are not indicated

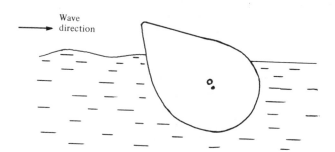

Fig. 12.13 Salter duck

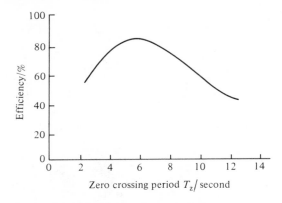

Fig. 12.14 Efficiency of 15 m diameter Salter duck fixed on spine

Further benefits and developments of the Salter design are:

(1) The duck can flip over and recover again after the impact of an unusually large wave; and
(2) A long line of ducks on a common flexible linkage or spine along the direction of the axis has little net translational motion, so mooring may not be difficult.

The scale of a duck is planned to be ~0.1 of a wavelength, i.e. ~10 m height for 100 m Atlantic waves. Strings of ducks several kilometers long have been proposed to be placed in the most intensive waves to the west of the Outer Islands of Scotland – to produce ~100 MW of power per string. Individual ducks, with the central axis fixed by stays under tension to the sea bottom, have been proposed for smaller scale operation in depths ~20 m.

All wave power devices have disadvantages, but the most serious of the ducks' disadvantages are

(1) The coupling of the slow oscillatory motion to electrical generators (Salter is working on a design involving gyroscopes, which will couple a slow perturbation of the gyroscope axis to electrical power takeoff); and
(2) The long 'spine' for each 'duck' to work against, as power is produced from a deep water floating array.

The Lanchester clam device has evolved from the Salter geometry. The system now incorporates a floating spine against which airbags are pressed by the waves. The compressed air is channelled from one bag to another as the wave passes at an angle to the spine. Electrical power is derived from the oscillating airstreams using Wells turbines. These turbines, once started, turn in the same direction to extract power from air flowing in either axial direction, i.e. the turbine motion is independent of the fluid direction.

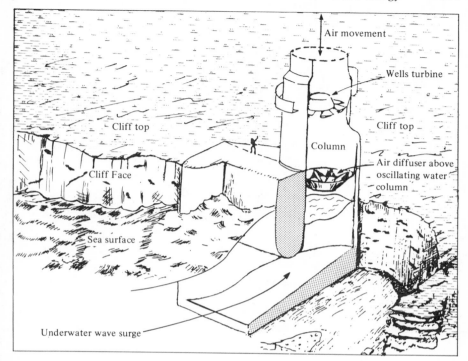

Fig. 12.15 Sketch of multi-resonant oscillating water column wave power system near Toftestallen in Norway

12.5.2 *Oscillating water column*

When a wave passes on to a partially submerged cavity open under the water (Fig. 12.16) a column of water oscillates up and down in the cavity. This can induce an oscillatory motion in the air above the column, which may be connected to the atmosphere through an air turbine. The ducting can be arranged so the air movement past the turbine is always in the same direction, or a Wells turbine can be used. At least two sets of commercial systems use the oscillating water column principle. A range of buoys with wave powered lighting are marketed by Masuda in Japan and the Queen's University in Belfast. More significantly, the first grid-linked wave power device has been constructed in Norway at Toftestallen by Kvaerner Brug A/S. (Fig. 12.15.) The basic principles of the oscillating water column are shown in Fig. 12.16. This has been used at Toftestallen for the 500 kW installation built into a cliff edge. The NEL proposal of Fig. 12.17 is for individual constructions sitting offshore on the sea bottom.

The main conceptual advantage of the oscillating water column is that the air velocity can be increased by reducing the cross-sectional area of the channel at

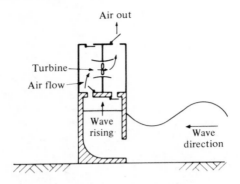

Unidirectional air flow drives turbine

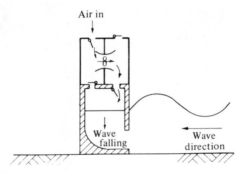

Fig. 12.16 Oscillating water column (NEL design for 'stand-on-the-bottom' sites). Printed with permission of NEL

the turbine. In this way the slow wave motions can be coupled to the high frequency turbine motion. There is also the opportunity to remove the generating equipment from the immediate saline water environment.

12.5.3 *Submerged devices*

Submerged devices have the benefit of avoiding the worst storm conditions, but have increased difficulties of power extraction and maintenance. One example is the proposed Bristol cylinder, consisting of submerged air filled cylinders held below the surface by 'legs' fixed to the sea bottom. Each cylinder is pressurized to move by the subsurface circular motion of the deep sea wave and the changing hydrostatic pressure. The 'legs' consist of hydraulic pumps that absorb energy as the cylinder moves. The pumped hydraulic fluid can be piped to a generating station near the array of cylinders.

Fig. 12.17 Proposed NEL design for power generation off the Western Isles of Scotland. Printed with permission of the UK National Engineering Laboratory

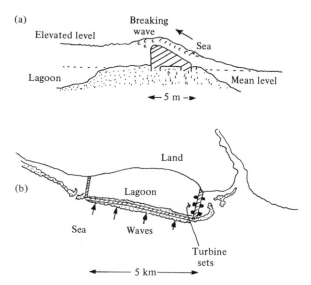

Fig. 12.18 Wave energy converter similar to the Tapered Wave Energy converter built near Toftestallen in Norway. (a) sea wall in section showing an impounded lagoon (b) plan of impounded lagoon

12.5.4 *Wave capture systems*

These schemes develop from a phenomenon often observed in natural lagoons. Waves break over a sea wall (equivalent to a natural reef) and water is impounded at a height above the mean sea level. This water may then return to the sea through a conventional low head hydraulic generator. Fig. 12.18 is a diagram of a scheme seriously considered for providing 20 MW of electrical power directly to the island of Mauritius. Overall wave to electrical busbar efficiency may be 30% from a delivered wave energy of 22 kW m^{-1} (i.e. the energy extracted from the wave at the sea wall and held in the form of potential energy in the reservoir). The length of the sea wall would be about 5 km. The world's first commercial power production system based on this principle has been built in Norway at a focussed wave site.

Problem

12.1 By considering elements of water lifted from depth z below the mean sea level to a height z above this level in a crest, show that the potential energy per unit length per unit width of wave front in the direction of the wave is

$$E_P = \cdot \tfrac{1}{4}\rho a^2 g$$

Solution

12.1 The wave surface is $h = a \sin kx$. Consider unit width of wave front, and a wavelength λ. Elements of water of mass $\rho\,dx\,dz$ are lifted a height of $2z$. The potential energy per wavelength is

$$E_{P,\lambda} = \int_{x=0}^{x=\lambda/2} \int_{z=0}^{z=h} (\rho\,dx\,dz)\,g(2z)$$

$$= \rho g \int_{x=0}^{x=\lambda/2} h^2\,dx$$

$$= \rho g a^2 \int_{x=0}^{x=\lambda/2} (1/2)(1 - \cos 2kx)\,dx$$

$$= \rho g a^2 \lambda/4$$

The potential energy per unit length is therefore

$$E_P = \rho g a^2/4$$

which equals the kinetic energy (12.21).

Bibliography

Barber, N. F. (1969) *Water Waves*, Wykeham, London.
An easily read but elementary introduction to water waves. Worth reading before tackling the full theory.

Bascom, W. (1977) 'Ocean waves' in Menard, H. W. (ed.) *Ocean Science*, W. H. Freeman, San Francisco.
An excellent introduction to ocean waves; most descriptive.

Bott, A. N., Hailey, J. S. M. and Hunter, P. D. (1978) 'Wave power prospects for Mauritius', *Water Power and Dam Construction*, December, 32–9.
An important techno-economic study for a project showing distinct possibilities of being attempted.

Coulson, C. A. and Jeffrey, A. (1977) *Waves*, Longman, London.
An excellent theoretical text, partly considering water waves.

Dawson, J. K. (1979) *Wave Energy*, Energy Paper no. 42, UK Dept of Energy, HMSO.
One of a series of excellent reviews of practical developments.

Glendenning, I. (1977) 'Energy from the sea', *Chemistry and Industry*, July, 592–9.
A very clear and concise review of practical propositions, with results of essential analysis and costing.

Glendenning, I. (1979) 'Wave power – some practical considerations', in *Future Energy Concepts*, IEE Conference Publication 171, IEE, Stevenage, UK, 109–13.

Lighthill, M. J. (1978) *Waves in Fluids*, Cambridge University Press.
Advanced level text with clear physical description.

NEL (1976) *The Development of Wave Power – A Techno-economic Study*, by Leishman, J. M. and Scobie, G. of the National Engineering Laboratory, East Kilbride, Glasgow, Report EAU M25.
A most valuable review of many suggested devices, with basic analysis and realistic assessment of practicability.

Salter, S. H. (1974) 'Wave power', *Nature*, **249**, 720–4.
Now seen as a classic paper. Later papers deal with duck developments.

Shaw, R. (1982) *Wave Energy – A Design Challenge*, Ellis Horwood, Chichester, and Halsted Press, New York.
Excellent analysis by a wave energy expert.

Tucker, M. J. (1963) 'Analysis of records of sea waves', *Proc. Inst. Civil. Engrs*, **26**, 305–16.
A careful analysis of sea wave data using conventional recording methods.

13 *Tidal power*

13.1 Introduction

The level of water in the large oceans of the earth rises and falls according to predictable patterns. The main periods τ of these tides are diurnal at about 24 h, and semidiurnal at about 12 h 25 min. The change in height between successive high and low tides is the range R. This varies between about 0.5 m in general and about 10 m at particular sites near continental land masses. The movement of the water produces tidal currents, which may reach speeds of $\sim$5 m s^{-1} in coastal and inter-island channels.

The sea water can be trapped at high tide in an estuarine basin of area A behind a dam or barrier. If the water of density ρ runs out through turbines at low tide, the maximum average power produced (13.35) is $\bar{P} = \rho A R^2 g/(2\tau)$. For example, if $A = 10\,\text{km}^2$, $R = 4\,\text{m}$, $\tau = 12\,\text{h}\,25\,\text{min}$, then $\bar{P} = 17\,\text{MW}$. Obviously sites of high range give the greatest potential for tidal power, but other vital factors are the need for the power, and the costs and secondary benefits of the construction. The civil engineering costs charged to a tidal range power scheme can be reduced if other benefits are included. Examples are the construction of roads on dams, flood control, irrigation improvement, pumped water catchments for energy storage, and navigation or shipping benefits. Thus the development of tidal power is very site specific.

Tidal current power may be harnessed in a manner similar to wind power. The power per unit area q in a current of maximum speed u_0 is $\bar{q} \approx 0.1\,\rho u_0^3$ (13.33). For $u_0 = 3\,\text{m}\,\text{s}^{-1}$, $q \approx 14\,\text{kW}\,\text{m}^{-2}$.

The harnessing of tidal range power (henceforward called tidal power) has been used for small mechanical power devices, e.g. in medieval England and in China. The best known large scale electricity generating plant is the 240 MW La Rance system at an estuary into the Gulf of St Malo in Brittany, France. Other key sites under review are the Severn estuary in England, and the Bay of Fundy on the eastern boundary between Canada and the United States. There is an important 400 kW capacity prototype tidal power plant at Kislayaguba on the Barents Sea in Russia.

The range, flow and periodic behavior of tides at most coastal regions is well documented and analyzed because of the demands of navigation and oceanography. The behavior may be predicted accurately, within an uncertainty of less than 4%, and so tidal power presents a very reliable and assured form of renewable power. The major drawbacks are:

(1) The mismatch of the principal lunar driven periods of 12 h 25 min and 24 h 50 min with the human (solar) period of 24 h, so that optimum tidal power generation is not in phase with demand
(2) The changing tidal range and flow over a two-week period, producing changing power production
(3) The requirement for large water volume flow at low head, necessitating many specially constructed turbines set in parallel
(4) The very high capital costs of most potential installations, and
(5) Potential ecological harm and disruption to extensive estuaries or marine regions.

For optimum electrical power generation from tides, the turbines should be operated in a regular and repeatable manner. The mode of operation will depend on the scale of the power plant, the demand and the availability of other sources. Very many variations are possible, but certain generalizations apply:

(1) If the tidal power is for local use, then other assured power supplies must exist when the tidal power is unavailable. Hydroelectric generation from rivers and water storage could provide such supplies.
(2) If the tidal power can feed into a large grid and so form a proportionately minor source within a national system, then the predictable tidal power variations can be submerged into the national demand.
(3) If the immediate demand is not fixed to the human (solar) period of 24 h, then the tidal power can be used whenever available. For example, if the electrical power is for transport by charging batteries or by electrolyzing water for hydrogen, then such a decoupling of supply and use can occur.

The final criterion for the success of a tidal power plant is the cost per unit kWh of the power produced. This can be lowered (1) if other advantages can be costed as benefit to the project, (2) if interest rates of money borrowed to finance the high capital cost are low, and (3) if the output power can be used to decrease consumption of expensive fuels such as diesel oil. With such economic complexity it may not be obvious that large scale (1000 MW) tidal power plants are the best. Smaller schemes for use in outlying regions may prove themselves more economic.

The following sections outline the physical background to tides and tidal power. Readers interested only in power generating devices should turn directly to Sections 13.4 and 13.5.

13.2 The cause of tides

The analysis of tidal behavior has been developed by many notable mathematicians and applied physicists, including Newton, Airy, Laplace, George

Darwin (son of Charles Darwin) and Kelvin. We shall use Newton's physical theory to explain the phenomena of tides. However, present day analysis and prediction depends on the mathematical method of harmonic analysis developed by Lord Kelvin in Glasgow. A complete physical understanding of tidal dynamics has not yet been attained owing to the topological complexity of the ocean basins.

The seas are liquids held on the solid surface of the rotating earth by gravity. The gravitational attraction of the earth with the moon and the sun perturbs these forces and motions so that tides are produced. Tidal power is derived from turbines set in this liquid, so harnessing the kinetic energy of the rotating earth. If all the world's major tidal power sites were utilized this would lead to a slowing of the earth's rotation by no more than one day in 2000 years: this is not a significant environmental problem.

13.2.1 *The lunar induced tide*

The moon and earth revolve about each other in space (Fig. 13.1), but since the mass of the earth is nearly 100 times greater than the moon's mass, the moon's motion is more apparent.

The center of revolution is at O, such that

$$ML = M'L'$$

So

$$L' = MD/(M' + M) \tag{13.1}$$

$$L' = 4670\,\text{km}$$

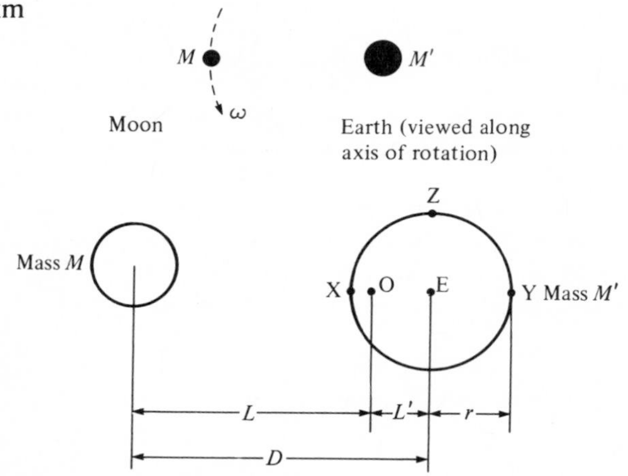

| Moon: | $M = 7.35 \times 10^{22}$ kg | $D = L + L' = 384 \times 10^6$ m |
| Earth: | $M' = 598 \times 10^{22}$ kg | $r = 6.38 \times 10^6$ m |

Rotation of earth and moon about O at frequency ω. $L' = 4670\,\text{km}$

Fig. 13.1 Motion of the moon and the earth

The earth's mean radius is 6371 km, so the point of revolution O is inside the surface of the earth.

The earth–moon separation is maintained by a balance of gravitational attraction and centrifugal force. If the gravitational constant is G,

$$\frac{GMM'}{D^2} = ML\omega^2 = M'L'\omega^2 \tag{13.2}$$

If all the mass of the earth could be located at the center of the earth E, then each element of mass would be at the equilibrium position with respect to the moon. However, the mass of the earth is not all at one point, and so is not all in this equilibrium. Material furthest from the moon at Y (Fig. 13.1) experiences an increased outward centrifugal force with distance of rotation $r + L'$ and a decreased gravitational force from the moon. Material nearest the moon at X has an increased gravitational force towards the moon, plus the centrifugal force, also towards the moon but reduced, that is due to the reduced rotation distance $r - L'$. The solid material of the earth experiences these changing forces as the moon revolves, but is held with only small deformation by the structural forces of the solid state. Liquid on the surface is however free to move, and it is this movement relative to the earth's surface that causes the tides. If the moon is in the equatorial plane of the earth, the water of the open seas attempts to heap together to form peaks at points X and Y, closest to and furthest from the moon. The solid earth would rotate with a period of one day underneath these two peaks (Fig. 13.2(a)). Thus with no other effect occurring, each sea covered position of the earth would experience two rises and two falls of the water level as the earth turns through the two peaks. This is the semidiurnal (half daily) tide. Note that the daily rotation of the earth on its own axis has no first order effect in producing the tidal height changes.

We may estimate the resultant force causing the tides from (1) the centrifugal force about O at the lunar frequency ω, and (2) the force of lunar gravitational attraction (Fig. 13.1). For a mass m of water, furthest from the moon at Y,

$$F_Y = m(L' + r)\omega^2 - \frac{GMm}{(D + r)^2} \tag{13.3}$$

Nearest to the moon at X

$$F_X = \frac{GMm}{(D - r)^2} + m(r - L')\omega^2 \tag{13.4}$$

At position E (Fig. 13.1), by definition of L',

$$\frac{GMm}{D^2} = mL'\omega^2 \tag{13.5}$$

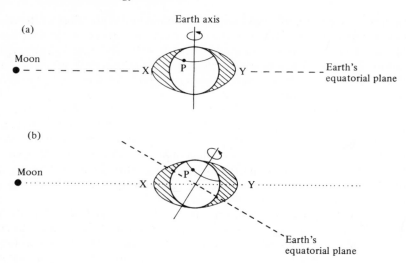

Fig. 13.2 Basic physical explanation of the semidiurnal and diurnal tide (a) Simple theory of equilibrium tide with the moon in the plane of the earth's equator, P experiences two equal tides each day (semidiurnal tide). (b) Normally the moon is not in the earth's equatorial plane, and so P may experience only one tide each day (diurnal tide)

However, since $r \ll D$,

$$\frac{1}{(D \pm r)^2} = \frac{1}{D^2}\left(1 \pm \frac{2r}{D}\right)$$

So by substituting (13.5) in (13.3) and (13.4),

$$F_X = F_Y = mr\omega^2(1 + 2L'/D) \tag{13.6}$$

We therefore expect two lunar tidal ranges each day of equal amplitude. This does indeed happen in the large oceans when the moon is in the equatorial plane of the earth.

At low tide on this equilibrium tide model the lunar related force is $mr\omega^2$, and so the tide raising force within (13.6) is $mr\omega^2 \, 2L'/D$. It can be shown (see Problem 13.1) that this would produce a maximum equilibrium tidal range of 0.36 m.

There are three principal reasons why actual tidal behavior is different from this simplistic 'equilibrium tide' explanation:

(1) The explanation of the forcing function producing the tides is physically correct, but the theory has failed to determine if the peaks of water can move at about 1600 km h^{-1} to keep up with the earth's rotation. In practice the

tidal wave cannot move fast enough to remain in the meridian of the moon (see Problem 13.3). Thus as the moon passes over an ocean, tidal motion is induced that propagates at a speed of about $500\,\mathrm{km\,h^{-1}}$ and lags behind the moon's position. The time of this lag is the 'age of the tide'. Each ocean basin tends to have its own system of moon induced tides that are not in phase from one ocean to another.

(2) The moon is not usually in the equatorial plane of the earth (Fig. 13.2(b)), and so a diurnal component of the tide occurs. Other minor frequency components of the tidal motion occur. For instance, the moon–earth distance oscillates slightly, from a maximum of $4.06 \times 10^8\,\mathrm{m}$ at apogee to a minimum of $3.63 \times 10^8\,\mathrm{m}$ at perigee in a period of 27.55 solar days (the anomalistic month). Also the moon's plane of motion moves about $2°$ in and out of the earth–sun ecliptic plane.

(3) A great many other complications occur, mostly associated with particular ocean basins. For instance resonances can occur, especially near continental shelves, that produce distinct enhancements of the tidal range. We will show in Section 13.3 that these resonant enhancements are of extreme importance for tidal power installations.

13.2.2 *Period of the lunar tides*

To calculate the period of the tides more precisely, we have to be more precise about what we mean by a 'day' (see Fig. 13.3). At a point A on the earth, a solar day is the interval between when the sun crosses the meridian at A today and when it does so tomorrow. This period actually varies through the year because of the irregularities in the earth's orbit, and so the common unit of time, the *mean solar day* t_S, is defined to be the interval averaged over a whole year. Its value is defined as exactly 24 hours, i.e.

$$t_S = 86\,400\,\mathrm{s} \tag{13.7}$$

The *sidereal day* t^* is similarly defined to be the average interval between successive transits of a 'fixed star', i.e. one so distant that its apparent motion relative to the earth is negligible. The sidereal day is therefore the 'true' period of rotation of the earth, as seen by a distant observer.

Fig. 13.3(a) shows how the difference between t_S and t^* is related to the revolution of the earth around the sun (period $T_S = 365.256\,t_S$). Suppose that at midday on a certain day the center of the earth E, point A on the earth's surface, the sun S and some fixed star are all aligned. One solar day later the sun, A and the earth's center are again aligned. In this time E has moved through an angle θ_1 around the sun to E'. Since t_S is the mean solar day, for time keeping purposes we can regard the earth as moving uniformly around a circular orbit so that

$$\frac{\theta_1}{2\pi} = \frac{t_S}{T_S} \tag{13.8}$$

(a)

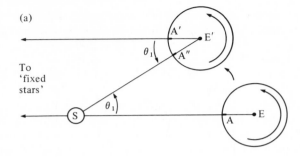

To 'fixed stars'

(b)

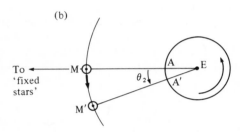

To 'fixed stars'

Fig. 13.3 Comparison of three different 'days' that can be observed from earth: (a) sidereal and solar day (b) sidereal and lunar. The solar day is 24 hours by definition, the sidereal is slightly shorter and the lunar slightly longer. The diagrams are not to scale

In this time A has rotated around E through an angle $(2\pi + \theta_1)$ to A''. Its time to rotate through an angle 2π (as seen by a distant observer) is just t^*, so that

$$\frac{\theta_1}{2\pi} = \frac{t_S - t^*}{t^*} \qquad (13.9)$$

Equating (13.8) and (13.9) gives

$$t^* = \frac{t_S}{1 + (t_S/T_S)}$$

$$= 86\,164\,\text{s}$$

$$= 23\,\text{h}\,56\,\text{min}\,4\,\text{s} \qquad (13.10)$$

Similarly the *mean lunar day* t_M is defined to be the mean interval between successive alignments of E, A and the moon's center. Fig. 13.3(b) shows the fictitious mean moon M moving uniformly in a circular orbit around the earth. In a time t_M, the moon moves through an angle θ_2 from M to M', while A on the earth rotates through $2\pi + \theta_2$. Thus as seen by a distant observer

$$\frac{\theta_2}{2\pi} = \frac{t_M}{T^*} = \frac{t_M - t^*}{t^*} \tag{13.11}$$

where $T^* = 27.32\,t_s$ (called the sidereal month, the 'true' lunar month) is the period of revolution of the moon about the earth's position as seen by a distant observer. This is shorter than the lunar month as recorded by an observer on earth ($T_m = 29.53$ days) owing to the earth moving around the sun. Equation (13.11) implies that

$$t_M = \frac{t^*}{1 - (t^*/T^*)}$$

$$= 89\,428\,\text{s} = 24\,\text{h}\,50\,\text{min}\,28\,\text{s} \tag{13.12}$$

13.2.3 *The solar induced tide and combined effects*

The same Newtonian theory that explains the major aspects of the twice daily lunar tide can be applied to the sun–earth system. A further twice daily tide is induced with a period of exactly half the solar day of 24 h. Other aspects being equal, the range of the solar tide will be 2.2 times less than the range of the lunar tide, which therefore predominates. This follows from considering that the tidal range is proportional to the *difference* of the gravitational force from the moon and the sun across the diameter d of the earth. If M_M and M_S are the masses of the moon and the sun at distances from the earth of D_M and D_S, then for either system

gravitational force $\alpha\ M/D^2$

difference in force $\alpha\ \dfrac{\partial F}{\partial D}d = -2Md/D^3$ \hfill (13.13)

The range of the lunar tide R_M and solar tide R_S are proportional to the difference, so

$$\frac{R_M}{R_S} = \frac{(M_M/D_M^3)}{(M_S/D_S^3)} = \left(\frac{D_S}{D_M}\right)^3 \frac{M_M}{M_S}$$

$$= \left(\frac{1.50 \times 10^{11}\,\text{m}}{3.84 \times 10^8\,\text{m}}\right)^3 \left(\frac{7.35 \times 10^{22}\,\text{kg}}{1.99 \times 10^{30}\,\text{kg}}\right) = 2.2 \tag{3.14}$$

The solar tide moves in and out of phase with the lunar tide. When the sun, earth and moon are aligned in conjunction, the lunar and solar tides are in phase, so producing tides of maximum range. These are the spring tides of maximum range occurring twice per lunar (synodic) month at times of both full and new moons (Fig. 13.8(c)).

When the sun/earth and moon/earth directions are perpendicular (in quadrature) the ranges of the tides are least. These are the neap tides that again occur twice per synodic month. If the spring tide is considered to result from the sum of the lunar and solar tides, and the neap tide from the difference, then the ratio of spring to neap ranges might be expected to be

$$\frac{R_s(\text{spring})}{R_n(\text{neap})} = \frac{1 + \dfrac{1}{2.2}}{1 - \dfrac{1}{2.2}} = 3 \tag{3.15}$$

In practice dynamical and local effects alter this rather naïve model, and the ratio of spring to neap range is more frequently about 2. Spring tides at the moon's perigee have greater range than spring tides at apogee, and a combination of effects including wind can occur to cause unusually high tides.

13.3 Enhancement of tides

The normal mid-ocean tidal range is less than one meter and of little use for power generation. However, near many estuaries and some other natural features, enhancement of the tidal range may occur by (1) funneling of the tides (as with sound waves in an old-fashioned trumpet shaped hearing aid), and (2) by resonant coupling to natural frequencies of water movement in coastal contours and estuaries.

Tidal movement of the sea has the form of a moving wave called a tidal wave. The whole column of water from surface to sea bed moves at the same velocity in a tidal wave, and the wavelength is very long compared with the sea depth (Fig. 13.4). Motion of a continuously propagating natural tidal wave has a velocity c related to the acceleration of gravity g and the sea depth h such that $c = \sqrt{(gh)}$.

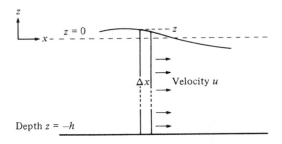

Fig. 13.4 Motion of water in a tidal wave

Normally the forced motion of the lunar and solar induced tides does not coincide with the requirements for a freely propagating tidal wave, and resonant enhancement of the forced motion does not occur in the open oceans. In certain estuaries and bays resonance can occur, however, and most noticeable changes in tidal motion therefore appear. Underwater volcanic or earthquake activity can induce a freely propagating 'tidal wave' in deep oceans called a tsunami, but this has no relationship to the tides other than its form of motion.

We consider a slab of water of depth h, width b, thickness Δx and surface level above the mean position z. The change in surface level over the thickness Δx is $(\partial z/\partial x)\Delta x$, and this is small compared with z. The side area of the slab is $A = hb$.

The form of the wave is obtained by considering Newton's equation of motion for the slab, and the requirement for conservation of water mass. The pressure difference across each side of the slab arises from the small change in height of the surface:

$$\Delta p = -\rho g \left(\frac{\partial z}{\partial x}\right) \Delta x \tag{13.16}$$

So the equation of motion of the slab of velocity u is

force = mass × acceleration

$$\left[-\rho g \left(\frac{\partial z}{\partial x}\right) \Delta x\right] hb = \rho hb \, \Delta x \frac{\partial u}{\partial t} \tag{13.17}$$

$$\frac{\partial u}{\partial t} = -g \frac{\partial z}{\partial x} \tag{13.18}$$

The difference between the flow of water into and out of the slab must be accounted for by a change in volume V of the slab with time, and this conservation of water mass leads to the equation of continuity

$$-\left[\frac{\partial}{\partial x}(A + bz)u\right] \Delta x = \frac{\partial V}{\partial t} = \frac{\partial[(A + bz)\Delta x]}{\partial t} \tag{13.19}$$

Since $A = bh$ is constant and much larger than bz,

$$-A \frac{\partial u}{\partial x} = b \frac{\partial z}{\partial t}$$

$$\frac{\partial u}{\partial x} = -\frac{1}{h}\frac{\partial z}{\partial t} \tag{13.20}$$

From (13.18) and (13.20),

$$\frac{\partial^2 u}{\partial t \, \partial x} = -g \frac{\partial^2 z}{\partial x^2} = \frac{\partial^2 u}{\partial x \, \partial t} = -\frac{1}{h}\frac{\partial^2 z}{\partial t^2} \tag{13.21}$$

So

$$\frac{\partial^2 z}{\partial t^2} = gh \frac{\partial^2 z}{\partial x^2} = c^2 \frac{\partial^2 z}{\partial x^2}$$ (13.22)

This is the equation of a wave of velocity c,

$$c = \sqrt{(gh)}$$ (13.23)

Resonant enhancement of the tides in estuaries and bays occurs in the same manner as the resonance of sound waves in open and closed pipes, for example as in Fig. 13.5.

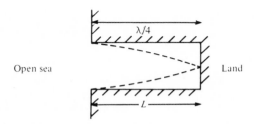

Fig. 13.5 Resonant enhancement of a tidal wave in an estuary, plan view. Idealized bay of constant depth h. Amplitude of tidal range indicated for a quarter wavelength resonance

Resonance with the open sea tide occurs when

$$L = j\lambda/4, \; j \text{ an odd integer}$$ (13.24)

The natural frequency of the resonance f_r and the period T_r is given by

$$f_r = \frac{1}{T_r} = \frac{c}{\lambda}$$ (13.25)

So

$$T_r = \frac{\lambda}{c} = \frac{4L}{jc} = \frac{4L}{j\sqrt{(gh)}}$$ (13.26)

Resonance occurs when this natural period equals the forced period of the tides in the open sea T_f, in which case

$$T_f = \frac{4L}{j\sqrt{(gh)}}; \; \frac{L}{\sqrt{h}} = \frac{j}{4} \sqrt{(g)} \, T_f$$ (13.27)

The semidiurnal tidal period is about 12 h 25 min (45 000 s), so resonance for $j = 1$ occurs when

$$\frac{L}{\sqrt{h}} = \frac{45\,000\,\text{s}}{4}\,\sqrt{(9.8\,\text{m}\,\text{s}^{-2})} = 36\,000\,\text{m}^{1/2} \tag{13.28}$$

Example 13.1
The River Severn estuary between Wales and England has a length of about 200 km and depth of about 30 m, so

$$\frac{L}{\sqrt{h}} \sim \frac{200 \times 10^3\,\text{m}}{\sqrt{(30\,\text{m})}} \sim 36\,400\,\text{m}^{1/2} \tag{13.29}$$

As a result there is close matching of the estuary's resonance frequency with the normal tidal frequency, and large amplitude tidal motions of 10–14 m range occur.

In practice estuaries and bays do not have the uniform dimensions implied in our calculations, and analysis is extremely complicated. It becomes necessary to model the conditions (1) in laboratory wave tanks using careful scaling techniques, and (2) by theoretical analysis. One dominant consideration for tidal power installations is to discover how barriers and dams will affect the resonance enhancement. For the Severn estuary some studies have concluded the barriers would reduce the tidal range and hence the power available: yet other studies have concluded the range will be increased! Tidal power schemes are too expensive to allow for mistakes to occur in understanding these effects.

13.4 Tidal flow power

Near coastlines and between islands, tides may produce strong water currents that can be considered for generating power. The total power produced may not be large, but generation at competitive prices for local consumption may be possible. The practical devices for tidal flow power will be similar to river flow power systems.

The theory of tidal flow power is similar to wind power (Chapter 9), with the advantage of predictable velocities of a fluid of 1000 times greater density than air, and the disadvantage of low fluid velocity and an aquatic environment.

The power density in the water current is, from (9.2),

$$q = \frac{\rho u^3}{2} \tag{13.30}$$

For a tidal or river current of velocity, for example, $3\,\text{m}\,\text{s}^{-1}$,

$$q = (1025\,\text{kg}\,\text{m}^{-3})(27\,\text{m}^3\,\text{s}^{-3})/2 = 13.8\,\text{kW}\,\text{m}^{-2}$$

Only a fraction η of the power in the water current can be transferred to useful power and, as for wind, η will not exceed about 60%. In practice η may approach a maximum of 40%.

Tidal current velocities vary with time approximately as

$$u = u_0 \sin(2\pi t/\tau) \tag{13.31}$$

where τ is the period of the natural tide, 12 h 25 min for a semidiurnal tide, and u_0 is the maximum velocity of the current.

Generation of electrical power per unit cross-section may therefore be on average (assuming 40% efficiency of tidal current power to electricity),

$$\bar{q} \approx \frac{0.4}{2} \rho u_0^3 \frac{\int_{t=0}^{t=\tau/4} \sin^3(2\pi t/\tau)\, dt}{\int_{t=0}^{\tau/4} dt} \tag{13.32}$$

$$= 0.2\rho u_0^3 (\tau/3\pi)(4/\tau)$$

$$\approx 0.1\rho u_0^3 \tag{13.33}$$

For a device that could generate power in the ebb (out) and flow (in) tidal currents, and with a maximum current of $3\,\mathrm{m\,s^{-1}}, q \sim 2.8\,\mathrm{kW\,m^{-2}}$. With a maximum current of 5 m s^{-1}, which occurs in a very few inter-island channels, $\bar{q} \sim 14\,\mathrm{kW\,m^{-2}}$. If the intercepted area was $1000\,\mathrm{m^2}$, then the total average power generation would be 14 MW.

The periodic nature of the power generation would lead to complications, but we note that tidal flow power lags about $\pi/2$ behind range power from a single basin, so the two systems could be complementary.

Few modern tidal flow power devices seem to have been developed, but Fig. 13.6 shows the design of one suggested device. Capital cost per unit of power produced would appear to be high. If tidal flow devices are to be used at all, the best opportunities would seem to be at remote locations where unusually high flow velocities occur, and where alternative sources are expensive.

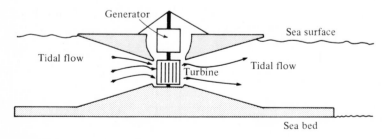

Fig. 13.6 Tidal current power device

13.5 Tidal range power

13.5.1 *Basic theory*

The basic theory of tidal power, as distinct from the tides themselves, is quite simple. Consider water trapped at high tide in a basin, and allowed to run out through a turbine at low tide (Fig. 13.7).

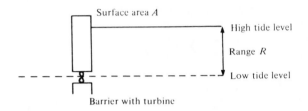

Fig. 13.7 Power generation from tides

The basin has a constant surface area A that remains covered in water at low tide. The trapped water, having a mass ρAR at a center of gravity $R/2$ above the low tide level, is all assumed to run out at low tide. The potential maximum energy available per tide if all the water falls through $R/2$ is therefore

$$\text{energy per tide} = (\rho AR)g\frac{R}{2} \tag{13.34}$$

If this energy is averaged over the tidal period τ, the average potential power for one tidal period becomes

$$\bar{P} = \frac{\rho AR^2 g}{2\tau} \tag{13.35}$$

The range varies through the month from a maximum R_s for the spring tides, to a minimum R_n for the neap tides. The envelope of this variation is sinusoidal, according to Fig. 13.8, with a period of half the lunar month.

At any time t after a mean high tide within the lunar month of period $T(T = 29.53$ days), the range is given by

$$\frac{R}{2} = \left(\frac{R_s + R_n}{4}\right) + \left(\frac{R_s - R_n}{4}\right)\sin\left(4\pi t/T\right) \tag{13.36}$$

If

$$R_n = \alpha R_s \tag{13.37}$$

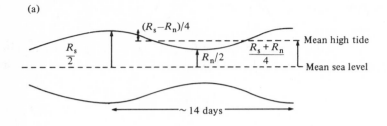

(a)

$(R_s - R_n)/4$

$\dfrac{R_s}{2}$ $R_n/2$ $\dfrac{R_s + R_n}{4}$ — Mean high tide

Mean sea level

~ 14 days

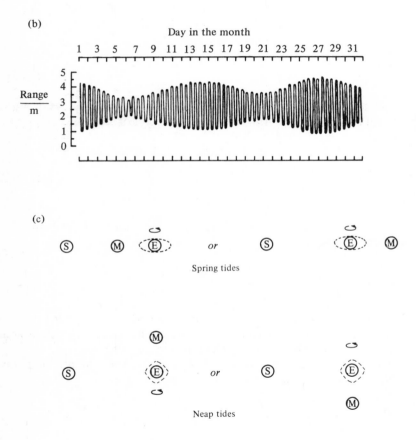

(b)

Day in the month

1 3 5 7 9 11 13 15 17 19 21 23 25 27 29 31

$\dfrac{\text{Range}}{\text{m}}$

5
4
3
2
1
0

(c)

Ⓢ   Ⓜ (Ⓔ)   *or*   Ⓢ   (Ⓔ)   Ⓜ

Spring tides

Ⓜ

Ⓢ   (Ⓔ)   *or*   Ⓢ   (Ⓔ)

Ⓜ

Neap tides

Fig. 13.8 (a) Sinusoidal variation of tidal range. (b) Tidal range variation for one month (from Bernstein, 1965) for a regular semidiurnal tide. Large range at spring tides, small range at neap tides. (c) Positions of the sun (S) moon (M) and earth (E) that produce spring and neap tides twice per month

then the range is given by

$$R = \frac{R_S}{2}[(1+\alpha)+(1-\alpha)\sin(4\pi t/T)] \tag{13.38}$$

The power is obtained from the mean square range:

$$\bar{R}^2 = \frac{R_s^2}{4}\frac{\int_{t=0}^{T}[(1+\alpha)+(1-\alpha)\sin(4\pi t/T)]^2\,dt}{\int_{t=0}^{T}dt} \tag{13.39}$$

Hence

$$\bar{R}^2 = \frac{R_s^2}{8}(3+2\alpha+3\alpha^2) \tag{13.40}$$

The mean power produced over the month is

$$\bar{P}_{\text{month}} = \frac{\rho A g}{2\tau}\frac{R_s^2}{8}(3+2\alpha+3\alpha^2) \tag{13.41}$$

where $R_n = \alpha R_s$ and τ is the intertidal period.

Since $\alpha \approx 0.5$, (13.41) differs little from the approximations often used in the literature, i.e.

$$\bar{P} \approx \frac{\rho A g}{2\tau}(\bar{R})^2 \tag{13.42}$$

where $\bar{R}$ is the mean range of all tides, and

$$\bar{P} \approx \frac{\rho A g (R_{\max}^2 + R_{\min}^2)}{2\tau}\frac{}{2} \tag{13.43}$$

where $R_{\max}$ and $R_{\min}$ are the maximum and minimum ranges.

Example 13.2
If $R_s = 5\,\text{m}$, $R_n = 2.5\,\text{m}$, $\alpha = 0.5$, $\bar{R} = 3.7\,\text{m}$, $R_{\max} = 5\,\text{m}$, $R_{\min} = 2.5\,\text{m}$, $A = 10\,\text{km}^2$, $\rho = 1.03 \times 10^3\,\text{kg}\,\text{m}^{-3}$ and $\tau = 12\,\text{h}\,25\,\text{min} = 4.47 \times 10^4\,\text{s}$, then in (13.41)

$$\bar{P} = 16.6\,\text{MW}$$

in (13.42)

$$\bar{P} = 15.4\,\text{MW} \tag{13.44}$$

and in (13.43)

$$\bar{P} = 16.1\,\text{MW}$$

13.5.2 *Application*

The maximum potential power of a tidal range system cannot be obtained in practice, although high efficiencies are possible. The complications are:

(1) Power generation cannot be maintained near to low tide conditions and so some potential energy is not harnessed.
(2) The turbines must operate at low head with large flow rates – a condition that is uncommon in conventional hydro-power practice. The turbines are least efficient at lowest head. The French have most experience of such turbines, having developed low head, large flow bulb turbines for generation from rivers and the Rance tidal scheme.
(3) The electrical power is usually needed at a near constant rate, and so there is a constraint to generate at times of less than maximum head.

Efficiency can be improved if the turbines are operated as pumps at high tide to increase the head. Consider a system where the range is 5 m. Water lifted 1 m at high tide can be let out for generation at low tide when the head becomes 6 m. Even if the pumps and generators are 50% efficient there will be a net energy gain of ~200% (see Problem 13.5).

In Fig. 13.7, note that power can be produced as water flows with the incoming and outgoing tide. Thus a carefully optimized tidal power system that uses reversible turbines to generate at both ebb and flow, and where the turbine can

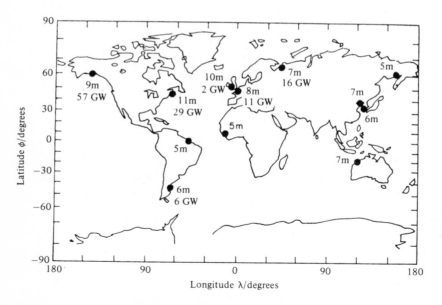

Fig. 13.9 Location of major world tidal power sites, showing the average tidal power range and power potential. From Sørensen (1979)

operate as pumps to increase the head, can produce energy of 90% of the potential given by (13.41).

13.6 World range power sites

The total dissipation of energy by water tides in the earth is estimated to be 3000 GW, including no more than about 1000 GW dissipated in shallow sea areas accessible for large civil engineering works. The sites of greatest potential throughout the world are indicated in Fig. 13.9 and detailed in Table 13.1. They have a combined total potential of about 120 GW. As Sorenson (1979) points out, this is about 10% of the total world hydro-power (river) potential.

Table 13.1 Major world tidal power sites (from Hubbert, 1969)

			Potential	
	Mean range	Basin area	Mean power	Annual production
	m	km^2	MW	1000 M Wh
North America				
Passamaquoddy	5.5	262	1800	15800
Cobscook	5.5	106	722	6330
Annapolis	6.4	83	765	6710
Minas-Cobequid	10.7	777	19900	175000
Amherst Point	10.7	10	256	2250
Shepody	9.8	117	520	22100
Cumberland	10.1	73	1680	14700
Petitcodiac	10.7	31	794	6960
Memramcook	10.7	23	590	5170
South America				
San Jose, Argentina	5.9	750	5870	51500
England				
Severn	9.8	70	1680	14700
France				
Aber-Benoit	5.2	2.9	18	158
Aber-Wrac'h	5.0	1.1	6	53
Arguenon	8.4	28	446	3910
Frenaye	7.4	12	148	1300
La Rance	8.4	22	349	3060
Rotheneuf	8.0	1.1	16	140
Mont St Michel	8.4	610	9700	85100
Somme	6.5	49	466	4090
Ireland				
Strangford Lough	3.6	125	350	3070
USSR				
Kislaya	2.4	2	2	22
Lumbouskii Bay	4.2	70	277	2430
White Sea	5.65	2000	14400	126000
Mezen Estuary	6.6	140	1370	12000
Australia				
Kimberley	6.4	600	630	5600
			~62000	~560000

Problems

13.1 (a) In Fig. 13.1, consider the lunar related force F_Z on a mass m of sea water along the earth's radius EZ. Since $D \gg r$, show that $F_Z = mr\omega^2$.

(b) Hence show that the difference in lunar related force on this mass between high and low tide is the tide raising force

$$F_t = F_X - F_Z = 2MmGr/D^3$$

(c) The tide raising force must equal the difference in the earth's gravitational attraction on m between low and high tide. Hence show that the tidal range R is 0.36 m and is given by

$$R = \frac{Mr^4}{M'D^3}$$

13.2 The sidereal month T^* is defined after (13.11). The synodic month T_m is defined as the average period between two new moons as seen by an observer on earth. T_m is greater than T^* because of the motion of the earth and moon together about the sun that effectively 'delays' the appearance of the new moon. What is the relation between T^* and T_m?

13.3 (a) The earth's oceans have an average depth of 4400 m. Show that the speed of a naturally propagating tidal wave is about $200 \, \text{m s}^{-1}$ $(750 \, \text{km h}^{-1})$.

(b) Compare this speed with the speed of the lunar tidal force passing round the earth's equator.

(c) What is the effect of the difference between these two speeds?

13.4 A typical ocean on the earth's surface has a depth of 4400 m.

(a) What is the speed of a freely traveling tidal wave for this depth?

(b) At this speed, how long would it take the tidal wave to circle the earth?

(c) If the tidal wave is started by the influence of the moon, can its motion be reinforced continually as the earth rotates?

13.5 Water is pumped rapidly from the ocean at high tide to give an increased water level in a tidal power basin of 1.0 m. If the tidal range is 5.0 m and if the pump/generator system is only 50% efficient, show that the extra energy gained can be nearly twice the energy needed for pumping.

Solutions

13.1 (a) The centrifugal force about O is $F_{OZ} = m(L'^2 + r^2)^{1/2}\omega^2$. Resolved along EZ, the radial (vertical) component

$$F_{EZ} = F_{OZ}\cos < \text{ZOE} = mr\omega^2$$

(b) From (13.6)

$$F_t = F_X - F_{EZ} = mr\omega^2 2L'/D$$

With (13.4)

$$F_t = (GMm/D^2)(2r/D) = 2MmGr/D^3$$

(c) The difference in gravitational attraction from the earth between r and $(r + R)$ is $2M'mR/r^3$. Equating this to F_t,

$$R = (M/M')(r^4/D^3)$$

Substituting the data from Fig. 13.1, $R = 0.36\,\text{m}$.

13.2 In figure 13.3(a), point A can be taken to represent the moon revolving around the earth E. In (13.10) we replace t^* by T^*, t_S by T_m, and hence

$$T_m = \frac{T^*}{1 - (T^*/T_s)}$$

$$= \frac{27.32 t_S}{1 - (27.32 t_S)/(365.256 t_S)} = 29.53 t_S$$

13.3 (a) $u = \sqrt{(gh)} = \sqrt{(9.8\,\text{m s}^{-2})(4400\,\text{m})}$

$$= 210\,\text{m s}^{-1} = 760\,\text{km h}^{-1}$$

(b) $v = 2\pi r/t_S = (2\pi)(6.38 \times 10^6\,\text{m})/(86\,400\,\text{s}) = 464\,\text{m s}^{-1}$
$$= 1670\,\text{km h}^{-1}$$

(c) The tidal wave cannot keep up with the tidal forcing function and so the equilibrium tide never materializes.

13.4 (a) $v = \sqrt{(gh)} = [(10\,\text{m s}^{-2})(4400\,\text{m})]^{1/2} = 210\,\text{m s}^{-1}$
(b) $t = 2\pi r/v = 2\pi(6.4 \times 10^6\,\text{m})/(210\,\text{m s}^{-1}) = 53\,\text{h}$ NB $53\,\text{h} \gg 24\,\text{h}$
(c) The freely traveling tidal wave propagates at less than half the speed necessary for continual reinforcement by the moon's tidal influence. Thus each ocean basin tends to have independent tidal properties from tides that dissipate their energy in shallow water and do not couple with neighboring oceans.

13.5 Consider a mass M of pumped water. Input energy to pump $1.0\,\text{m}$ is

$$2 \times [Mg(0.5)] = Mg$$

Output energy at low tide is

$$\tfrac{1}{2}[Mg(5.5)] = 2.7\,Mg$$

Energy gain/energy input is $(2.7 - 1)/1 = 1.7$.

Bibliography

Bernshtein, L. B. (1965) *Tidal Energy for Electric Power Plants*, Israel Program for Scientific Translations (translated from the Russian), Jerusalem.
A careful comprehensive text considering the important properties of tides and generating systems. Comparison of 13 different basin arrangements in a model tidal bay. Discussion of low head generation and dam construction.
Cotillon, J. (1979) 'La Rance tidal power station, review and comments', in Severn, R. T. *et al.* (1979).
Useful review of the world's only operating installation.
Darwin, G. H. (1898) *The Tides*, John Murray, London.
Clear explanations by an authority on tides. G. H. Darwin was the son of Charles Darwin.
Davey, N. (1923) *Studies in Tidal Power*, Constable, London.
A clear exposition of principles and applications, with many examples of potential sites, including Scotland.
Hubbert, M. K. (1969) *Resources and Man*, US National Academy of Sciences and National Science Foundation, W. Freeman, San Francisco.
Classic global assessment of energy resources.
Hubbert, M. K. (1971) *Scientific American*, September, 60–87.
Global tidal estimates.
Severn, R. T., Dineley, D. and Hawker, L. E. (eds) (1979) *Tidal Power and Estuary Management*, Colston Research Society 30th Meeting 1978, Colston Research Society, Bristol.
A set of useful research and development reviews, including economic and ecological aspects. Summaries of French (La Rance) and Russian experience, and suggested plans for the UK (Severn estuary).
Sørensen, B. (1979) *Renewable Energy*, Academic Press, London.
Useful but short summary of tidal power potential.
Struben, A. M. A. (1921) *Tidal Power*, Pitman, London.
An interesting, but dated, survey of principles. Do not despise such older work.
Webb, D. J. (1982) 'Tides and tidal power', *Contemp. Phys.*, **23**, 419–42.
Excellent review of tidal theory and resonant enhancement.
Wilson, E. M. (1973) 'Energy from the sea – tidal power', *Underwater Journal*, 175–85.

14 Ocean thermal energy conversion (OTEC)

14.1 Principles

The ocean is the world's largest solar collector. Temperature differences of 20 °C between warm, solar absorbing surface water and cooler 'bottom' water can occur. This can provide a continually replenished store of thermal energy, which is in principle available for conversion to other forms. The term *ocean thermal energy conversion* (abbreviated OTEC) refers to the conversion of some of this thermal energy into work and thence into electricity.

Fig. 14.1 shows one system for doing this. Essentially it is a heat engine operating between the 'cold' temperature T_c of the water at some substantial depth, and the 'hot' temperature, $T_h = T_c + \Delta T$, of the surface water. A working fluid circulates in a closed cycle and takes up heat from the warm water through a heat exchanger. As the fluid expands, it drives a turbine, which in turn drives a generator. The working fluid is cooled by the cold water, and the cycle continues. This *closed cycle* is the system which has been examined in most detail and is the only type discussed in this chapter. Other types are conceivable, e.g. with ocean water as the working fluid. Nevertheless the fundamental physical and geographical principles outlined in this and the next section apply to all OTEC systems.

We begin by defining P_0. This is the power given up from the warm water in an ideal system. For this we assume that a volume flow Q of warm water passes into the system at temperature T_h and leaves at T_c (the cold water temperature of lower depths). In defining P_0 we are obviously assuming perfect heat exchangers. In such an idealized system, if

$$\Delta T = T_h - T_c$$

then

$$P_0 = \rho c Q \Delta T \tag{14.1}$$

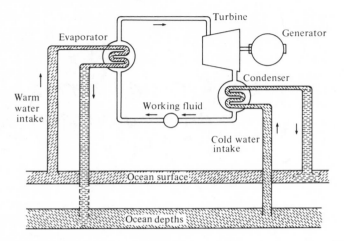

Fig. 14.1 Schematic diagram of an OTEC system. A heat engine operates between the warm water from the ocean surface and the cold water from the ocean depths

The second law of thermodynamics dictates that the *maximum* output of mechanical power we can obtain from the heat flow P_0 is

$$P_1 = \eta_{\text{Carnot}} P_0 \tag{14.2}$$

where

$$\eta_{\text{Carnot}} = \Delta T / T_{\text{h}} \tag{14.3}$$

is the efficiency of an ideal Carnot engine operating between T_{h} and $T_{\text{c}} = T_{\text{h}} - \Delta T$. Of course the output from a real system will be substantially less than P_1. Real engines will not follow the Carnot cycle but may operate closer to the ideal Rankine cycle of vapor turbines. Nevertheless these equations suffice to illustrate the promise and limitations of OTEC. From (14.1)–(14.3) the ideal mechanical output power is

$$P_1 = (\rho c Q / T_{\text{h}})(\Delta T)^2 \tag{14.4}$$

Example 14.1 Required flow rate
For $\Delta T = 20\,°C$ the flow rate required to yield $1\,MW$ from an ideal heat engine is (from (14.4))

$$Q_1 = \frac{(10^6\,J\,s^{-1})(300\,K)}{(10^3\,kg\,m^{-3})(4.2 \times 10^3\,J\,kg^{-1}\,K^{-1})(20\,K)^2}$$

$$= 0.18\,m^3\,s^{-1}$$

$$(\equiv 650\,t\,h^{-1})$$

Thus a substantial flow is required to give a reasonable output, even at the highest ΔT available in the world. This requires large, and therefore expensive, machinery.

Since P_1 depends *quadratically* on ΔT, experience shows that only sites with $\Delta T \gtrsim 15°C$ have much chance of being economically attractive. Fig. 14.2 indicates that such sites are confined to the tropics, and Fig. 14.3 illustrates the effect of depth on ΔT.

Sites under active investigation include Hawaii (20°N, 160°W), Nauru (0°S, 166°E) and the Gulf Stream off Florida (30°N, 80°E). Tropical sites have the

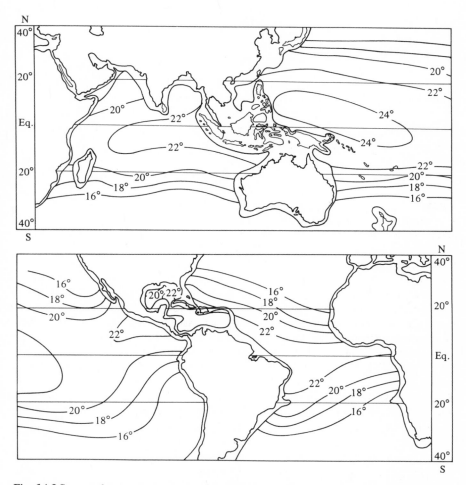

Fig. 14.2 Seasonal average of temperature difference ΔT between sea surface and a depth of 1000 m. Zones with $\Delta T \geq 20°C$ are most suitable for OTEC. These zones all lie in the tropics. Source: US Department of Energy

added advantage that T_h and T_c hardly vary from season to season, so that the potential output of the system is steady through the year.

Indeed steadiness and independence of the vagaries of weather is a major advantage of OTEC as a renewable source of energy. Its other major advantages are:

(1) At a suitable site, the resource is essentially limited only by the size of the machinery.
(2) The machinery to exploit it economically requires only marginal improvements in such well-tried engineering devices as heat exchangers and turbines. No dramatically new or physically impossible devices are required.

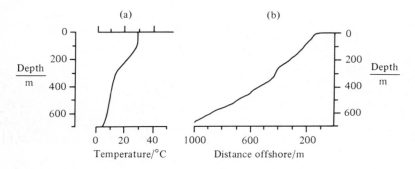

Fig. 14.3 Ocean conditions offshore from the island of Nauru, in the Central Pacific Ocean (0°S, 166°E): (a) water temperature (b) cross-section of sea bottom. The water temperatures are typical of those at good OTEC sites, and the steeply sloping sea floor allows a land based system. Data from Tokyo Electric Power Services Co. Ltd

The major disadvantages are cost and scale. If the power obtainable were in fact P_1, the costs would be lower, but resistances to the flow of heat and to fluid motion lower the output considerably. Sections 14.2 and 14.3 estimate the losses due to imperfect heat exchangers and to pipe friction. The installed cost of the experimental OTEC plants so far built have been as high as $40000 per kW of electricity capacity. However, the theory of Sections 14.2–14.4 suggests that larger systems will be more economical, which makes OTEC worth pursuing. Active development is under way in France, the USA and Japan.

One factor increasing the cost of OTEC systems is the expense of maintaining them at sea and of bringing the power ashore (Section 14.4). There are however a few especially favorable coastal sites where the sea bottom slopes down so steeply that the machinery can all be placed on dry land. The island of Nauru in the South Pacific has such a topography. Fig. 14.3 shows a section of the sea bottom there, and Fig. 14.4 is a photograph of an experimental OTEC installation on the shore.

Fig. 14.4 Experimental land based OTEC plant on Nauru, built by Tokyo Electric Power Services Company in 1981. It is a 'closed cycle' system, rated at 100 kW output. The large horizontal cylinder is the evaporator. Vertical framework at rear houses the condenser. Turbine 'house' is at left. The cold water pipe runs out to sea (in the background). Cylinders in foreground hold spare working fluid

14.2 Heat exchangers

In calculating the ideal output power P_1 we have assumed that there is perfect heat transfer between the ocean waters and the working fluid. In practice this is far from being the case, even for transfer through the best available heat exchangers.

14.2.1 *General analysis*

A heat exchanger transfers heat from one fluid to another, while keeping the fluids apart. Many different designs are described in engineering handbooks, but a typical and common type is the shell-and-tube design (Fig. 14.5). Water flows one way through the tubes, while the working fluid flows through the shell around the tubes.

Fig. 14.6 shows some of the resistances to heat transfer. The most fundamental of these arises from the relatively low thermal conductivity of water. As in Section 3.4, one can think of heat being carried by blobs of water to

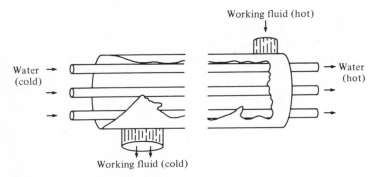

Fig. 14.5 Shell-and-tube heat exchanger (cutaway view)

within a fraction of a millimeter of the metal surface, but the last transfer from liquid to solid has to be by pure conduction. Similarly the heat flow through the metal, and any scum or biological growth adhering to it, is also by pure conduction. A temperature drop δT is required to drive the heat flow across these conductive resistances.

Let P_{wf} be the heat flow from water (w) to working fluid (f). Then

$$P_{wf} = \delta T / R_{wf} \qquad (14.5)$$

where R_{wf} is the thermal resistance between water and fluid. If it is assumed that there will be a similar temperature drop δT in the other heat exchanger, the temperature difference actually available to drive the heat engine is not ΔT but

$$\Delta_2 T = \Delta T - 2\delta T \qquad (14.6)$$

With an idealized Carnot engine the mechanical power output would be

$$P_2 = \left(\frac{\Delta T - 2\delta T}{T_h} \right) \frac{\delta T}{R_{wf}} \qquad (14.7)$$

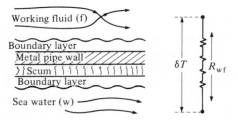

Fig. 14.6 Resistances to heat flow across a heat exchanger wall

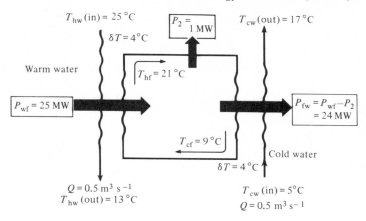

Fig. 14.7 Temperatures and heat flows in the OTEC system of Example 14.2. The other quantities are calculated from $T_{cw}^{(in)}, T_{hw}^{(in)}, \delta T, P_2$

Equation (14.7) implies that if $\delta T/R_{wf}$ is small there will be little output power. However, δT must be small to obtain maximum engine efficiency, so it is *crucial* to minimize the transfer resistance R_{wf}, i.e. to make the heat exchanger as efficient as possible. Therefore the tubes must be made of metal (good conductor) and there must be many of them – perhaps hundreds – in order to provide a large surface area. Other refinements may include fins or porous surfaces on the tubes, and baffles within the flow. With such an elaborate construction, it is not surprising that the heat exchangers constitute one of the major expenses of an OTEC system. This is the more so since the tube material has to be resistant to corrosion by sea water and the working fluid, and all joints must be leakproof.

The overall thermal resistance can be analyzed in terms of the thermal resistivity of unit area r_{wf} and the total wall area A_{wf} (see Section 3.2 for the definition and explanation of these terms):

$$R_{wf} = r_{wf}/A_{wf} \qquad (14.8)$$

Much of the development work in OTEC concerns improvements in the design of existing heat exchangers. The aim is to decrease r_{wf}, and thereby decrease the area A_{wf}. Having smaller heat exchangers with less metal can lead to substantial cost reductions. Values for r_{wf} of $3 \times 10^{-4} m^2 K W^{-1}$ (i.e. $h = 1/r = 3000 \, W \, m^{-2} K^{-1}$) can be obtained by the best of existing technology.

The flow rate required through the heat exchanger is determined by the power P_{wf} removed from the water, and by the heat transfers and temperatures involved. These are indicated in Fig. 14.7, which shows a counterflow heat exchanger on each side of the working fluid circuit. At each point along the heat exchanger, the temperature difference between the working fluid and the water is δT. Thus the hottest point in the working fluid is at

$$T_{hf} = T_{hw}^{(in)} - \delta T$$

and the coldest is at

$$T_{cf} = T_{cw}^{(in)} + \delta T$$

Therefore the power given up by the hot water is

$$P_{wf} = \rho c Q (T_{hw}^{(in)} - T_{hw}^{(out)}) \tag{14.9}$$

with the temperature drop

$$T_{hw}^{(in)} - T_{hw}^{(out)} = \Delta T - 2\delta T \tag{14.10}$$

14.2.2 *Size*

Example 14.2 Heat exchanger dimensions
Find a set of working dimensions for a shell-and-tube heat exchanger suitable for an OTEC system set to produce 1 MW. Assume a Carnot cycle for the working fluid, but allow for temperature reductions in nonperfect heat exchangers.

Assume $r_{wf} = 3 \times 10^{-4} \, m^2 \, K \, W^{-1}$, $\Delta T = 20°C$, $\delta T = 4°C$ etc., as in Fig. 14.7.

Solution
(1) *Surface area*
From (14.8),

$$A_{wf} = r_{wf}/R_{wf}$$

From (14.7),

$$1/R_{wf} = \frac{P_2 T_h}{(\Delta T - 2\delta T)\delta T}$$

so

$$A_{wf} = \frac{(1 \times 10^6 \, W)(300 \, K)(3 \times 10^{-4} \, m^2 \, K \, W^{-1})}{(20 - 8) \, K \, (4 \, K)}$$

$$= 1.9 \times 10^3 \, m^2$$

This is a very large area of transfer surface.

(2) *Flow rate*

$$\eta'_{Carnot} = \frac{(21 - 9)°C}{(273 + 21) \, K} = 12/294$$

$$P_{wf} = P_2/\eta'_{Carnot} = (1 \, MW)(294/12) = 25 \, MW$$

Therefore, from (14.9), and (14.10), the flow rate is

$$Q = \frac{(25 \times 10^6 \, \text{W})}{(10^3 \, \text{kg m}^{-3})(4.2 \times 10^3 \, \text{J K}^{-1} \text{kg}^{-1})(12 \, \text{K})}$$

$$= 0.50 \, \text{m}^3 \text{s}^{-1}$$

(3) *Thermal resistance of the boundary layers*
We suppose that each fluid boundary layer of Fig. 14.6 contributes about half of r_{wf}. In particular, assume that the thermal resistivity of the boundary layer (of water) on the inside of the pipe is given by

$$r_\text{v} = 1.5 \times 10^{-4} \, \text{m}^2 \text{K W}^{-1}$$

Let d be the diameter of each tube in the heat exchanger. The convective heat transfer to the inside wall of a smooth tube is given by (C.14) (see Section 3.4):

$$\mathcal{N} = 0.027 \mathcal{R}^{0.8} \mathcal{P}^{0.33}$$

By definition of the Nusselt number, $\mathcal{N} = d/(r_\text{v} k)$. Thus the Reynolds number in each tube is

$$\mathcal{R} = [d/(0.027 \, r_\text{v} k \, \mathcal{P}^{0.33})]^{1.25}$$

$$= [(0.027)(0.6 \, \text{W m}^{-1} \text{K}^{-1})(7.0)^{0.33}]^{-1.25} (d/r_\text{v})^{1.25}$$

$$= a d^{1.25}$$
where $a = 4.67 \times 10^6 \, \text{m}^{-1.25}$.

(4) *Diameter of tube*
As an initial estimate, suppose $d = 0.02 \, \text{m}$. Then

$$\mathcal{R} = 3.5 \times 10^4$$

Hence, flow speed in each tube is

$$u = \mathcal{R} v/d$$

$$= \frac{(3.5 \times 10^4)(1.0 \times 10^{-6} \, \text{m}^2 \text{s}^{-1})}{(0.02 \, \text{m})} = 1.7 \, \text{m s}^{-1}$$

Since the total flow through n tubes is

$$Q = nu \pi d^2/4$$

the number of tubes required is

$$n = \frac{(0.50\,\text{m}^3\,\text{s}^{-1})(4)}{(1.7\,\text{m}\,\text{s}^{-1})(3.14)(0.01\,\text{m})^2}$$

$$= 3600$$

(5) *Length of tubes*
To make up the required transfer area

$$A = n\pi d l$$

each tube must have length

$$l = \frac{(1.9 \times 10^3\,\text{m}^2)}{(3600)\pi(0.02\,\text{m})} = 32\,\text{m}$$

This example makes it clear that large heat exchangers, with substantial construction costs, are required for OTEC systems. Indeed the example under-estimates the size involved because it does not allow for imperfections in the heat engines etc. (which increase the required Q to achieve the same power output). Also the example assumes that the pipe is clean and smooth.

14.2.3 *Biofouling*
The inside of the pipe is vulnerable to encrustation by marine organisms, which will increase the resistance to heat flow (Fig. 14.6) and thereby lower the performance. Such *biofouling* is one of the major problems in OTEC design, since increasing the surface area available for heat transfer also increases the opportunity for organisms to attach themselves. Among the methods being tried to keep this fouling under control are mechanical cleaning by continual circulation of close fitting balls, and chemical cleaning by additives to the water.

The effect of all these complications is that the need for cost saving will encourage the use of components working at less than optimal performance (e.g. undersized heat exchangers).

14.3 Pumping requirements

Work is required to move large quantities of hot water, cold water, and working fluid around the system against friction. This will have to be supplied from the gross power output of the OTEC system, i.e. it constitutes yet another loss of energy from the ambient flow P_0. The work may be estimated numerically using the methods of Section 2.6, although analytic calculations are difficult. The effect of cooling the water in the hydrostatic 'circuit' is small, but does encourage circulation.

Example 14.3 Friction in the cold water pipe
The OTEC system of Example 14.2 (Fig 14.7) with $P_2 = 1\,\text{MW}$, $\Delta T = 20\,°\text{C}$ has a cold water pipe with $L = 1000\,\text{m}$, diameter $D = 1\,\text{m}$. Calculate the power required to pump water up the pipe.

Solution
The mean speed is

$$u = Q/A$$
$$= \frac{(0.50\,\text{m}^3\text{s}^{-1})}{\pi(0.5\,\text{m})^2} = 6.3\,\text{m}\,\text{s}^{-1}$$

Therefore the Reynolds number is

$$\mathscr{R} = uD/\nu$$
$$= \frac{(0.63\,\text{m}\,\text{s}^{-1})(1\,\text{m})}{(1.0 \times 10^{-6}\,\text{m}^2\text{s}^{-1})} = 6.3 \times 10^5$$

In practice, many varieties of marine organisms brought up from the depths will adhere to the pipe, giving an equivalent roughness height $\xi \sim 20\,\text{mm}$, i.e. $\xi/D = 0.02$. Thus, from Fig. 2.6, the pipe friction coefficient is

$$f = 0.012$$

From (2.12), the head loss is

$$H_f = 2fLu^2/Dg = 1.0\,\text{m}$$

To overcome this requires the same power as to lift a mass ρQ per second through a height H_f, i.e.

$$P_f = \rho Q g H_f = 4.7\,\text{kW}$$

We see that the cold water pipe can be built large enough to avoid major friction problems. However, because the head loss varies as (pipe diameter)$^{-4}$, or even as (diameter)$^{-5}$, friction loss can become appreciable in the smaller piping between the cold water pipe and the heat exchanger, and in the heat exchanger itself. Indeed, because the same turbulence carries both heat and momentum from the heat exchanger surfaces, all attempts to increase heat transfer by increasing the surface area necessarily increase fluid friction in the heat exchangers.

In addition, the flow rate required in practice to yield a given output power is greater than that calculated in Example 14.2, because a real heat engine is less efficient than a Carnot engine in converting the input heat into work. This increases the power lost to fluid friction. Fouling of the heat exchanger tubes makes the situation worse, both by further raising the Q required to yield a certain power output, and by decreasing the tube diameter. As a result, in some systems over 50% of the input power may be lost to fluid friction.

14.4 Other practical considerations

The calculations of the previous sections confirm that there are no fundamental thermodynamic difficulties that prevent an OTEC system from working success-fully. Although there remain a number of practical, engineering and environ-mental difficulties, we shall see that none of these appears insuperable.

14.4.1 *The platform*

American designers are aiming for large systems (yielding about 400 MW (electric)) which would be based on a massive floating offshore platform, similar to those used in oil drilling. Since such a platform will be heavy and unwieldy, there will be a major problem in connecting it to the cold water pipe (CWP), because of the stresses from surface waves and currents.

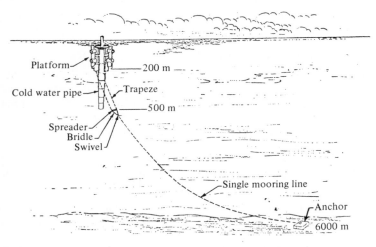

Fig. 14.8 Underwater platform for 400 MW(e) system, proposed by Lockheed for the US Department of Energy. The platform can be moored in position in any depth of water

One response to this problem is to make the platform neutrally buoyant and moor it underwater (Fig. 14.8), thereby avoiding the major stresses at the surface.

14.4.2 *Construction of the cold water pipe*

The pipe is subject to many forces in addition to the stresses at the connection. These include drag by currents, oscillating forces due to vortex shedding, forces due to harmonic motion of the platform, forces due to drift of the platform, and the dead weight of the pipe itself. It is debatable whether a rigid material (e.g. steel) or a flexible one (e.g. polythene) would better withstand these forces. In addition, there are substantial problems involved in assembling and positioning the pipe. Some engineers favor bringing out a prefabricated pipe and slowly sinking it into place. But transporting an object of several meters in diameter and perhaps a kilometer long is not easy.

14.4.3 *Link to the shore*

High voltage, high power submarine cables are standard components of electrical power transmission systems. They are expensive (like all marine engineering), but a cable about 50 km long is quite practicable.

Alternatively it has been suggested that large OTEC plants, which might be hundreds of kilometers away from energy demand, could use the electricity on board to produce a chemical store of energy (e.g. H_2, Section 16.3).

Land based systems, like that of Fig. 14.4, are possible at certain favorable locations, where the sea bottom slopes sharply downward. Their main advantage is lower cost, since the link to shore, the assembly and the maintenance are much simplified. The cold water pipe is also not so subject to stress, since it rests on the sea bottom. But it is still vulnerable to storm damage from wave motion to a depth of about 20 m.

14.4.4 *The turbine*

Apart from its physical size, the turbine presents no great difficulty and existing designs can be used. As with all practical heat engines the efficiency will not be greater than 0.5 relative to an ideal Carnot engine with the same heat input *to the working fluid*.

14.4.5 *Choice of working fluid*

There are many common fluids having an appropriate boiling point, e.g. ammonia, freon or water. Indeed, by applying a partial vacuum (i.e. lowering the pressure) the boiling point of water can be lowered to the temperature of the warm water intake. This is the basis of the *open cycle* system, in which the warm water itself is used as the working fluid. Such a system provides not only power, but also substantial quantities of distilled water.

14.4.6 *Environmental effects*

The richness of marine life poses the problem of biofouling (Section 14.2.3). It also offers the possibility of fish farming. Sea water from the depths is rich in nutrients, and these would be discharged near the surface around an OTEC plant. This would certainly encourage the growth of algae, which in turn would attract other marine creatures higher up the food chain. This could possibly provide a basis for commercial fish farming. However, the total biological effects of releasing large quantities of cool water into the warmer surface environment are not yet known. The effects may or may not be desirable, and will have to be estimated from small scale trials.

Problems

14.1 Calculate the dimensions of a shell-and-tube heat exchanger to produce an output power $P_2 = 10\,\text{MW}$. Assume $r_v = 3 \times 10^{-4}\,\text{m}^2\text{K W}^{-1}$, $\delta T = 4\,°\text{C}$, and tube diameter $D = 5\,\text{cm}$.
Hint: follow Example 14.2.

14.2 Calculate the power lost to fluid friction in the heat exchanger of Example 14.2.

14.3 *Heat engine for maximum power*   As shown in textbooks of thermo-dynamics, no heat engine can be more efficient than the ideal concept of the Carnot engine. Working between temperatures T_h and T_c, its efficiency is

$$\eta_{\text{Carnot}}(\Delta T) = (T_h - T_c)/T_h$$

However, the power output from a Carnot engine is zero. Why?
   Use (14.7) to show that the engine which produces the greatest *power*, for constant thermal resistance of pipe, has $\delta T = \frac{1}{4}\Delta T$, i.e. it 'throws away' half the input temperature difference. What is the efficiency of this engine as an energy converter compared with an ideal Carnot engine?

Solutions

14.1 $A_{\text{wf}} = 19 \times 10^3\,\text{m}^2$, $Q = 5\,\text{m}^3\text{s}^{-1}$, $\mathscr{R} = 1.1 \times 10^5$, $u = 2.2\,\text{m s}^{-1}$, $n = 1200, l = 100\,\text{m}$

14.2 $f = 0.007$ (when clean), $P_f = 35\,\text{kW}$

14.3 $\eta_{\text{Carnot}}/4$

Bibliography

Popular articles

Ford, G., Niblett, C. and Walker, L. (1981) *Ocean Thermal Energy: Prospects and Opportunities*, Occasional Paper no. 9, Policy Research Unit in Engineering, Science and Technology, University of Manchester, UK.
   Focuses on policy issues.
Kamogawa, H. (1980) 'OTEC research in Japan', *Energy*, **5** (6), 481–92.
   Describes Japanese research.
Lavi, A. (1980) 'Ocean thermal energy conversion: a general introduction', *Energy*, **5** (6), 469–80.
   Part of a special issue on OTEC. Neatly presents the basic physics.
McGowan, J. G. (1976) 'Ocean thermal energy conversion – a significant solar resource', *Solar Energy*, **18**, 81–92.
   Reviews US design philosophy.

Marchaud, Ph. (1981) 'Travaux français sur l'énergie thermique des mers', *La Houille · Blanche*, 4/5, 315–21.
  French work on OTEC (in French).
Whitmore, W. F. (1978) 'OTEC: electricity from the sea', *Technology Review*, **81** (1), 2–7.
  Includes some useful illustrations.
Zener, C. (1974) 'Solar sea power', in *Physics and the Energy Problem – 1974*, American Institute of Physics Conference Proceedings no. 19, pp. 412–19.
  Useful focus on heat exchanger thermodynamics. (This whole volume makes interesting reading.)

Technical reports

Very few technical reports have appeared in established journals. Most are in the form of contractor's reports to government agencies, or conference papers, e.g. US Department of Energy (1981) *Proceedings of the 8th Ocean Energy Conference*, Washington DC, report no. conf-810622 (available through US National Technical Information Service)

Thermodynamics of real engines

Curzon, F. L. and Ahlborn, B. (1975) 'Efficiency of a Carnot engine at maximum power output', *Amer. J. Phys.*, **43**, 22–4.

15 *Geothermal energy*

15.1 Introduction

The inner core of the earth reaches a maximum temperature of about 4000 °C. Heat passes out through the solid submarine and land surface mostly by conduction – geothermal heat – and occasionally by active convective currents of molten magma or heated water. The average geothermal heat flow at the earth's surface is only $0.06 \, \text{W m}^{-2}$, with a temperature gradient of $<30 \, °\text{C km}^{-1}$. This continuous heat current is trivial compared with other renewable supplies in the above surface environment that will in total average about $500 \, \text{W m}^{-2}$. However, at certain specific locations increased temperature gradients occur, indicating significant geothermal resources. These may be harnessed over areas of the order of square kilometers and depths of $\sim 5 \, \text{km}$ at fluxes of 10 to $20 \, \text{W m}^{-2}$ to produce $\sim 100 \, \text{MW}$ (thermal) km^{-2} in commercial supplies for at least 20 years of operation.

Geothermal heat is generally of low quality, and is best used directly for building or process heat at about 50 to 70 °C, or for preheating of conventional high temperature energy supplies. Such supplies are established in several parts of the world and many more projects are planned. Occasionally geothermal heat is available at temperatures above about 150 °C, so electrical power production from turbines can be contemplated. Several important geothermal electric power complexes are fully established, especially in Italy, New Zealand and the USA.

It is common to use ground heat as the input to a heat pump. Although this is strictly a 'geothermal' source, we do not include such systems as geothermal supplies for the purposes of this chapter.

In Chapter 1 renewable energy was defined as currents of energy occurring naturally in the environment. By this definition some sources of geothermal energy can be classed as renewable, because the energy would otherwise be dissipated in the environment, e.g. from hot springs or geysers. In other geothermal sites, however, the current of heat is increased artificially (e.g. by fracturing and actively cooling hot rocks, or by drilling into hot aquifers), and so the supply is not renewable at the extraction rate on a long time scale. Such finite supplies are included in this text only because they are usually included with other 'alternative' supplies. We recognize, however, that the processes for extracting and using geothermal power are more akin to those for centralized fossil fuels than the dispersed renewable supplies of the natural environment. For this reason we only present a short chapter on geothermal energy.

15.2 Geophysics

A section through the earth is shown in Fig. 15.1. Heat transfer from the semifluid mantle maintains a temperature difference across the relatively thin crust of 1000°C, and a mean temperature gradient of $\sim 30°C\,km^{-1}$. The crust solid material has a mean density $\sim 2700\,kg\,m^{-3}$, specific heat capacity $\sim 1000\,J\,kg^{-1}\,K^{-1}$ and thermal conductivity $\sim 2\,W\,m^{-1}\,K^{-1}$. Therefore the

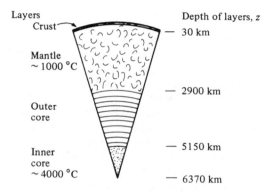

Fig. 15.1 Section through the earth, showing average lower depths of named layers. The crust has significant variation in composition and thickness over a local scale of several kilometers

average geothermal flux is $\sim 0.06\,W\,m^{-2}$, and $\sim 10^{20}\,J\,km^{-2}$ of heat is stored in the crust above surface temperature. If just 0.1% of this heat was removed in 30 years, the heat power available would be $100\,MW\,km^{-2}$. These calculations give the order of magnitude of the quantities involved and show that geothermal sources are a large potential energy supply.

Heat passes from the crust by (1) natural cooling and friction from the core, (2) radioactive decay of elements such as uranium and thorium, and (3) chemical reactions. The time constants of such processes over the whole earth are so long that it is not possible to know whether the earth's temperature is presently increasing or decreasing. The radioactive elements are concentrated in the crust by fractional recrystallization from molten material, and are particularly pronounced in granite. However, the production of heat by radioactivity or chemical action is only significant over many millions of years (see Problem 15.2), and the use of geothermal heat usually relies on draining heat stored in the thermal capacity of solid material and water in the crust.

Through uniform material with heat flowing outward from the mantle there will be a constant temperature gradient if conduction is the only heat transfer mechanism. The temperature gradient is high in poorly conducting solid strata, and low in regions of increased heat transfer where convection occurs usually by water. If radioactive or exothermic chemical heat sources occur there will be anomalous temperature gradients.

The earth's crust consists of large plates (Fig. 15.2). At the plate boundaries there is active convective thermal contact with the mantle, evidenced by seismic activity, volcanoes, geysers, fumaroles and hot springs. The geothermal energy potential of these regions is very great, owing to increased temperature gradients (to $\sim 100\,°C\,km^{-1}$) and to active release of water as steam or superheated liquid, often at considerable pressure when tapped by drilling.

Moderate increases in temperature gradient to $\sim 50\,°C\,km^{-1}$ occur in localized regions away from plate boundaries, owing to anomalies in crust composition and structure. Heat may be released from such regions naturally by deep penetration of water in aquifers and subsequent convective water flow. The resulting hot springs, with increased concentrations of dissolved chemicals, are often famous as health spas. Deep aquifers can be tapped by drilling to become sources of heat at temperatures from ~ 50 to $\sim 200°C$. If the anomaly is associated with material of low thermal conductivity, i.e. dry rock, then a high temperature gradient occurs with a related increase in stored heat.

Geothermal information has been obtained from mining, oil exploration and geological surveys. Only since about 1975 has there been a determined effort to obtain direct geothermal information, and now at least 80 countries have active geothermal surveying programs. The most important parameter is temperature gradient, and accurate measurements depend on leaving the drill hole

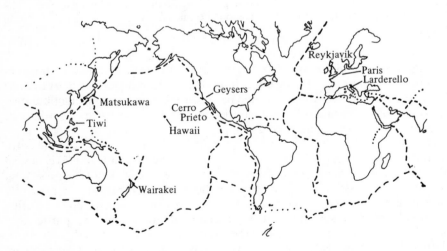

Fig. 15.2 World geothermal activity. Some well-known generation and thermal fields are indicated.   _ _ _ _ _ main plate boundaries . . . . regions of strain

undisturbed so that temperature equilibrium is re-established after drilling. Deep drilled survey wells commonly reach depths of 6 km, and the technology is available to drill to 15 km. The principal components of a geothermal energy plant are the boreholes, and so heat extraction from depths to 15 km can also be contemplated.

There are three classes of geothermal region:

(1) *Hyperthermal* Temperature gradient $\gtrsim$ 80 °C km^{-1}. These regions are usually on tectonic plate boundaries. The first such region to be tapped for electricity generation was in 1904 at Larderello in Tuscany, Italy. Nearly all geothermal power stations are in such areas.
(2) *Semithermal* Temperature gradient ~ 40 °C km^{-1} to 80 °C km^{-1}. Such regions are associated generally with anomalies away from plate boundaries. Heat extraction is from harnessing natural aquifers or fracturing dry dock. A well-known example is the geothermal district heating system for houses in Paris.
(3) *Normal* Temperature gradient < 40 °C km^{-1}. These remaining regions are associated with average geothermal conductive heat flow at ~ 0.06 W m^{-2}. It is unlikely that these areas can ever supply heat at competitive prices to present finite, or future renewable, supplies.

In each class it is, in principle, possible for heat to be obtained by:

(1) *Natural hydrothermal circulation*, in which water percolates to deep aquifers to be heated to dry steam, vapor/liquid mixtures, or hot water. Emissions of each type can be observed in nature. If pressure increases by steam formation at deep levels, spectacular geysers may occur, as at the Geysers near Sacramento in California and in the Wairakei area near Rotorua in New Zealand. Note, however, that liquid water is ejected, and not steam.
(2) *Hot igneous systems* associated with heat from semimolten magma that solidifies to lava. The first power plant using this source was the 3 MW(e) station in Hawaii, completed in 1982.
(3) *Dry rock fracturing* Poorly conducting dry rock, e.g. granite, stores heat over millions of years with a subsequent increase in temperature. Artificial fracturing from boreholes enables water to be pumped through the rock to extract the heat.

In practice geothermal energy plants in hyperthermal regions are associated with natural hydrothermal systems; in semithermal regions both hydrothermal and hot rock extraction is developed; and normal areas have too small a temperature gradient for commercial interest.

15.3 Dry rock and hot aquifer analysis

15.3.1 *Dry rock*

We consider a large mass of dry material extending from near the earth's surface to deep inside the crust (Fig. 15.3). The rock has density ρ_r, specific heat capacity

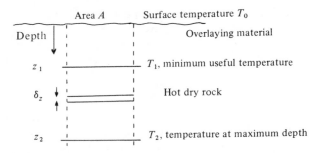

Fig. 15.3 Profile of hot dry rock system for calculating the heat content. Density ρ, specific heat capacity c, temperature gradient $dT/dz = G$

c_r and cross-section A. With uniform material and no convection, there will be a linear increase of temperature with depth. If z increases downward from the surface at $z = 0$,

$$T = T_0 + \frac{dT}{dz}z = T_0 + Gz \qquad (15.1)$$

Let the minimum useful temperature be T_1 at depth z_1, so

$$T_1 = T_0 + Gz_1; \; z_1 = \frac{T_1 - T_0}{G} \qquad (15.2)$$

The useful heat content δE, at temperature T greater than T_1, in an element of thickness δz at depth z is

$$\delta E = (\rho_r A \delta z)c_r(T - T_1) = (\rho_r A \delta z)c_r G(z - z_1) \qquad (15.3)$$

The total useful heat content of the rock to depth z_2 becomes

$$E_0 = \int_{z=z_1}^{z_2} \rho_r A c_r G(z - z_1)\, dz$$

$$= \rho_r A c_r G[z^2/2 - zz_1]_{z_1}^{z_2}$$

$$= \rho_r A c_r G(z_2 - z_1)^2/2 \qquad (15.4)$$

Alternatively, let the average available temperature greater than T_1 be θ:

$$\theta = (T_2 - T_1)/2 = G(z_2 - z_1)/2 \qquad (15.5)$$

Then $E_0 = C_r \theta$

where C_r is the thermal capacity of the rock between z_1 and z_2,

$$C_r = \rho_r A c_r (z_2 - z_1) \tag{15.6}$$

So as (15.4),

$$E_0 = \rho_r A c_r G (z_2 - z_1)^2 / 2 \tag{15.7}$$

Assume heat is extracted from the rock uniformly in proportion to the temperature greater than T_1 by a flow of water; volume flow rate $\dot{V}$, density ρ_w, specific heat capacity c_w. The water will be heated through a temperature difference of θ in the near perfect heat exchange process.

Thus

$$\dot{V} \rho_w c_w \theta = - C_r \frac{d\theta}{dt} \tag{15.8}$$

$$\frac{d\theta}{\theta} = - \frac{\dot{V} \rho_w c_w}{C_r} dt = - \frac{dt}{\tau} \tag{15.9}$$

and

$$\theta = \theta_0 e^{-t/\tau} \tag{15.10}$$

The useful heat content $E = C_r \theta$. So

$$E = E_0 e^{-t/\tau} \tag{15.11}$$

and

$$\frac{dE}{dt} = - \frac{E_0}{\tau} e^{-t/\tau} \tag{15.12}$$

where the time constant τ is given by

$$\tau = \frac{C_r}{\dot{V} \rho_w c_w} = \frac{\rho_r A c_r (z_2 - z_1)}{\dot{V} \rho_w c_w} \tag{15.13}$$

Example 15.1 (after Garnish, 1976)

(1) Calculate the useful heat content per square kilometer of dry rock granite to a depth of 7 km. Take the geothermal temperature gradient at 40°C km^{-1}, the minimum useful temperature as 140 K above the surface temperature, $\rho_r = 2700 \, \text{kg m}^{-3}$, $c_r = 820 \, \text{J kg}^{-1} \text{K}^{-1}$.

(2) What is the time constant for useful heat extraction using a water flow rate of $1\,\mathrm{m^3\,s^{-1}\,km^{-2}}$?

(3) What is the useful heat extraction rate initially and after 10 years?

Solution

(1) At $7\,\mathrm{km}$ the temperature $T_2 = 280\,\mathrm{K}$ above T_0. The minimum useful temperature of $140\,\mathrm{K}$ above T_0 occurs at $3.5\,\mathrm{km}$. By (15.7),

$$E_0/A = \rho_r c_r (z_2 - z_1)(T_2 - T_1)/2$$

$$= (2.7 \times 10^3\,\mathrm{kg\,m^{-3}})(0.82 \times 10^3\,\mathrm{J\,kg^{-1}\,K^{-1}})(3.5\,\mathrm{km})(70\,\mathrm{K})$$

$$= 5.42 \times 10^{17}\,\mathrm{J\,km^{-2}} \tag{15.14}$$

(2) Substituting in (15.13)

$$\tau = \frac{\rho_r c_r A (z_2 - z_1)}{\dot{V} \rho_w c_w}$$

$$= \frac{1}{1\,\mathrm{m^3\,s^{-1}\,km^{-2}}} \frac{2.7}{1} \frac{0.82}{4.2} (1\,\mathrm{km^{-2}})(3.5\,\mathrm{km})$$

$$= 1.84 \times 10^9\,\mathrm{s} = 58\,\mathrm{y} \tag{15.15}$$

(3) By (15.12),

$$\left(\frac{dE}{dt}\right)_{t=0} = \frac{5.42 \times 10^{17}\,\mathrm{J\,km^{-2}}}{1.84 \times 10^9\,\mathrm{s}} = 294\,\mathrm{MW\,km^{-2}} \tag{15.16}$$

$$\left(\frac{dE}{dt}\right)_{t=20y} = 294\exp(-10/58) = 247\,\mathrm{MW\,km^{-2}} \tag{15.17}$$

15.3.2 *Hot aquifers*

In a hot aquifer the heat resource lies within a layer of water deep beneath the ground surface (Fig. 15.4). We assume that the thickness of the aquifer (h) is much less than the depth (z_2) below ground level, and that consequently the water is all at temperature T_2. The fraction of the aquifer containing water is the porosity p', with the remaining space of rock of density ρ_r. The minimum useful temperature is T_1. The characteristics of the resource are calculated similarly to those for dry rock in Section 15.3.1.

$$T_2 = T_0 + \frac{dT}{dz}z = T_0 + Gz \tag{15.18}$$

$$\frac{E_0}{A} = C_a(T_2 - T_1) \tag{15.19}$$

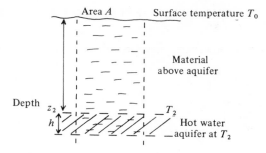

Fig 15.4 Profile of hot aquifer system for calculating the heat content

where

$$C_a = [p' \rho_w c_w + (1-p') \rho_r c_r] h \tag{15.20}$$

As with (15.8) onwards, we calculate the removal of heat by a water volume flow rate $\dot{V}$ at θ above T_1:

$$\dot{V} \rho_w c_w \theta = -C_a \frac{d\theta}{dt} \tag{15.21}$$

So

$$E = E_0 \exp(-t/\tau_a) \tag{15.22}$$

$$\frac{dE}{dt} = -(E_0/\tau_a) \exp(-t/\tau_a) \tag{15.23}$$

and

$$\tau_a = \frac{C_a}{\dot{V}\rho_w c_w} = \frac{[p' \rho_w c_w + (1-p')\rho_r c_r]h}{\dot{V}\rho_w c_w} \tag{15.24}$$

Example 15.2 (after Garnish, 1976)

(1) Calculate the initial temperature, and heat content per square kilometer above 40 °C, of an aquifer of thickness 0.5 km, depth 3 km, porosity 5%, under sediments of density 2700 kg m^{-3}, specific heat capacity 840 J kg^{-1} K^{-1}, temperature gradient 30 °C km^{-1}. Suggest a use for the heat if the average surface temperature is 10 °C.

(2) What is the time constant for useful heat extraction with a pumped water extraction of $100 \, 1\text{s}^{-1}\text{km}^{-2}$?

(3) What is the thermal power extracted initially and after 10 years?

Solution
(1) Initial temperature

$$T_2 = 10\,°\text{C} + (30 \times 3)\,\text{K} = 100\,°\text{C} \tag{15.25}$$

From (15.20),

$$C_a = [(0.05)(1000)(4200) + (0.95)(2700)(840)](\text{kg}\,\text{m}^{-3}\,\text{J}\,\text{kg}^{-1}\,\text{K}^{-1})(0.5\,\text{km})$$

$$= 1.18 \times 10^{15}\,\text{J}\,\text{K}^{-1}\,\text{km}^{-2} \tag{15.26}$$

With (15.19),

$$E_0 = (1.18 \times 10^{15}\,\text{J}\,\text{K}^{-1}\,\text{km}^{-2})\,(100 - 40)\,°\text{C}$$

$$= 0.71 \times 10^{17}\,\text{J}\,\text{km}^{-2} \tag{15.27}$$

The quality of the energy is suitable for factory processes or household district heating.

(2) In (15.24),

$$\tau_a = \frac{(1.2 \times 10^{15}\,\text{J}\,\text{K}^{-1}\,\text{km}^{-2})}{(0.1\,\text{m}^3\text{s}^{-1}\text{km}^{-2})(1000\,\text{kg}\,\text{m}^{-3})(4200\,\text{J}\,\text{kg}^{-1}\,\text{K}^{-1})}$$

$$= 2.8 \times 10^9\,\text{s} = 90\,\text{y} \tag{15.28}$$

(3) From (15.23),

$$\left(\frac{dE}{dt}\right)_{t=0} = \frac{(0.71 \times 10^{17}\,\text{J}\,\text{km}^{-2})}{(2.8 \times 10^9\,\text{s})}$$

$$= 25\,\text{MW}\,\text{km}^{-2} \tag{15.29}$$

Check:

$$\left(\frac{dE}{dt}\right)_{t=0} = \dot{V}\rho_w c_w (T_2 - T_1)$$

$$= (0.1\,\text{m}^3\text{s}^{-1}\text{km}^{-2})(1000\,\text{kg}\,\text{m}^{-3})(4200\,\text{J}\,\text{kg}^{-1}\,\text{K}^{-1})(60\,\text{K})$$

$$= 25\,\text{MW}\,\text{km}^{-2}$$

From (15.23),

$$\left(\frac{dE}{dt}\right)_{t\,=\,10y} = 25\,\text{MW km}^{-2}\exp(-10/90)$$

$$= 22\,\text{MW km}^{-2} \tag{15.30}$$

15.4 Harnessing geothermal resources

Geothermal power arises from heat sources having a great range of temperatures and local peculiarities. In general, available temperatures are much less than from furnaces, and although much energy is accessible the thermodynamic quality is low. The sources share many similarities with industrial waste heat processes and ocean thermal energy conversion (Chapter 14). In this chapter we shall review the strategy for using geothermal energy.

15.4.1 *Matching supply and demand*

With a geothermal source it is always sensible to attempt electricity generation since this is a valued product, and the rejected heat can be used in a combined heat and power mode. Electricity can be distributed on a widely dispersed grid and integrates with other national power supplies. Nevertheless, the energy demand for heat at $< 100\,°C$ is usually greater than that for electricity, and so the use of geothermal energy as heat is important. Electricity generation will probably be attractive if the source temperature is $> 300\,°C$, and unattractive if $< 150\,°C$.

Heat cannot be distributed easily over distances greater than $\sim 30\,\text{km}$, and so concentrated uses near to the point of supply are needed. In cold climates, household and business district heating schemes make sensible loads if the population density is $\gtrsim 350$ people km^{-2} (> 100 premises km^{-2}). Thus a $100\,\text{MW(th)}$ geothermal plant might serve an urban area $\sim 20\,\text{km} \times 20\,\text{km}$ at $\sim 2\,\text{kW}$ per premises. Such geothermal schemes have been long established in Iceland and, on a smaller scale, in New Zealand. Other heating loads are for glasshouse heating (at $60\,\text{MW(thermal)}\,\text{km}^{-2}$ in one installation in northern Europe), fish farming, food drying, factory processes etc.

Several factors fix the scale of geothermal energy use. The dominant costs are capital costs, especially for the boreholes, whose costs increase exponentially with depth. Since temperature increases with depth, and the value of the energy increases with temperature, most schemes settle on optimum borehole depths of $\sim 5\,\text{km}$. Consequently, the scale of the energy supply output is usually $\gtrsim 100\,\text{MW}$ (electricity and heat for high temperatures, heat only for low temperatures), as shown in Examples 15.1 and 15.2.

The total amount of heat extracted from a geothermal source can be increased by reinjecting the partially cooled water from the above ground heat exchanger.

This has the extra advantage of disposing of this water, which may have about $25\,kg\,m^{-3}$ of solute and be a substantial pollutant. Nevertheless a substantial extra cost is incurred.

15.4.2 *Extraction techniques*

The most successful geothermal projects have boreholes sunk into natural water channels in hyperthermal regions (Fig. 15.5). This is the method used at the Geysers, California, and Wairakei, New Zealand, where there is a considerable overpressure in the boreholes. Similar methods are used for extraction from hot aquifers in semithermal regions, where natural convection can be established from the borehole without extra pumping.

Recent interest has concentrated on extraction from dry hot rocks since these may be more extensive than wet aquifer sources. The leading development group (at the Los Alamos Scientific Laboratory, USA) has pioneered methods

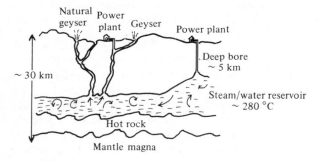

Fig. 15.5 Schematic diagram, not to scale, of hydrothermal power stations in a hyperthermal region, e.g. the Geysers geothermal field, California

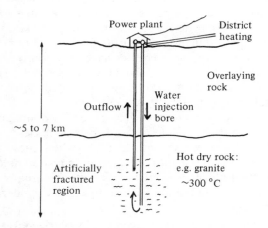

Fig 15.6 Schematic diagram of heat extraction from a hot dry rock system

of fracturing the rock with pressurized cold water around the end of the injection borehole (Fig. 15.6). After the initial fracturing, water is pumped down the injection bore to percolate through the hot rock at depths of ~5km and temperatures ~250°C before returning through shallower return pipes. Complex arrays of injection and return boreholes might enable gigawatt supplies of heat to be obtained.

15.4.3 *Electricity generating systems*

The choice of the heat exchange and turbine system for a particular geothermal source is complex, requiring specialist experience. Milora and Tester (1977) provided one of the first extensive reviews of the subject. Fig. 15.7 outlines some of the possible arrangements for generating plant.

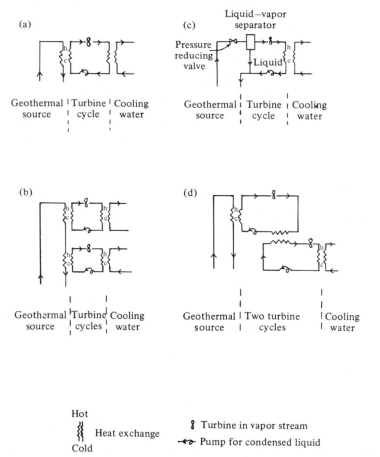

Fig. 15.7 Example of geothermal turbine cycles for electricity generation. After Milora and Tester (1977). (a) Single fluid cycle, e.g. water, freon. (b) Two fluid series cycles, e.g. first water, second freon. (c) Direct steam flashing cycle. (d) Topping/bottoming cycles

Table 15.1 Electricity generating capacity using only geothermal sources. The countries with major experience are listed, but about 50 other countries have plans for generation using geothermal sources in whole or part (after Goodman and Love, 1980, and other sources)

Country	Major fields	National installed capacity 1980/MW (e)	Realistic further development /MW(e)
USA	Geysers	700	~ 10 000
Italy	Larderello	420	—
New Zealand	Wairakei	250	~ 100
Mexico	Cerro Prieto	150	~ 5 000
Japan	Matsukawa	250	~ 10 000
Philippines		250	1 000
USSR	Pauzhetsk	6	—
El Salvador		100	—
Iceland	Namafjall	40	500

The low source temperature requires that compounds other than water may be considered to drive the turbines (e.g. freons or toluene), and novel techniques are needed to improve efficiency. Particular difficulties may occur with heat exchangers owing to the high concentration of chemicals in the borehole water. In Chapter 14, similar problems with heat exchangers are discussed for ocean thermal energy conversion.

Table 15.1 lists major geothermal electricity generating systems, with an indication of the very substantial increase expected in future years. Capital costs are now \$1500 to \$2500 per installed kilowatt (electric) capacity, which are competitive with those of nuclear and hydro power stations.

Problems

15.1 (a) A cube of 'hot rock' of side h has its top surface at a depth d below the earth's surface. The rock has a density ρ and specific heat capacity c. The material above the cube has thermal conductivity k. If the rock is treated as an isothermal mass at temperature T above the earth's surface with no internal heat source, show that the time constant for cooling is given by

$$\tau = \frac{\rho h c d}{k}$$

(b) Calculate τ for a cubic mass of granite (of side 10 km, density $2.7 \times 10^3 \, kg \, m^{-3}$, specific heat capacity $0.82 \times 10^3 \, J \, kg^{-1} \, K^{-1}$), that is 10 km below ground under a uniform layer material of thermal conductivity $0.40 \, J \, m^{-1} \, s^{-1} \, K^{-1}$.

(c) Compare the natural conductive loss of heat from the granite with commercial extraction at 100 MW from the whole mass.

15.2 (a) Calculate the thermal power produced from the radioactive decay of ^{238}U in $5\,km^3$ of granite. (^{238}U is 99% of the uranium in granite, and is present in average at a concentration of $4 \times 10^{-3}\%$. The heat produced by pure ^{238}U is $3000\,J\,kg^{-1}y^{-1}$.)

(b) ^{238}U radioactivity represents about 40% of the total radioactive heat source in granite. Is the total radioactive heat a significant continuous source of energy for geothermal supplies?

Solutions

15.1 (a) Heat flow through the surface material with a temperature difference T is

$$P = kAT/d$$

Heat lost by granite mass is

$$P = \rho V c\, dT/dt$$

Thus

$$dT/dt = [kA/\rho Vc]T/d$$

So

$$T = T_0 e^{-t/\tau}$$

where

$$\tau = \rho Vcd/kA = \rho hcd/k$$

(b) $\tau \approx 5 \times 10^{11}\,s \approx 10\,000\,y$

15.2 (a) Mass of granite $= (5 \times 10^9\,m^3)(2700\,kg\,m^{-3}) = 13.5 \times 10^{12}\,kg$

Mass of $^{238}U = (4 \times 10^{-5})(13.5 \times 10^{12}\,kg) = 54.0 \times 10^7\,kg$

$$\text{Heat produced} = \frac{(54 \times 10^7\,kg)(3000\,J\,kg^{-1})}{(3.1 \times 10^7\,s)} = \frac{162}{3.1}\frac{10^{10}}{10^7}\,W$$

$$= 50\,kW$$

(b) Total thermal power produced $= 50/0.4\,kW = 120\,kW$. This is insignificant compared with geothermal power extraction of $\sim 100\,MW(th)$ from such a mass of dry rock.

Bibliography

Armstead, H. C. H. (1978) *Geothermal Energy*, E. and F. N. Spon, London. Fundamental text, often quoted.

Garnish, J. D. (1976) *Geothermal Energy: The Case for Research in the UK*, Department of Energy paper no. 9, HMSO, London.
   Succinct evaluation with basic analysis.
Goodman, L. J. and Love, R. N. (1980) *Geothermal Energy Projects*: *Planning and Management*, Pergamon Press, New York.
   Useful for case studies of complete projects, especially Wairakei in New Zealand.
Milora, S. L. and Tester, J. W. (1977) *Geothermal Energy as a Source of Electric Power*, MIT Press, Cambridge, Mass.
   Evaluates and analyzes the varied and specialized turbine cycles and systems suitable for geothermal heat sources.
Rowley, J. C. (1977) Geothermal Energy, *Physics Today*, 36–45.

16 *Energy storage and distribution*

16.1 The importance of energy storage and distribution

Energy is useful only if available when and where it is wanted. Carrying energy to *where* it is wanted is called *distribution*; keeping it available until *when* it is wanted is called *storage*.

As discussed in Chapter 1, renewable energy supplies have different requirements for storage and distribution than do fossil or nuclear energy supplies. In particular the low intensity and wide distribution of renewable sources favor decentralized end-use. Therefore energy from renewable sources will often not require much further distribution since the sources are already distributed. However, some renewable sources can fruitfully be harnessed in a moderately centralized manner (e.g. large hydroelectric schemes), and we therefore discuss some mechanisms of large scale distribution in Section 16.8.

Since use of renewable energy supplies constitutes a diversion of a continuing natural *flow* of energy, there are problems in matching supply and demand in the time domain, i.e. in matching the *rate* at which energy is used. This varies with time on scales of months (e.g. house heating in temperate climates), days (e.g. artificial lighting) and even seconds (e.g. starting motors). In contrast to fossil fuels and nuclear power, the initial input power of renewable energy sources is outside our control. As discussed more fully in Chapter 1, we have the choice of either matching the load to the availability of renewable energy supply, or storing the energy for future use. Energy can be stored in many forms, i.e. chemical energy, heat, electric, potential energy, gravitational potential energy or kinetic energy. This chapter discusses these different storage forms, in sufficient detail for users to choose the most suitable system for a particular application.

To help in this selection, Table 16.1 and Fig. 16.1 summarize the performance of various storage mechanisms. 'Performance' can be measured in $MJ\$^{-1}$, $MJ\,m^{-3}$ and $MJ\,kg^{-1}$. Of these, the first is usually the deciding factor but is the hardest to estimate. The second is important when space is at a premium (e.g. in buildings of fixed size). The third is considered when weight is vital (e.g. in aircraft). In this chapter we show how these performance figures are estimated.

Energy storage is not a new concept in energy planning. Fossil fuels are

Table 16.1 Storage devices and their performance[a]

Store	Energy density MJ kg^{-1}	Energy density MJ l^{-1}	Operating temp/°C	Likely commercial Development time/y	Likely commercial Operating value (MJ US\$$^{-1}$)	Conversion Type	Conversion Efficiency (%)
Conventional fuels							
Diesel oil	45	39	Ambient	In use	200[b]	Chem. → work	30
Coal	29	45	Ambient	In use	1000[b]	Chem. → work	30
Wood	15	7	Ambient	In use	200–∞[b]	Chem. → heat	60
Other chemicals							
Hydrogen gas	140	1.7[d]	(−253)–30	10	0.1–10[b]	Elect. → chem.	60
Ammonia (to $N_2 + H_2$)	2.9	0.3[d]	0–700	5	~1[b]	Heat → chem.	70
FeTiH$_{1.7}$ (releases H_2)	1.8	20	100	10	1[b]	Chem. → chem.	90
Sensible heat							
Water	0.2	0.2	20–100	In use	3–100[c]	Heat → heat	50–100[e]
Cast iron	0.05	0.4	20–400	In use	0.2[c]	Heat → heat	50–90[e]
Heat (phase change)							
Steam	2.2	0.02[f]	100–300	In use	10	Heat → heat	70[d]
Na$_2$SO$_4 \cdot$ 10H$_2$O	0.25	0.29	32	In use	2[c]	Heat → heat	80
Electrical							
Capacitors	—	10^{-6}		Unlikely			
Electromagnets	—	10^{-3}		Unlikely			
Batteries (in practice)							
Lead acid	0.10	0.29	Ambient	In use	0.1[c]	Elect. → elect.	75
Sodium sulfur	0.65		350	10	0.1[c]	Elect. → elect.	75
Li/TiS$_2$	0.48		100	10	0.1[c]	Elect. → elect.	75
Fuel cell	—	—	150	10	10[b]	Chem. → elect.	38
Mechanical							
Pumped hydro	0.001	0.001	Ambient	In use	1[c]	Elect. → elect.	80
Flywheel (steel disk)	0.05	0.4	Ambient	In use	0.05[c]	Elect → elect.	80
Flywheel (composite)	0.05	0.15	Ambient	In use	0.05	Elect. → elect	80
Compressed air	0.2–2	2[f]	20–1000	In use	0.3	Elect. → elect.	50

(a) These figures are for 'typical' operation and are only approximations for a particular application. This is especially true for those relating to commercial operation, e.g. costs. Most of the data are based on Jensen (1980).

(b) Energy throughput per unit cost.

(c) Energy capacity per unit cost; e.g. if a battery cycles 10 times to 50% discharge, its throughput per unit cost would be $(0.1 \, \text{MJ} \, \$^{-1})(10)(50\%)$ because it may be used repeatedly.

(d) At 150 atmospheres pressure.

(e) Depends on time and heat leaks.

(f) At 20 atmospheres pressure.

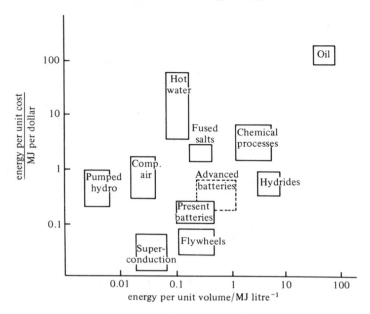

Fig. 16.1 Energy per unit cost versus energy per unit volume of storage methods (1980 prices, $US). Note the superiority of oil. For more details see Table 16.1

effectively energy stores, whose energy density is high. However, as these conventional sources become increasingly unavailable and expensive, there is a need to develop other storage methods, and indeed to make renewable fuels.

16.2 Biological storage

The growth of plants by photosynthesis, and the storage of the solar energy in oxygen, is fully discussed in Chapter 10. This energy is released in the combustion of biological and fossil fuel material. We may therefore consider the energy as stored in the material itself, although strictly this is not correct. The heats of combustion of many of these materials are listed in Appendix B, Table B.6. The many chemical paths for producing biofuels are described in Chapter 11 (see Fig. 11.2).

Some of the biofuels are liquids and gases that may be used in internal combustion engines, and are therefore important for transport in place of conventional petroleum fuels. The generation of electricity by such fuels in small diesel engines is also important. These and other aspects of biological energy storage are fully discussed in Chapter 11.

16.3 Chemical storage

16.3.1 *Introduction*

Energy can be held in the bonds of many chemical compounds and released by exothermic reactions, notably combustion. Sometimes it is necessary to apply heat or other catalysts (e.g. enzymes) to promote the desired reaction. Biological compounds are a special case. Here we discuss the most important inorganic compounds which have been suggested as practical energy stores by means of their combustion in air.

16.3.2 *Hydrogen*

Hydrogen can be made from water by electrolysis, using any source of DC electricity. The gas can be stored, distributed and burnt to release heat. The only product of combustion is water, so no pollution results. The enthalpy change is $\Delta H = -242\,\mathrm{kJ\,mol^{-1}}$; i.e. 242 kJ are released for every mole (18g) of H_2O formed.

Hydrogen (with CO in the form of 'town gas' made from coal) was used for many years as an energy store and supply, and there is no overriding technical reason why hydrogen based systems could not come into wide use again.

Electrolysis is a well-established commercial process yielding pure hydrogen, but at efficiencies of only $\sim 60\%$. Some of this loss is due to electrical resistance in the circuit, especially around the electrodes where the evolving bubbles of gas block the current carrying ions in the water. New electrodes with bubble removing mechanisms have been tested. These electrodes are also porous, giving a greater effective area, and thus allowing a higher current density. Higher current densities imply fewer cells and therefore lower cost for a given gas output. Efficiencies $\sim 80\%$ have been obtained, and can be increased further by using (expensive) catalysts.

High temperatures also promote the decomposition of water. The change in Gibbs free energy associated with a reversible electrochemical reaction at absolute temperature T is

$$\Delta G = -nF\xi$$

$$= \Delta H - T\Delta S \tag{16.1}$$

where ξ is the electrical potential, ΔH is the enthalpy change and ΔS is the entropy change, $F = 96\,500\,\mathrm{C\,mol^{-1}}$ is Faraday's constant, and n is the number of moles of reactant.

The decomposition reaction

$$H_2O \rightarrow H_2 + \tfrac{1}{2}O_2 \tag{16.2}$$

has ΔG, ΔH, ΔS all positive. Then (16.1) shows that as T increases the electric potential ξ required for decomposition decreases. Problem 16.1 shows that $\xi = 0$

for $T \approx 2700\,K$, so it is impracticable to decompose water by straightforward heating.

A more promising idea is to replace some of the input electrical energy by heat from a cheaper source. Heat at $T \approx 1000\,K$ from solar concentrators may be cheaper than electricity, and this may be the cheapest route to hydrogen.

A technical difficulty in the electrolysis of sea water is that chlorine may also be evolved at the 'oxygen' electrode. Approximate chemical calculations suggest that the O_2 can be kept pure if the applied voltage per cell is kept below $1.8\,V$ (Bockris, 1975). Unfortunately this limits the current density, so that electrodes of large surface area would be needed.

Several other methods of producing hydrogen without using fossil fuels have been tried in the laboratory – even including the use of special algae which 'photosynthesize' H_2 (see Section 10.7) – but none has yet shown worthwhile efficiencies.

To store hydrogen in large quantities is not trivial. Most promising is the use of underground caverns, such as those from which natural gas is now extracted. But storage of gas – even if compressed – is bulky. Hydrogen can be liquefied, but since its boiling point is $20\,K$ these stores are awkward to maintain. Chemical storage as metal hydrides, from which the hydrogen can be released by heating, is more manageable and allows large volumes of H_2 to be stored (see Table 16.1). For example

$$FeTiH_{1.7} \xrightarrow{\ T \sim 50°C\ } FeTiH_{0.1} + 0.8\,H_2 \tag{16.3}$$

This reaction is reversible, so that a portable hydride store can be replenished with hydrogen at a central 'filling station'. The heat released in this process can be used for district heating, and the portable hydride store can be used as the 'fuel tank' of a vehicle. The main difficulty is the weight and cost of the metals used (Table 16.1). Hydrogen can also be distributed through the extensive pipeline networks already used for natural gas in many countries. It is also possible to convert H_2 very efficiently to electricity, by means of fuel cells (Section 16.6).

16.3.3 *Ammonia*

Unlike water, ammonia can be dissociated at realizable temperatures:

$$N_2 + 3\,H_2 \rightleftharpoons 2\,NH_3 \tag{16.4}$$

In conjunction with a heat engine, these reactions form the basis of what may be the most efficient way to generate continuous electrical power from solar heat. The system proposed by Carden is described in Section 6.9.

16.4 Heat storage

A substantial fraction of world energy use is as low temperature heat. For example, Fig. 16.2 shows the demand in Britain for total energy and for space

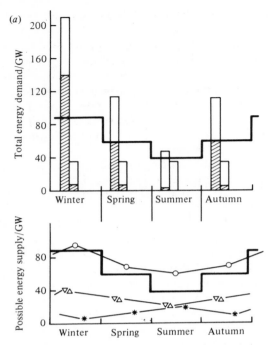

Fig. 16.2 (a) Seasonal changes in total demand for energy for all purposes in Great Britain. All figures are in gigawatts of electrical equivalent. The two bars for each season correspond to demand on (1) the coldest day of the season, and (2) during sleeping hours. Hatched areas represent demand for space heat and hot water (for public buildings, housing and industry). The heavy line shows what the demand would be if 7 day storage of heat was available to smooth out the diurnal pattern and weather fluctuations. Note that the hatched areas show when heat is *wanted*, so that even (existing) 12 hours storage substantially reduces the day/night fluctuations shown. (b) Firm power that could *potentially* be obtained in Great Britain from solar heat (*), wind (▽) and waves (△), allowing for 7 day storage of heat. Note that the total power available from these three renewable sources (O) exceeds the smoothed demand (heavy line). For details of assumptions, see Problem 16.4 and Ryle (1977)

heating. We see that in winter over half of the national energy consumption is for space heating at temperatures of about 18 ± 3 °C.

It is usually not sensible to meet this demand for heat from the higher grade energy sources, which are best saved for higher grade uses (Section 1.3.3). It would be better to use passive solar heat (Section 6.4), and to keep the house comfortably warm overnight using heat storage.

Likewise only when energy is to be used at 'low' temperatures, as for space conditioning, is it worth storing it as heat. Heat storage also provides a way of fruitfully using 'waste' energy shed from other processes, e.g. by load control devices (Section 1.4.4).

In the higher latitudes, solar heat is significantly greater in summer than winter (see Figs 4.7 and 4.10), whereas the demand for heat is greatest in winter. Therefore to have the maximum benefit from solar heating requires a way of storing heat for at least 3 months.

Let us therefore estimate the time, t_{loss}, it takes for a heat store to lose 50% of its heat while maintaining a uniform temperature T_s, placed in an environment of temperature T_a. Its heat balance is

$$mc\frac{\partial T_s}{\partial t} = -\frac{T_s - T_a}{R} \tag{16.5}$$

where mc is its heat capacity, and R is the thermal resistance between the store and the surroundings. The solution of (16.5) is

$$\frac{T_s - T_a}{T_s(0) - T_a} = \exp\left(-\frac{t}{mcR}\right) \tag{16.6}$$

from which it follows that

$$t_{loss} = 1.3\,mcR \tag{16.7}$$

If the store is a sphere of radius a, the thermal resistance is $R = r/4\pi a^2$, where r is the thermal resistivity of unit area, and $m = 4\pi a^3 \rho/3$, so for a sphere,

$$t_{loss} = 0.43\,\rho cra \tag{16.8}$$

Example 16.1 Size and insulation of a domestic heat store
A small well-insulated passive solar house requires an average internal heat supply of 1 kW. Together with the free gains of lighting etc., this will maintain an internal temperature of 20°C. It is decided to build a hot water store in a rectangular tank whose top forms the floor of the house, of area 200 m². The heating must be adequate for 100 days as all the heat loss from the tank passes by conduction through the floor, and as the water cools from an initial 60°C to a final 40°C.

(1) Calculate the volume of the tank and
(2) the thermal resistivity of the heat path from the tank to the floor.
(3) Suggest how the tank should be enclosed thermally.
(4) What is the energy density of storage?

Solution
(1) Heat required = $(1\,\text{kW})(100\,\text{day})(24\,\text{h}\,\text{day}^{-1})(3.6\,\text{MJ}\,\text{kWh}^{-1})$
$= 8640\,\text{MJ}$

Volume of water $= \dfrac{(8640\,\text{MJ})}{(1000\,\text{kg}\,\text{m}^{-3})(4200\,\text{J}\,\text{kg}^{-1}\,\text{K}^{-1})(20\,\text{K})} = 103\,\text{m}^3$

Depth of tank $= (103\,\text{m}^3)/(200\,\text{m}^2) = 0.5\,\text{m}$

(2) Assume the heat only leaves through the top of the tank.
From (16.7),

$$R = \frac{(100 \text{ day})(86\,400 \text{ s day}^{-1})}{(1.3)(103 \text{ m}^3)(1000 \text{ kg m}^{-3})(4200 \text{ J kg}^{-1}\text{K}^{-1})}$$

$$= 0.0154 \text{ K W}^{-1}$$

From (3.6) the thermal resistivity $r = R$ (area)
$$= (0.0154)(200 \text{ m}^2 \text{ K W}^{-1})$$
$$= 3.1 \text{ m}^2 \text{ K W}^{-1}$$

(3) Insulating material (e.g. expanded polystyrene) has a thermal conductivity $k \sim 0.04 \text{ W m}^{-1}\text{K}^{-1}$. A satisfactory layer on top of the tank, protected against excess pressure, would have a depth

$$d = (3.1 \text{ m}^2 \text{ K W}^{-1})(0.04 \text{ W m}^{-1}\text{K}^{-1}) = 12 \text{ cm}$$

To avoid unwanted heat loss, the base and sides should be insulated by the equivalent of 50 cm of expanded polystyrene.
(4) Energy density of the used storage above 40 °C $= (8640 \text{ MJ})/(103 \text{ m}^3)$
$$= 84 \text{ MJ m}^{-3}$$
Energy density above ambient house temperature at 20 °C $= 168 \text{ MJ m}^{-3}$
Note: an active method of extracting the heat by forced convection through a heat exchanger would enable better control, a lower initial temperature, and/or a smaller tank.

Example 16.1 shows that 3 month heat storage is realistic if this forms part of the initial design criteria, and if other aspects of the construction are considered. These include high standards of thermal insulation with damp proof barriers, controlled ventilation (possibly with recycling of heat), and the inclusion of free gains from lighting, cooking and metabolism. Examples exist of such high technology houses, and the best also have imaginative architectural features so that they are pleasant to live in. It is more usual to have rock bed storage than the water system of the example.

It follows from Example 16.1 that short term heat storage of about 4 days should be possible, with the fabric of the building used as the store. Similarly thermal capacity and cold storage can have important implications for building design in hot weather conditions.

Ryle (1977) argues cogently that the widespread use of such storage would allow much of the increased demand for heat in high latitude marine countries to be met by wind and wave power. Both these sources are strongest in winter, and their supply, though fluctuating hour by hour, rarely drops for more than a few days (see Fig. 16.2(b)).

Materials which change phase offer a much larger heat capacity, over a limited temperature range, than systems using sensible heat. For example, Glauber's salt ($Na_2SO_4 \cdot 10H_2O$) has been used as a store for room heating. It decomposes at 32 °C to a saturated solution of Na_2SO_4 plus an anydrous residue of Na_2SO_4. This reaction is reversible and evolves $250\,kJ\,kg^{-1} \approx 650\,MJ\,m^{-3}$. Since much of the cost of a store for house heating is associated with the construction, such stores may be cheaper overall than simple water tanks of lower energy density per unit volume. Unfortunately, some practical difficulties still remain. In particular, the phases often separate too much to allow them to fully recombine, so that after many cycles, the system becomes inefficient.

16.5 Electrical storage: the lead acid battery

Electricity is a high grade form of energy, and therefore great effort is made to find cheap and efficient means for storing it. A device that has electricity both as input and output is called an (electrical) storage battery or (electrical) accumulator. Usually the combination of electrolyzer and fuel cell is not included as 'electrical storage', however. Batteries form an essential component of almost all photovoltaic and small wind electric systems, and there is steady development of battery powered vehicles.

Although many electrochemical reactions are reversible in theory, few are suitable for a practical storage battery, which will be required to cycle hundreds of times between charging and discharging currents of 1–100 A. The most widely used storage battery is the lead acid battery, invented by Planté in 1860.

A battery is built up from cells, one of which is shown schematically in Fig. 16.3. As in all electrochemical cells, there are two solid plates immersed in a conducting solution (electrolyte). In this case the plates are in the form of grids holding pastes of lead and lead dioxide respectively. The electrolyte is sulfuric acid, which ionizes as follows:

$$H_2SO_4 \rightarrow H^+ + HSO_4^- \qquad (16.9)$$

During *discharge*, the reaction at the negative electrode is

$$Pb + HSO_4^- \rightarrow PbSO_4 + H^+ + 2e^- \qquad (16.10)$$

Lead (Pb) is oxidized to Pb^{2+} which is deposited as $PbSO_4$. The sulfate takes the place of the Pb paste in the plate. The electrons so liberated travel through the external circuit to the positive electrode, where they contribute to the reaction:

$$PbO_2 + HSO_4^- + 3H^+ + 2e^- \rightarrow PbSO_4 + 2H_2O \qquad (16.11)$$

This $PbSO_4$ likewise replaces the PbO_2 in that plate. The electrical current through the solution is carried by H^+ and HSO_4^- ions from the sulfuric acid, which themselves take part in the plate reactions.

Knowing the reactions involved and the corresponding standard electrode potentials (given in chemical tables), the theoretical energy density of any proposed battery can be calculated.

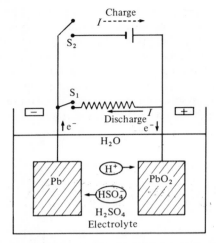

Fig. 16.3 Schematic diagram of lead acid cell. The charge carriers move in the direction shown during the discharge reaction. The reactions and carrier movements are reversed during charging (switch S1 open and S2 closed)

Example 16.2 Theoretical energy density of lead acid battery
The reactions (16.9), (16.10) show that to transfer 2 mol of electrons requires

1 mol	Pb	=	207 g
1 mol	PbO_2	=	239 g
2 mol	H_2SO_4	=	196 g

Total active material   642 g

But 2 mol of electrons represent a charge:

$$(2\,\text{mol})(-1.60 \times 10^{-19}\,C)(6.02 \times 10^{23}\,\text{mol}^{-1})$$

$$= -(2)(9.6)(10^4)\,C = -1.93 \times 10^5\,C$$

The standard electrode potential for (Pb/PbSO$_4$) is   0.30 V
and for (PbSO$_4$/Pb^{4+}) is -1.62 V
So the theoretical cell EMF for (Pb/PbSO$_4$/H$_2$SO$_4$/PbSO$_4$/PbO$_2$) is
$$\xi_{cell} = +1.92\,V$$
with the PbO$_2$ plate positive, according to the IUPAC sign convention.

The actual cell EMF depends on the concentration of reagents, and can be calculated by standard electrochemical methods. In general, the voltage of a cell operating at low currents differs by only a few per cent from the theoretical cell voltage (Fig. 16.4). In particular, lead acid batteries are usually set to give 2.0 V per cell.

Therefore the work done in moving 2 mol of electrons is

$$(1.93 \times 10^5\,C)(2.0\,V) = 0.386 \times 10^6\,J$$

Thus the energy stored in 1 kg of active ingredients is, in theory,

$$W_m{}^{(0)} = (0.386 \times 10^6 \, J)/(0.642 \, kg) = 0.60 \, MJ \, kg^{-1}$$

Unfortunately, the energy density W_m of any practical battery is always much less than the theoretical value $W_m^{(0)}$, if the total mass of the whole battery is considered. Most commercial batteries have $W_m \sim 0.15 \, W_m^{(0)}$, although more careful (and more expensive!) designs can reasonably be expected to achieve energy densities up to 25% of the theoretical values.

In the specific case of the lead acid battery, the main reasons for this 'under-achievement' are

(1) A working battery necessarily contains nonactive materials, e.g. the case, the separators (which prevent the electrodes short circuiting) and the water in which the acid is dissolved. (The acid concentration must not be too high, or the battery will discharge itself.) Since the mass of an actual battery exceeds the mass of the active ingredients, the energy density is less than the theoretical value calculated from the active mass alone.
(2) The reactions cannot be allowed to go to completion. If all the lead was consumed by reaction (16.10) there would be no electrode left for the reverse reaction to operate at, i.e. the battery could not be cycled. Similarly, if the concentration of H_2SO_4 is allowed to fall too low, the electrolyte ceases to be an adequate conductor. In practice, the battery cannot be allowed to discharge more than about 50% of its stored energy, or it will be ruined. Such a discharge is called a 'deep discharge'.

A further limitation of real batteries is familiar to all car owners: they do not last forever. Solid Pb is almost twice as dense as the $PbSO_4$ found in the discharge reaction (16.10). Therefore it is difficult to fit the $PbSO_4$ crystals into the space originally occupied by the Pb paste in the negative electrode. In practice, some $PbSO_4$ falls to the bottom of the cell in every discharge. This constitutes an irreversible loss of active material. This loss is worse if the battery is allowed to fully discharge; indeed, it may rapidly become impossible to recharge the battery.

The other main factor limiting the life of even a well-maintained battery is self-discharge of the positive electrode. This is particularly acute in vehicle (SLI) batteries in which the grid is not pure Pb but a lead–antimony alloy, which is stronger and better able to stand the mechanical stresses during motion. Unfortunately antimony promotes the reaction

$$5PbO_2 + 2Sb + 6H_2SO_4 \rightarrow (SbO_2)_2SO_4 + 5PbSO_4 + 6H_2O \tag{16.12}$$

which also slowly, but irreversibly, removes active material from the battery.

Batteries for stationary applications (e.g. photovoltaic lighting systems) can use Sb-free plates, and have longer life (up to 7 years) if not excessively discharged.

The performance of a battery depends on the current at which it is charged and discharged, and the depth to which it is regularly discharged. Figure 16.4(a) shows the *discharge characteristics* of a typical lead acid car battery. Its nominal capacity is $Q_{20} = 100$ Ah, which is the charge which can be extracted if it is discharged at a constant current over 20 hours. The voltage per cell during this discharge drops only slightly from 2.07 V to 1.97 V as the first 60% of Q_{20} is discharged. If the same battery is discharged over 1 hour, its voltage drops much more sharply, and the total charge which can be removed from it is only about $0.5 Q_{20}$. This is because the rate of reaction of the electrodes is limited by the rate at which the reactants can diffuse into contact with each other. A rapid buildup of reaction products ($PbSO_4$ in particular) can block this contact. Moreover the internal resistance across this $PbSO_4$ layer lowers the voltage available from the cell.

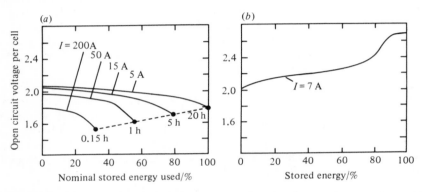

Fig. 16.4 Operating characteristics of a typical lead acid battery (SLI type of about 100 Ah nominal capacity). (a) Discharge. The curves are labeled by the discharge current (assumed steady), and by the time taken to 'fully' discharge at that current. (b) Charge. The curve is for charging at a constant low current. After Weissman (1980)

A set of *charging characteristics* for the same battery is shown in Fig. 16.4 (b). To commence charging, an EMF of at least 2.1 V per cell is required. The voltage required initially increases slowly but increases rapidly to about 2.6 V per cell as the battery nears full charge (if constant charging current is maintained). This is because the water in the cell begins to electrolyze. If the cell is overcharged, considerable H_2 gas may be released, possibly mechanically damaging the cell, and certainly raising the concentration of acid to the point where the ions are not mobile enough to allow the battery to work. Charging at constant voltage alleviates this problem.

16.6 Fuel cells

A fuel cell converts chemical energy of a fuel into electricity directly, with no intermediate combustion cycle. Since there is no intermediate heat → work conversion, the efficiency of fuel cells is not limited by the second law of thermodynamics, unlike conventional fuel → heat → work → electricity systems. The efficiency of conversion from chemical energy to electricity by a fuel cell could be (theoretically) 100%. Although not strictly 'storage' devices, fuel cells are treated in this chapter because of their many similarities to batteries (Section 16.5) and their possible use with H_2 stores (Section 16.3). Therefore we shall discuss only fuel cells using H_2, although other types exist.

Like a battery, a fuel cell consists of two electrodes separated by an electrolyte, which transmits ions but not electrons. In the fuel cell, hydrogen (or another reducing agent) is supplied to the negative electrode and oxygen (or air) to the positive (Fig. 16.5). A catalyst on the porous anode causes hydrogen molecules to dissociate into hydrogen ions and electrons. The H^+ ions migrate

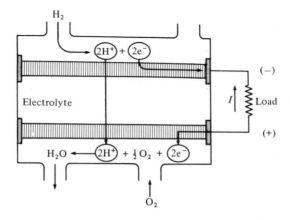

Fig. 16.5 Schematic diagram of a fuel cell. Hydrogen and oxygen are combined to give water and electricity. The porous electrodes allow hydrogen ions to pass

through the electrolyte – usually an acid – to the cathode, where they react with electrons (supplied through the external circuit) and oxygen to form water.

The efficiency of practical fuel cells is well below the theoretical 100%, for much the same reasons as for batteries. However, the efficiency – perhaps 40% for the conversion of chemical energy to electricity – is not dependent on whether the cell is working at its full rated power or not (in contrast to diesel engines, gas turbines etc.).

Since the efficiency of an assembly of fuel cells is nearly equal to that of a single cell, there are few economies of large scale. Therefore small localized plants (1 to 100 kW) are a promising proposition. A single building could be supplied with both electric and thermal energy (from the waste heat of the cells), for the same amount of fuel ordinarily required for the thermal demand alone. The main reason why fuel cells are not in wide use for such applications is their cost ($>$2000 \, k \, W^{-1}$).

16.7 Mechanical storage

16.7.1 *Water*

Hydro-power systems draw on a natural flow of energy $P_0 = \rho g Q_0 H$, where Q_0 is the natural flow rate of water at the site and H is the vertical distance through which it falls (Section 8.2). Since the natural flow Q_0 depends on rainfall, which has a different rhythm to the demand for power, all large hydroelectric systems incorporate energy storage by means of a dam (Figs 1.5(b), 8.7). Water is stored at the high elevation H, and released to the turbine below at a controlled flow rate Q. The potential energy stored in a dam at 100 m head has an energy density $W_v = 1.0\,\text{MJ m}^{-3}$. Though this is a relatively low energy density, the total energy stored in a hydro dam can still be very large.

A *pumped hydro* system uses two reservoirs, an upper and a lower. When power is available but not in demand, water is pumped uphill. When demand increases the water is allowed to fall again, driving a hydroelectric turbine at the bottom and thereby generating power. In practice, the same machine (usually a Francis turbine, Fig. 8.5(b)) is used as both pump and turbine. Some very large systems of this type have been built to smooth the fluctuating demand on conventional power stations, thereby allowing them to run at constant load and greater overall efficiency.

Since about 15% of the input power is used to keep the turbines/pumps spinning (to allow quick response) and a further 15% is lost in friction and distribution, it is arguable that the high capital cost of such schemes would have been better spent on automatic load control (Section 1.4.4).

16.7.2 *Flywheels*

The kinetic energy of rotation of an object is

$$E = \tfrac{1}{2}I\omega^2 \tag{16.13}$$

where I is the moment of inertia of the object about its axis, and ω is its angular velocity (rad s^{-1}). In the simplest case, where the mass m is concentrated in a rim of radius $r = a$, then $I = ma^2$. For a uniform disk of the same mass, I is lower ($ma^2/2$) because the mass nearer the shaft contributes less to I.

From (16.13) we find that the energy density of a uniform disk is

$$W_m = E/m = \tfrac{1}{2}a^2\omega^2 \tag{16.14}$$

For a flywheel to be a useful store of energy (and not just a smoothing device), it follows from (16.14) that it must rotate as fast as possible. However, its angular velocity is limited by the strength of the material which has to resist the centrifugal forces tending to fling it apart. For a uniform wheel of density ρ the maximum tensile stress is

$$\sigma^{max} = \rho\omega^2 a^2 \tag{16.15}$$

In general $I = Kma^2/2$ for a particular shape, where K is a constant ~ 1. So

$$W_m = Ka^2\omega^2/2 \tag{16.16}$$

and

$$W_m^{max} = \frac{K\sigma^{max}}{2\rho} \tag{16.17}$$

Conventional materials such as steel give rather low energy densities.

Example 16.3 Maximum energy density of a rotating steel disk
For a fairly strong steel, (16.17) gives, with $K = 1$,

$$W_m^{(max)} = \frac{(1000 \times 10^6\,\mathrm{N\,m^{-2}})}{(2)(7800\,\mathrm{kg\,m^{-3}})}$$

$$= 0.06\,\mathrm{MJ\,m^{-3}}$$

Much higher energy densities can be obtained by using lightweight fiber composite materials, such as fiberglass in epoxy resin, which have higher tensile strength σ^{max} and lower density ρ. To make the best use of these materials, flywheels should be made in unconventional shapes with the strong fibers aligned in the direction of maximum stress. Such systems can have energy densities of $0.5\,\mathrm{MJ\,kg^{-1}}$ (better than lead acid batteries) or even higher (Problem 16.3).

For use in smoothing demand in large electricity networks, flywheels have the advantage over pumped hydro-systems that they can be installed anywhere and take up little land area. Units with a 100 tonne flywheel would have a storage capacity of about 10 MWh. Larger storage demands would probably best be met by cascading many such 'small' units.

Flywheels also offer an interesting alternative to storage batteries for use in electrically powered vehicles (Problem 16.2), especially since the energy in a flywheel can be replenished more quickly than in a battery.

16.7.3 *Compressed air*

Air can be rapidly compressed and slowly expanded, and this provides a good way of smoothing large pressure fluctuations in hydraulic systems. An example is the hydraulic ram pump of Section 8.7.

The energy densities available are moderately large. Consider, for example, the slow compression of $V_1 = 1\,\mathrm{m^3}$ of air, at pressure $p_1 = 2\,\mathrm{atmos} = 2 \times 10^5$ $\mathrm{Nm^{-2}}$, to $V_2 = 0.4\,\mathrm{m^3}$, at constant temperature. For n moles of this perfect gas.

$$pV = nR_0T \tag{16.18}$$

from which it follows that the work done (energy stored) is

$$E = \int_{V_1}^{V_2} p \, dV = nR_0 T \int_{V_1}^{V_2} \frac{dV}{V}$$

$$= p_1 V_1 \log(V_1/V_2) \tag{16.19}$$

$$= 0.19 \, \text{MJ}$$

for the figures quoted.

In the compressed state, $W_v = E/V_2 = 0.48 \, \text{MJ} \, \text{m}^{-3}$. For a system operating under less idealized conditions, W_v will be less but of similar magnitude. A major difficulty is to decrease the losses of heat production during the compression.

16.8 Distribution of energy

16.8.1 *Introduction*

Table 16.2 sets out the major methods by which commercial energy is distributed to its point of use.

The methods are categorized by whether they involve continuous flow (e.g. pipelines) or batch movement (e.g. lorries), and by their suitability for use over long, medium or short distances. Also shown are numerical values representing the energy flows in a typical unit of the type in question (e.g. an individual gas pipeline). Although these flows vary greatly in scale, the energy flow through each system *per end-user per day* is in practice remarkably similar, i.e. $\sim 10 \, \text{MJ}$ user^{-1} day$^{-1} \approx 100 \, \text{W user}^{-1}$. The following subsections, and Problems 16.6–16.8, give the basis of the numerical values in the table.

For renewable energy supplies, short distance distribution is more important than long distance, because the sources themselves are usually widespread and of low intensity. In particular, short haul carriage of biomass, and movement of heat (e.g. through a building), are most significant. The renewable energy supplies that are mechanical in origin, e.g. hydro, wave and wind, are usually best distributed by electricity. In this way electricity is a carrier or vector of energy, and not necessarily the main end-use requirement. Movement of gas, perhaps on a large scale, would be required if hydrogen becomes a common store of energy.

16.8.2 *Gas pipelines*

In the pipelines usually used for carrying fuel gases, the flow is turbulent but not supersonic. Therefore the theory of Section 2.7 applies, although its interpretation is affected by the compressibility of the gas, as follows.

According to (2.12) the pressure gradient along a small length of pipe of diameter D is

$$\frac{dp}{dx} = -2f \frac{\rho u^2}{D} \tag{16.20}$$

Table 16.2 Summary of major means and flows for distributing energy.

	Long distance (>1000 km)	Flow (MW per unit)	(MJ user⁻¹ day⁻¹)	Medium distance (1–1000 km)	Flow (MW per unit)	(MJ user⁻¹ day⁻¹)	Short distance 10 m–1 km	Flow (MW per unit)	(MJ user⁻¹ day⁻¹)
Continuous	Oil pipeline	15 000	60	Oil pipeline	10 000	60			
	Gas pipeline (high pressure)	500	20	Gas pipeline (high pressure)	500	20	Gas pipeline (low pressure)	500	7
				Electricity (high voltage line)	100	20	Electricity (low voltage)	100	10
							Heat in gas, vapor or liquid		
Batch	Oil tanker	1 200		Oil (or substitute, e.g. ethanol) In vehicle					
				As cargo	200				
				As fuel in tank		28			
	Coal in ships	15		Coal in trains	15				
				Biomass on lorry	15		Biomass on lorry	15	
							Wood 'by hand'	0.03	15

where f is the friction coefficient, and ρ and u are respectively the density and mean speed of the fluid.

In a steady flow of *gas*, both ρ and u vary along the length of a long pipe, but the mass flow rate

$$\dot{m} = \rho u A \tag{16.21}$$

is constant. $A = \pi D^2/4$ is the cross-sectional area.

Also the density ρ varies with the pressure p:

$$p = \left(\frac{RT}{M}\right)\rho \equiv K\rho \tag{16.22}$$

with K more or less constant for a given gas. R is the universal gas constant, T is absolute temperature, and M is the gram molecular mass/1000, so as to have units of kg mol^{-1}. If the Reynolds number ($\mathcal{R} = uD/\nu$) is large, f will not vary appreciably along the pipe, and we can integrate (16.20) between stations x_1 and x_2 to

$$p_1^2 - p_2^2 = \frac{64f\,RT\dot{m}^2(x_2-x_1)}{\pi^2 MD^5} \tag{16.23}$$

Thus the pressure falls off rapidly along the length of the pipe, and frequent pumping (recompression) stations are needed to maintain the flow. As a numerical example, a pipe of diameter 30 cm, carrying methane at a mean pressure about 40 times atmospheric, holds an energy flow of about 500 MW, which is very substantial (see Problem 16.7).

According to (16.23), larger pipes (bigger D) will require much less pumping. The most economical balance between pipe size (capital cost) and pump separation (running cost) depends largely on the accessibility of the pipe. Construction costs are very variable, but are unlikely to be less than US\$0.2 GJ^{-1}(1000 km)$^{-1}$.

The compressibility of the gas offers another benefit. The pipe itself can be used as a temporary store, simply by pumping gas in faster than it is taken out so that compressed gas accumulates in the pipe. For the pipe considered above, the energy 'stored' in the 100 km length might be

$$(32\,\text{kg m}^{-3})(50\,\text{MJ kg}^{-1})(10^5\,\text{m})\pi(0.3\,\text{m})^2$$

$$= 44 \times 10^6\,\text{MJ per 100 km}$$

Such storage is very substantial.

16.8.3 *Electricity transmission*

Consider two alternative systems transmitting the same useful power P to a load R_L at different voltages V_1, V_2 in wire of the same resistance per unit length

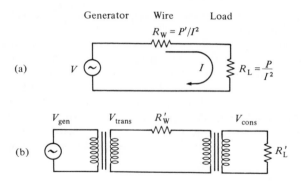

Fig. 16.6 Electrical transmission. (a) Power transmission to a load of resistance R_L, through a wire of finite resistance R_W. (b) More likely realization, with generated voltage transformed up for transmission and then down for consumption

(Fig. 16.6). The corresponding currents are $I_1 = P/V_1$, $I_2 = P/V_2$, and therefore the ratio of power lost in the two systems is

$$\frac{P_1'}{P_2'} = \frac{I_1^2 R_w}{I_2^2 R_w} = \left(\frac{P}{V_1}\right)^2 \left(\frac{V_2}{P}\right)^2 = \frac{V_2^2}{V_1^2} \tag{16.24}$$

Significantly less power is dissipated in the system working at high voltage. The low voltage system can have the same loss as the high voltage system only with thick, and therefore expensive, cable. For instance if electricity is to be transmitted at domestic mains voltage ($\sim 110\,V$ or $\sim 220\,V$), the cost of cable becomes prohibitive for distances greater than about 200 m. The difficulty becomes even greater at very low voltage, $\sim 12\,V$.

These factors govern the design of all electrical power networks. Normal rotating generators work best at voltages $\lesssim 10\,kV$. The ease with which AC can be transformed up and down in voltage explains why AC transmission systems are now standard for all but the smallest networks. As indicated in Fig. 16.6 (b), power is generated at low voltage, stepped up for transmission, and then down again to a safer level for consumption. The transmission voltage is limited by dielectric breakdown of the air around the overhead cables and by the insulation of the cables from the metal towers (at earth potential). Improvements in insulation have allowed transmission voltages for long lines to increase from 6000 V in the year 1900 to over 200 000 V today.

Grids using even higher voltages are now being constructed, but will probably make only a marginal improvement in costs. The same is true for very high voltage DC systems, which have certain advantages in transmission but require more expensive step-up equipment. Superconducting lines of zero resistance are attractive but can operate only at very low temperatures ($\lesssim - 260\,°C$).

Maintaining such low temperatures is difficult over large distances and such lines are not yet a commercial proposition.

Electrical power companies usually link power stations into a common national network (grid). Thus if one station shuts down for maintenance, repair or economy, its demand can be met by the others in the network. Power from variable renewable sources can be fed directly into such a grid. Since all major grid systems can cater for rapid demand fluctuations of $\sim 20\%$, it is also possible for the same systems to cater for rapid supply fluctuations of $\sim 20\%$. Thus the inclusion of unpredictable supply, e.g. wind power, is possible. If hydro-power is available, then rapid control of the grid is easy whatever the supply.

16.8.4 *Batch transport*

Biomass and coal can be carried in suitable containers by hand, road, rail or boat. However, the low density and bulky nature of most biomass means that it is rarely economic to distribute it over large distances ($> 1000\,$km). Even over medium distances (1–$1000\,$km), it is unlikely to be economic to distribute biomass for its energy value alone. Economic, and ecological, use of biomass energy sources therefore depends on harnessing energy from a flow of biomass which is already occurring for some other purpose, e.g. the extraction of sugar from sugarcane after which the spent cane is used to fuel the factory. In this case the transport of fuel may be regarded as 'free', or nearly so (Section 11.1 amplifies this argument). Biofuels can however be transported over medium to long distances after conversion from raw biomass (e.g. by pyrolysis: see Section 11.5).

Alternatively, the biomass can be used very close to its source. This is usually the case with firewood, which remains a fuel of major importance in most developing countries.

16.8.5 *Heat*

The movement of heat within a building through hot air 'ducts' (whether specially built as such or not) and through steam pipes is a major means of distributing energy over short distances. This is especially true in cold climates, where space heating dominates energy use (e.g. Fig. 16.2). Heat transport by steam is also used in many industrial processes.

Heat transport is limited to short distances by heat losses from the sides of the duct.

Example 16.4 Heat loss from a steam pipe

A pipe 5 cm in diameter is to deliver heat over a distance of 100 m. It is insulated with glass wool of thickness $\Delta x = 1\,$cm. Estimate the heat loss along the path. (Take $T_a = 10\,°$C.)

Solution

As a first approximation, assume the steam is at $T_v = 100\,°$C along the whole pipe. (Steam at higher pressure will actually be at higher temperature: see books on engineering thermodynamics.) The conductivity of mineral wool is

$k = 0.04\,\text{Wm}^{-1}\text{K}^{-1}$ (similar to that of other insulators, using trapped air). The major resistance to heat loss is by conduction through the insulation, so from (3.9)

$$P_{\text{loss}} = -kA\,\Delta T/\Delta x$$

$$= (-0.04\,\text{Wm}^{-1}\text{K}^{-1})(100\,\text{m})\,\pi\,(0.06\,\text{m})(100 - 10)\text{°C}/(0.01\,\text{m})$$

$$= 6.8\,\text{kW}$$

As calculated here, this loss is independent of the flow rate in the pipe. Obviously, very large heat flows ($\geq 10\,\text{MW}$) are needed if the losses are to be proportionately small. *District heating* of this kind operates successfully in many cities.

The *heat pipe* offers another way to move large quantities of heat over very short distances. It is a tube containing vapor with the condensate recycled by a wick, and offers an effective conductivity much greater than that of copper.

Problems

16.1 The changes in enthalpy, free energy, and entropy in the formation of water

$$H_2 + \tfrac{1}{2}O_2 \rightarrow H_2O_{(gas)}$$

are respectively

$$\Delta H = -242\,\text{kJ}\,\text{mol}^{-1}$$

$$\Delta G = -228\,\text{kJ}\,\text{mol}^{-1}$$

$$\Delta S = -64\,\text{J}\text{K}^{-1}\text{mol}^{-1}$$

Estimate the temperature above which H_2O is thermodynamically unstable. *Hint*: consider (16.1).

16.2 A passenger bus used in Switzerland derived its motive power from the energy stored in a large flywheel. The flywheel was brought up to speed, when the bus stopped at a station, by an electric motor which could be attached to the electric power lines. The flywheel was a solid steel cylinder of mass 1000 kg, diameter 180 cm, and would turn at up to 3000 rev min^{-1}.

(a) At its top speed, what was the kinetic energy of the flywheel?
(b) If the average power required to operate the bus was 20 kW, what was the average time between stops?

16.3 A flywheel of three uniform bars, rotating about their central points as spokes of a wheel, is made from fibers of E glass with density $\rho = 2200\,\text{kg}\,\text{m}^{-3}$, and tensile strength $3500\,\text{MN}\,\text{m}^{-2}$. The fibers are aligned along the bars and are held together by a minimal quantity (10%) of resin of negligible tensile strength and similar density. Calculate the maximum energy density obtainable. If $a = 1.0\,\text{m}$, what is the corresponding angular velocity?

16.4 Verify that the data plotted in Fig. 16.2 (taken from Ryle, 1977) are in fact 'reasonable'.

(a) The population of Great Britain is about 50 million. How does the total energy demand per person compare with the world average of 2.0 t oil equivalent (TOE) per person per year (UN statistic, 1980)?
(b) How does the nonheat demand vary with season? What types of industrial usage does this correspond to?
(c) The peak demand shown in winter is for a daytime temperature of $-3\,°\text{C}$. How much heat is being used per household? Does this seem likely? (The figure quoted for 'heating' in Fig. 16.2 includes non-household use.)
(d) Use the data in Chapter 4 (especially Fig. 4.16) to estimate the solar heat input on $1\,\text{m}^2$ of horizontal surface, and on $1\,\text{m}^2$ of (south-facing) vertical surface in each season. (The latitude of Britain is about $50\,°\text{N}$.) What is a typical efficiency of a solar heater? What collector area would be required to supply the power required for heating indicated in Fig. 16.2? How many m^2 per house does this represent? Is this reasonable? Would passive solar energy techniques, combined with thermal insulation, ventilation control, and the use of free gains, be of significance?
(e) Approximately what is the electrical power obtainable from $1\,\text{m}^2$ of swept area in a mean wind of $8\,\text{m}\,\text{s}^{-1}$, see (9.73). The land and shallow sea waters of Britain can be (very roughly) approximated by two rectangles $1000\,\text{km} \times 200\,\text{km}$, with the longer sides facing the prevailing wind. Consider large $100\,\text{m}$ diameter aerogenerators with mean wind speed $8\,\text{ms}^{-1}$ at hub height. How many aerogenerators would be needed to produce an average power of $30\,\text{GW}$ for the whole country? What would be the average spacing between them if half were on land and half at sea?
(f) Use the wave power map (Fig. 12.12) to estimate how long a barrage would be needed to generate a mean power of $30\,\text{GW}$ off the north west coast of Britain. How does this length compare with the length of the coast?

16.5 The largest magnetic field that can be routinely maintained by a conventional electromagnet is $B_0 \sim 1 \, \text{Wb m}^{-2}$. The energy density in a magnetic field is $W_v = \frac{1}{2} B^2 / \mu_0$. Calculate W_v for $B = B_0$.

16.6 Calculate the energy flows in the following cases:

(a) In 1984 about 30 million barrels of oil per day were shipped out of the Persian Gulf area (1 barrel = 160 l).

(b) The TAP crude oil pipeline from Iraq to the Mediterranean carries about 10 million tons of oil per year.

(c) A family of four in a household uses (for cooking) one cylinder of LPG (gas) (13 kg) per month.

(d) The same family runs a car which covers $8000 \, \text{km y}^{-1}$, with a petrol consumption of $7 \, \text{km l}^{-1}$.

(e) A villager in Papua New Guinea takes 2 hours to bring one load of 20 kg wood from the bush, carrying it on her back.

(f) A 3 t lorry carries fuelwood into town at a speed of $30 \, \text{km h}^{-1}$.

16.7 A steel pipeline of diameter 30 cm carries methane gas (CH_4). Recompression stations are sited at 100 km intervals along the pipeline. The gas pressure is boosted from 3 to $6 \, \text{MN m}^{-2}$ at each station. (These are typical commercial conditions.) Calculate (a) the mass flow and (b) the energy flow. (c) What volume per day of gas at STP would this correspond to?
Hint: Make a first estimate of f, assuming $\mathscr{R}$ is 'high enough'. Then find $\dot{m}$, and check for consistency. Iterate if necessary. Viscosity of methane at these pressures is

$$\mu = 10 \times 10^{-6} \, \text{N s m}^{-2}.$$

16.8 An electrical transmission line links a 200 MW hydroelectric installation A to a city B, 200 km away, at 220 kV. The cables are designed to dissipate 1% of the power carried. Calculate the dimensions of wire required, and explain why losses of 1% may be economically preferable to losses of 10% or 0.1%.

Solutions

16.1 3800 K (assuming ΔH etc. independent of T)

16.2 (a) 20 MJ (b) 16 minutes

16.3 $0.5 \, \text{MJ kg}^{-1}$; $1.8 \times 10^3 \, \text{rad s}^{-1} = 17\,000 \, \text{rpm}$

16.4 (a) $2 \, \text{TOE y}^{-1} \times 50 \, \text{M people} = 150 \, \text{GW}$

(b) Even, 60 GW; manufacturing etc.

(c) 14 kW per household. No, household really $\sim 1 \, \text{kW}$.

(d) Use Fig. 4.16. $25 \, \text{m}^2$ per house in summer, $360 \, \text{m}^2$ in winter. Possible, yet unlikely without the added use of passive solar design in substantial quantities of new housing stock.

(e) $90 \mathrm{W\,m^{-2}}$. If blade diameter $= 100\,\mathrm{m}$, need about 30000 aero-generators to yield $30\,\mathrm{GW}$. Area for each $= 2 \times (2 \times 10^{11}\mathrm{m^2})/30\,000$. Average spacing about $3\,\mathrm{km}$.

(f) $400\,\mathrm{km}$ at $70\,\mathrm{kW\,m^{-1}}$. Fits.

Note: these 'ballpark' answers show that renewable energy supplies are of significant potential, even for a country with independent supplies of oil and coal.

16.5 $0.4\,\mathrm{MJ\,m^{-3}}$

16.6 See Table 16.2.

16.7 (a) $11\,\mathrm{kg\,s^{-1}}$ (b) 540 MW (c) $1.3 \times 10^6\,\mathrm{m^3\,day^{-1}}$

16.8 For copper wire, total cross-sectional area $\approx 1500\,\mathrm{mm^2}$ (e.g. four wires each $22\,\mathrm{mm}$ diameter).

Bibliography

General

Jensen, J. (1980) *Energy Storage*, Newnes-Butterworth, London.
  One of the few books specifically on this topic. Good coverage at about same level as this book.
Kalhammer, F. R. (1979) 'Energy storage', *Sci. Am.*, **241** (Dec.), 42–51.
  Concentrates on large scale devices.

Most books on particular renewable energy sources (referred to in the appropriate chapters) include some discussion of storage media applicable to that source (e.g. heat stores for solar, batteries for wind).

Chemical storage

Bockris, J. O'M. (1975) *Energy: The Solar-Hydrogen Alternative*, ANZ Publishers, Sydney.
  Both Bockris and Gregory have published extensively on hydrogen. These are two of their more accessible publications. Their economic analyses are usually more favorable than those of their critics!
Carden, P. O. (1977) 'Energy corradiation using the reversible ammonia reaction', *Solar Energy*, **19**, 365–78.
  Sets out the main features of a solar/ammonia system using distributed collectors. Many later papers elaborate on details, e.g.
Carden, P. O. and Williams, O. M. (1978) 'The efficiencies of thermochemical energy transfer', *Energy Research*, **2**, 389–406.
Gregory, D. P. (1973) 'The hydrogen economy', *Sci. Am.*, **228** (Jan.), 13–21.
Any book on physical chemistry will give a thermodynamic analysis of the heat release in chemical reactions, e.g. Barrow, G. M. (1973) *Physical Chemistry*, 3rd edn, McGraw-Hill.

Heat storage

Duffie, J. A. and Beckman, W. A. (1980) *Solar Engineering of Thermal Processes*, Wiley, New York.
  Chapter 9 is specifically concerned with heat storage.
Ryle, M. (1977) 'Economics of alternative energy sources', *Nature*, **267**, 111–16.

Cogently argues that storage for about seven days enables wind/wave/solar to match most fluctuations in UK demand.

Electrical storage

Denaro, A. R. (1971) *Elementary Electrochemistry*, 2nd edn, Butterworths, London.
    Easily read but accurate account. See also Barrow (1973), but note that many older (or American) books use a different sign convention.
Rand, D. A. J. (1979) 'Battery systems for electric vehicles: a state-of-the-art review', *J. Power Sources*, **4**, 101–43.
    Covers all types of storage batteries.
Smith, R. (1980) *Batteries*, Pitman, London.
    Useful account, emphasizes operating characteristics.
Weizmann, E. (1977) 'Lead-acid storage batteries', in Kordesch, K. V. (ed.) *Batteries*, vol. 2, Marcel Dekker, New York.
    An exhaustive review of technology and limitations.

Almost any textbook on electricity and magnetism discusses the energy density of electric and magnetic fields, e.g. Bleaney, B. I. and Bleaney, B. *Electricity and Magnetism*, Oxford University Press.

Fuel cells

Fickett, A. P. (1978) Fuel cell power plants', *Sci. Am.*, **239** (Dec.), 70–6.

Flywheels

Post, R. F. and Post, S. F. (1973) 'Flywheels', *Sci. Am.*, **239** (Dec.), 17–23.
    New shapes and materials, and their possibilities.

Distribution

British Petroleum (1977) *Our Industry: Petroleum.*
    Full coverage of the oil industry at a not too technical level. Chapter 10 is about pipelines.
Hughes, E. (1969) *Electrical Technology*, 4th edn, Longmans, London.
    Alternating current, generators etc. (first-year engineering text).
Luten, D. B. (1971) 'The economic geography of energy', *Sci. Am.*, Sept. (reprinted in *Energy and Power*, Freeman, 1975). Still a useful overview.
Starr, A. T. (1957) *Generation, Transmission and Utilization of Electrical Power*, 4th edn, Pitman, London.
    Technical details of cables, transformers, switches etc.

Appendix A Units

A.1 Names and symbols for the SI units

Base units

Physical quantity	Name of SI unit	Symbol for SI unit
Length	meter	m
Mass	kilogram	kg
Time	second	s
Electric current	ampere	A
Thermodynamic temperature	kelvin	K
Amount of substance	mole	mol
Luminous intensity	candela	cd

Supplementary units

Physical quantity	Name of SI unit	Symbol for SI unit
Plane angle	radian	rad
Solid angle	steradian	sr

A.2 Special names and symbols for SI derived units

Physical quantity	Name of SI unit	Symbol for SI unit	Definition of SI unit	Equivalent form(s) of SI unit
Energy	joule	J	$m^2 kg s^{-2}$	$N m$
Force	newton	N	$m kg s^{-2}$	$J m^{-1}$
Pressure	pascal	Pa	$m^{-1} kg s^{-2}$	$N m^{-2}, J m^{-3}$
Power	watt	W	$m^2 kg s^{-3}$	$J s^{-1}$
Electric charge	coulomb	C	$s A$	$A s$
Electric potential difference	volt	V	$m^2 kg s^{-3} A^{-1}$	$J A^{-1} s^{-1}, J C^{-1}$
Electric resistance	ohm	Ω	$m^2 kg s^{-3} A^{-2}$	$V A^{-1}$
Electric capacitance	farad	F	$m^{-2} kg^{-1} s^4 A^2$	$A s V^{-1}, C V^{-1}$
Magnetic flux	weber	Wb	$m^2 kg s^{-2} A^{-1}$	$V s$
Inductance	henry	H	$m^2 kg s^{-2} A^{-2}$	$V A^{-1} s$
Magnetic flux density	tesla	T	$kg s^{-2} A^{-1}$	$V s m^{-2}, Wb m^{-2}$
Frequency	hertz	Hz	s^{-1}	

A.3 Examples of SI derived units and unit symbols for other quantities

Physical quantity	SI unit	Symbol for SI unit
Area	square meter	m^2
Volume	cubic meter	m^3
Wave number	1 per meter	m^{-1}
Density	kilogram per cubic meter	$kg m^{-3}$
Speed; velocity	meter per second	$m s^{-1}$
Angular velocity	radian per second	$rad s^{-1}$
Acceleration	meter per second squared	$m s^{-2}$
Kinematic viscosity	square meter per second	$m^2 s^{-1}$
Amount of substance concentration	mole per cubic meter	$mol m^{-3}$

A.4 Other units

Physical quantity	Unit	Unit symbol	Alternative representation
Energy	electron volt	eV	$1\,eV \approx 1.602\,189\,2 \times 10^{-19}\,J$
Time	year	y	$365.26\,d = 3.16 \times 10^7\,s$
Time	minute	min	$60\,s$
Time	hour	h	$60\,min = 3600\,s$
Time	day	d	$24\,h = 86\,400\,s$
Angle	degree	°	$(\pi/180)\,rad$
Angle	minute	'	$(\pi/10\,800)\,rad$
Angle	second	"	$(\pi/648\,000)\,rad$
Volume	liter	l	$10^{-3}\,m^3 = dm^3$
Volume	gallon (US)		$3.785 \times 10^{-3}\,m^3$
	gallon (Brit.)		$4.546 \times 10^{-3}\,m^3$
Mass	tonne	t	$10^3\,kg = Mg$
*Celsius temperature	degree Celsius	°C	K, kelvin
Area	acre (Brit.)		$4.047 \times 10^3\,m^2$
	hectare		$10^2 \times (10^2\,m^2) = 10^4\,m^2$

*The Celsius temperature is the excess of the thermodynamic temperature over 273.15 K.

A.5 SI prefixes

Multiple	Prefix	Symbol	Multiple	Prefix	Symbol
10^{-1}	deci	d	10	deca	da
10^{-2}	centi	c	10^2	hecto	h
10^{-3}	milli	m	10^3	kilo	k
10^{-6}	micro	μ	10^6	mega	M
10^{-9}	nano	n	10^9	giga	G
10^{-12}	pico	p	10^{12}	tera	T
10^{-15}	femto	f	10^{15}	peta	P
10^{-18}	atto	a	10^{18}	exa	E

A.6 Energy equivalents

$1\,kWh = 3.6\,MJ$
$1\,Btu = 1055.79\,J$
$1\,therm = 10^5\,Btu = 105.6\,MJ = 29.3\,kWh$
$1\,calorie = 4.18\,J$
$1\,tonne\ coal\ equivalent = 29.3\,GJ$ (UN standard)
$1\,tonne\ oil\ equivalent = 42.6\,GJ$ (UN standard)

A.7 Power equivalents

$1\,\mathrm{Btu\,s^{-1}} = 1.06\,\mathrm{kW}$
$1\,\mathrm{Btu\,h^{-1}} = 0.293\,\mathrm{W}$
$1\,\mathrm{horse\,power} = 746\,\mathrm{W}$

Appendix B Data

The following tables give sufficient physical data to follow the examples and problems in this book. They are not intended to take the place of the standard handbooks listed in the following bibliography, from which the data have been extracted.

Only two or three significant figures are given except in the few cases where the data and their use in this book justify more accuracy.

Bibliography

Kaye, G. W. C. and Laby, T. H. (1973) *Tables of Physical and Chemical Constants*, 14th edn, Longman, London.
  Useful range of data for physicists, with good attention to accuracy. Coverage sometimes better for special materials than for common engineering materials. SI units.
Ede, A. J. (1967) *An Introduction to Heat Transfer Principles and Calculations*, Pergamon Press, Oxford.
Monteith, J. (1973) *Principles of Environmental Physics*, Edward Arnold, London.
  Extremely useful set of tables for data on air and water vapor, and on heat transfer with elementary geometrical shapes.
Tennent, R. M. (ed.) (1971) *Science Data Book*, Oliver & Boyd, Edinburgh.
  Low priced book for students, with surprisingly wide range of carefully selected data. SI units.
Toulonkian, Y. S., de Witt, D. P. and Herncicz, R. S. (eds) (1972 onwards) *Thermophysical Properties of Matter*, 12 volumes, Plenum Press, New York.
International Critical Tables, McGraw-Hill, New York, 1929.
  Older, but more wide ranging than Toulonkian *et al.* (1972).
Rohsenow, W. M. and Hartnett, J. P. (eds) (1973) *Handbook of Heat Transfer*, McGraw-Hill, New York.
  Chapter 2 by W. Ibele is an extensive compilation of thermophysical data, but mostly in US units.
Handbook of Physics and Chemistry, Chemical Rubber Company.
  Chemical emphasis, but useful for all scientists. Mostly not in SI, and often disappointing for practical information.
Wong, H. Y. (1977) *Handbook of Essential Formulae and Data on Heat Transfer for Engineers*, Longman, London.
  Student priced book. Has a useful 20 pages of thermophysical data (the rest is like Appendix C). Highly recommended.

Table B.1 Physical Properties of dry air at atmospheric pressure
(adapted from Ede, 1967). $\mathscr{A}$ is the Raleigh number, X the characteristic dimension

Temperature T	Density ρ	Specific heat $c_{(P)}$	Kinematic viscosity $\nu = \mu/\rho$	Thermal diffusivity κ	Thermal conductivity k	Prandtl number $\mathscr{P}$	$\mathscr{A}/X^3\Delta T$
°C	kg m^{-3}	$10^3\,\text{J kg}^{-1}\text{K}^{-1}$	$10^{-6}\text{m}^2\text{s}^{-1}$	$10^{-6}\text{m}^2\text{s}^{-1}$	$10^{-2}\text{W m}^{-1}\text{K}^{-1}$ —		$10^8\text{m}^{-3}\text{K}^{-1}$
0	1.30	1.01	13.3	18.4	2.41	0.72	1.46
20	1.20	1.01	15.1	20.8	2.57		1.04
40	1.13	1.01	16.9	23.8	2.72		0.78
60	1.06	1.01	18.8	26.9	2.88	0.70	0.58
80	1.00	1.01	20.8	29.9	3.02		0.45
100	0.94	1.01	23.0	32.8	3.18	0.69	0.34
200	0.75	1.02	34.6	50	3.85	0.68	0.12
300	0.62	1.05	48.1	69	4.50		0.052
500	0.45	1.09	78	115	5.64		0.014
1000	0.28	1.18	174	271	7.6	0.64	0.0016

Other properties of air:

Velocity of sound in air $= 340\,\text{m s}^{-1}$ (15 °C)
Coefficient of diffusion of water vapor in air $= 25 \times 10^{-6}\,\text{m}^2\text{s}^{-1}$ (15 °C)
Coefficient of self-diffusion of N_2 or O_2 in air $= 18 \times 10^{-6}\,\text{m}^2\text{s}^{-1}$ (15 °C)
Coefficient of thermal expansion $\beta = (1/T) = 0.0033\,\text{K}^{-1}$ at 27 °C

Table B.2 Physical properties of water (at moderate pressures)
(a) *Liquid*

Temperature T	Density ρ	Kinematic viscosity $\nu = \mu/\rho$	Thermal diffusivity κ	Thermal conductivity k	Prandtl number $\mathscr{P}$	$\mathscr{A}/X^3\Delta T$	Expansion coefficient β
°C	$10^3\,\text{kg m}^{-3}$	$10^{-6}\text{m}^2\text{s}^{-1}$	$10^{-6}\text{m}^2\text{s}^{-1}$	$\text{W m}^{-1}\text{K}^{-1}$	—	$10^{10}\text{m}^{-3}\text{K}^{-1}$	10^{-4}K^{-1}
0	0.9998*	1.79	0.131	0.55	13.7	−0.24*	Changes
20	0.9982	1.01	0.143	0.60	7.0	+1.44	sign*
40	0.9922	0.66	0.151	0.63	4.34	3.81	3.0†
60	0.9832	0.48	0.155	0.65	3.07	6.9	4.5†
80	0.9718	0.37	0.164	0.67	2.23	10.4	5.7†
100	0.9584	0.30	0.168	0.68	1.76	14.9	6.7†

*The maximum density of water occurs at 3.98 °C and is 1000.0 kg m^{-3}. Therefore β is negative in the range 0 °C < T < 4 °C.

† These values of β apply to the range from the line above, e.g. $3.0 \times 10^{-4}\,\text{K}^{-1}$ is the mean value between 20 and 40°C.

(b) *Water vapor in air*

Temperature T $°C$	(Saturated) vapor pressure P_v $kN m^{-2}$	Mass of H_2O in $1 m^3$ of saturated air χ $g m^{-3}$
0	0.61	4.8
10	1.23	9.4
20	2.34	17.3
30	4.24	30.3
40	7.38	51.2
50	12.34	82.9
60	19.9	130
70	31.2	197
80	47.4	291
90	70.1	
100	101.3	

Note: $\chi = (2.17 \times 10^{-3} kg K m^2 N^{-1}) p_v/T$.
Other properties of water:

Specific heat capacity $c = 4.2 kJ kg^{-1} K^{-1} (0°C < T < 100°C)$
Latent heat of freezing $\Lambda_1 = 334 kJ kg^{-1}$
Latent heat of vaporization $\Lambda_2 = 2.45 MJ kg^{-1} (20°C)$
$= 2.26 MJ kg^{-1} (100°C)$
Surface tension (against air) $= 0.073 N m^{-1} (20°C)$

Table B.3 Density and conductivity of solids (at room temperatures)

Material	Density ρ $kg m^{-3}$	Thermal conductivity k $W m^{-1} K^{-1}$
Copper	8795	385
Steel	7850	47.6
Aluminum	2675	211
Glass (standard)	2515	1.05
Brick building	2300	0.6
Concrete (1:2:4)	2400	1.73
Ice (−1°C)	918	2.26
Gypsum plaster (dry, 20°C)	881	0.17
Oak wood (14% m.c.)	770	0.16
Pine wood (15% m.c.)	570	0.138
Pine fiberboard (24°C)	256	0.052
Asbestos cement, sheet (30°C)	150	0.319
Cork board (dry, 18°C)	144	0.042
Mineral wool, batts	32	0.035
Polyurethane (rigid foam)	24	0.025
Polystyrene, expanded	16	0.035
Still air (27°C, 1 atmos.)	1.18	0.026

Table B.4 Emittances of common surfaces

Material	Temperature	Emittance ϵ
	°C	%
Aluminum		
Polished	100	9.5
Roughly polished	100	18
Iron, roughly polished	100	17
Tungsten filament	1500	33
Brick (rough, red)	0–90	93
Concrete (rough)	35	94
Glass (smooth)	25	94
Wood (oak, planed)	90	90

Table B.5 Miscellaneous physical fundamental constants

Speed of light in vacuum	$c = 3.00 \times 10^8 \, \mathrm{m \, s^{-1}}$
Permeability of free space	$\mu_0 = 4\pi \times 10^{-7} \, \mathrm{H \, m^{-1}}$
Permittivity of free space	$\epsilon_0 = 8.85 \times 10^{-12} \, \mathrm{F \, m^{-1}}$
Gravitational constant	$G = 6.67 \times 10^{-11} \, \mathrm{N \, m^2 \, kg^{-2}}$
Elementary charge	$e = 1.60 \times 10^{-19} \, \mathrm{C}$
Avogadro constant	$N_0 = 6.02 \times 10^{23} \, \mathrm{mol^{-1}}$
Gas constant	$R = 8.31 \, \mathrm{J \, K^{-1} \, mol^{-1}}$
Stefan-Boltzmann constant	$\sigma = 5.67 \times 10^{-8} \, \mathrm{W \, m^{-2} \, K^{-4}}$

Table B.6 Calorific values of various fuels

Fuel	Gross calorific value[a]		Remarks
	$MJ\,kg^{-1}$	$MJ\,l^{-1}$ [b]	
Crops			
Wood			
Green	~8	~6	Varies more with moisture
Seasonal	~13	~10	content than species of wood
Oven dry	~16	~12	
Vegetation: dry	~15		Examples: grasses, hay
Crop residues			
Rice husk			For dry material.
Bagasse (sugarcane solids)	12–15		In practice residues may be
Cow dung			very wet
Peat			
Secondary biofuels			
Ethanol	30	25	C_2H_5OH: $789\,kg\,m^{-3}$
Methanol	23	18	CH_3OH
Biogas	28	20×10^{-3}	50% methane + 50% CO_2
Producer gas	5–10	$(4–8) \times 10^{-3}$	Depends on composition
Charcoal			
Solid pieces	32	11	
Powder	32	20	
Coconut oil	39	36	
'Cocohol'	39	33	Ethyl esters of coconut oil
Fossil fuels			
Methane	55	38×10^{-3}	Natural gas
Petrol	47	34	Motor spirit, gasoline
Kerosene	46	37	
Diesoline	46	38	Automotive distillate, derv
Crude oil	44	35	
Coal	27		Black, coking grade

(a) *Gross calorific value* (also called *heat of combustion*) is the heat evolved in a reaction of the type

$$CH_2O + O_2 \rightarrow CO_2(gas) + H_2O(liquid)$$

Some authors quote instead the *net calorific value*, which is the heat evolved when the final H_2O is gaseous.

(b) At 15 °C.

Appendix C Some heat transfer formulas

For notation, definitions and sources see Chapter 3. X is the characteristic dimension for calculation of Nusselt Number $\mathcal{N}$, Reynolds number $\mathcal{R}$ and Rayleigh number $\mathcal{A}$. These formulas represent averages over the range of conditions likely to be met in solar engineering. In particular $0.01 < \mathcal{P} < 100$, where $\mathcal{P}$ is the Prandtl number.

Table C.1 General

			Text references
Heat flow	$P = \Delta T / R$		(3.2)
Heat flux density	$q = P/A = \Delta T/r$		(3.4), (3.5)
Thermal resistance of unit area, thermal resistivity	$r = 1/h = RA$		(3.6), (3.8)
Conduction	$r_n = \Delta x/k$	$R_n = \Delta x/kA$	(3.11), (3.10)
Convection	$r_v = X/\mathcal{N}k$	$R_v = X/A\mathcal{N}k$	(3.16)
Radiation: in general	$r_r = (T_1 - T_2)/q$	$R_r = (T_1 - T_2)/P_{12}$	(3.3), (3.4)
	where P_{12} is given in Table C.5		
Nusselt number	$\mathcal{N} = \dfrac{XP_v}{kA\,\Delta T}$		(3.14)
Reynolds number	$\mathcal{R} = uX/\nu$		(2.10), (3.18)
Rayleigh number	$\mathcal{A} = \dfrac{g\beta X^3 \Delta T}{\kappa \nu}$		(3.22)
Prandtl number	$\mathcal{P} = \nu/\kappa$		(3.19)
Grashof number	$\mathcal{G} = \mathcal{A}/\mathcal{P}$		End of Section 3.4.4
Thermal diffusivity	$\kappa = k/\rho c$		(3.12)

Table C.2 Free convection. Comparative tables in other texts may refer to Grashof number $\mathscr{G} = \mathscr{A}/\mathscr{P}$

Shape	Case	Overall Nusselt number	Equation no.
Horizontal flat plate	Laminar ($10^2 < \mathscr{A} < 10^5$)	$\mathscr{N} = 0.54 \mathscr{A}^{0.25}$	(C.1)
	Turbulent ($\mathscr{A} > 10^5$)	$\mathscr{N} = 0.14 \mathscr{A}^{0.33}$	(C.2)
Horizontal cylinder	laminar ($10^4 < \mathscr{A} < 10^9$) turbulent ($\mathscr{A} > 10^9$)	$\mathscr{N} = 0.47 \mathscr{A}^{0.25}$ $N = 0.10 \mathscr{A}^{0.33}$	(C.3) (C.4)
Vertical flat plate or Vertical cylinder	If laminar, ($10^4 < \mathscr{A} < 10^9$)	$\mathscr{N} = 0.56 \mathscr{A}^{0.25}$	(C.5)
	If turbulent, ($10^9 < \mathscr{A} < 10^{12}$)	$\mathscr{N} = 0.20 \mathscr{A}^{0.40}$	(C.6)
Parallel plates (slope <50°)	Turbulent ($\mathscr{A} > 10^5$)	$\mathscr{N} = 0.062 \mathscr{A}^{0.33}$	(C.7)

Table C.3 Forced convection

Shape	Case	Overall Nusselt number	Equation no.
Flow over flat plate	Laminar ($\mathcal{R} < 5 \times 10^5$)	$\mathcal{N} = 0.664 \mathcal{R}^{0.5} \mathcal{P}^{0.33}$	(C.8)
	Turbulent ($\mathcal{R} > 5 \times 10^5$)	$\mathcal{N} = 0.37 \mathcal{R}^{0.8} \mathcal{P}^{0.33}$	(C.9)
Flow over circular cylinder	Laminar ($0.1 < \mathcal{R} < 1000$)	$\mathcal{N} = (0.35 + 0.56 \mathcal{R}^{0.52}) \mathcal{P}^{0.3}$	(C.10)
	Turbulent ($10^3 < \mathcal{R} < 5 \times 10^4$)	$\mathcal{N} = 0.26 \mathcal{R}^{0.6} \mathcal{P}^{0.3}$	(C.11)
Flow *inside* a circular pipe: from Wong, 1977 (see Appendix B, Bibliography)	Graetz number $\mathcal{G}_1 = \mathcal{R}\mathcal{P}(D/L)$ $= 4Q/\kappa \pi L$		
	Laminar flow, short pipe ($\mathcal{R} < 2300$, $\mathcal{G}_1 > 10$)	$\mathcal{N} = 1.86 \mathcal{G}_1^{0.33}$	(C.12) (C.13)
	Turbulent flow ($\mathcal{R} > 2300$)	$\mathcal{N} = 0.027 \mathcal{R}^{0.8} P^{0.33}$	(C.14)
	General	$P = \rho c Q(T_2 - T_1)$	(2.6)

〰〰〰 Indicates section of an extended shape

Table C.4 Mixed convection (forced and free together)

Shape	Case	Formula	
Air over flat plate	$X > 0.1\,\text{m}$ $u < 20\,\text{m s}^{-1}$	$h = a + bu$ $[a = 5.7\,\text{Wm}^{-2}\text{K}^{-1}$ $b = 3.8\,(\text{Wm}^{-2}\text{K}^{-1})/(\text{ms}^{-1})]$	(C.15)
General		$\mathcal{N}_1 = \max(\mathcal{N}_{\text{forced}}, \mathcal{N}_{\text{free}})$	
		$\mathcal{N}_1 < \mathcal{N}_{\text{mixed}} < \mathcal{N}_{\text{forced}} + \mathcal{N}_{\text{free}}$	(C.16)

Table C.5 Net radiative heat flow between two diffuse gray surfaces. For definitions and notation, see Chapter 3. NB In these formulas T is the *absolute temperature* (i.e. in kelvin)

System	Schematic presentation	Net radiative heat flow	Equation no.
Gray surface to surroundings $(A_1 << A_2)$		$P_{12} = \epsilon_1 \sigma A_1 (T_1^4 - T_2^4)$	(C.17)
Two closely spaced parallel planes $(L/D \to \infty)$		$P_{12} = \dfrac{\sigma A_1 (T_1^4 - T_2^4)}{(1/\epsilon_1) + (1/\epsilon_2) - 1}$	(C.18)
Closure formed by two surfaces (surface 1 convex or flat)		$P_{12} = \dfrac{\sigma A_1 (T_1^4 - T_2^4)}{\dfrac{1}{\epsilon_1} + \left(\dfrac{A_1}{A_2}\right)\left(\dfrac{1}{\epsilon_2} - 1\right)}$	(C.19)
General two-body system (neither surface receives radiation from a third surface)		F_{12} = shape factor $P_{12} = \dfrac{\sigma(T_1^4 - T_2^4)}{\dfrac{1-\epsilon_1}{\epsilon_1 A_1} + \dfrac{1}{A_1 F_{12}} + \dfrac{1-\epsilon_2}{\epsilon_2 A_2}}$	(3.36) (C.20)

Index